Cures Out of Chaos

Cures Out of Chaos

How Unexpected Discoveries Led to Breakthroughs in Medicine and Health

M. Lawrence Podolsky, MD

Foreword by Daniel E. Koshland Jr., MD

harwood academic publishers
Australia • Canada • China • France • Germany • India •
Japan • Luxembourg • Malaysia • The Netherlands • Russia •
Singapore • Switzerland • Thailand • United Kingdom

Copyright © 1997 OPA (Overseas Publishers Association) Amsterdam B.V. Published under license under the Harwood Academic Publishers imprint, part of The Gordon and Breach Publishing Group.

All rights reserved.

No part of this book may be reproduced or utilized in any form or by any means, electronic or mechanical, including photocopying and recording, or by any information storage or retrieval system, without permission in writing from the publisher. Printed in India.

Amsteldijk 166
1st Floor
1079 LH Amsterdam
The Netherlands

British Library Cataloguing in Publication Data

Podolsky, M. Lawrence
   Cures out of choas : how unexpected discoveries led to breakthroughs in medicine and health
   1. Medical innovations – History
   I.Title
   610.9

ISBN 90-5702-556-6

To my mentors

Russell J. Blattner, MD
Baylor University College of Medicine
Houston, Texas

and

John J. Sheinin, MD
Chicago Medical School
Chicago, Illinois

All the inventions that the world contains,
Were not by reason first found out, nor brains;
But pass for theirs who had the luck to light
Upon them by mistake or over sight.

—Samuel Butler

CONTENTS

Foreword   ix

Preface   xi

Acknowledgments   xv

1	Dyes and Destiny	1
2	A Lesson from "The Wizard of Oz"	43
3	Elegy for a Planet	71
4	Afflictions of Affluence	89
5	Too Much of a Good Thing	109
6	Foresight, Nearsight and Insight	133
7	Captain of the Men of Death	153
8	The Oxford Incidents	177
9	Orphans and Their Relatives	225
10	Janus and the Cancer Connection	271
11	Concept Out of Chaos	295
12	Bing, Bang and Levin and the Horseshoe Crab	327
13	"Mighty Like a Rose"	349
14	Backyard Full of Diamonds	375

Sources and References   395

Index   415

FOREWORD

Discovery is an exciting process in itself: from the discovery of a beautiful landscape to the finding of gold nuggets that indicate a gold mine may be nearby. Nothing, however, equals the discovery of an advance in medicine because this revelation will benefit thousands and, in many cases, millions. This book describes the incredible discoveries of not just one but many of the most important advances in modern medicine. Individual chapters read like a mystery novel. But unlike mystery stories, they are relevant not only to one person or one place but to all of humankind. Discoveries dealt with here are very important; they are also great fun to read about.

Daniel E. Koshland Jr., MD
Dept. of Molecular and Cell Biology
University of California, Berkeley

PREFACE

In his book *The Structure of Scientific Revolutions*, Thomas Kuhn proved that progress in medicine, as in general science, arises two ways: from normal science, which is the slow, steady accumulation of knowledge, and from unexpected discoveries that entail paradigm shifts—learning to see nature in a different way. Discovery, Kuhn said, commences with the awareness of anomaly.... And it closes only when the paradigm theory has been adjusted so that the anomalous becomes the expected.

A paradigm, in its broad sense, is a set of rules that define or exemplify something; in science it characterizes a natural phenomenon. When new facts or an unexpected discovery undermines an established paradigm, it is replaced by a new one—one that provides a better fit for the evidence. Chaos, however, precedes the formation of an original paradigm and reappears when the paradigm is compelled to change, hence the title, *Cures Out of Chaos*. Special features of particular paradigms are discussed in the Paradigm Pointers sections in each chapter.

The seeds for this book were sown at the very same time that Professor Kuhn was formulating his ideas about science, in the 1950s and 1960s. I was a fledgling young physician then and regularly attended the lectures of Dr. Julius Hiram Comroe at the University of California in San Francisco. As the foremost authority on heart and lung physiology and director of the university's cardiovascular institute, Dr. Comroe generously dispensed pearls of medical wisdom throughout his lectures but, more often than not, he would drift off into his favorite topic: the nature of medical discoveries. Dr. Comroe always maintained that: "Research is really a systematic way of looking for a needle in a haystack. The truly great investigator," he would emphasize, "is not necessarily the one who directly finds the needle but rather the one who finds an ugly toad in that haystack and has enough wisdom, luck, pluck, and intuition to consider that it may be a beautiful princess—and then goes on to prove it."

Dr. Comroe's anecdotes about the vagaries of medical research intrigued me and I found myself collecting similar episodes throughout my career in teaching and practicing medicine. But not until Thomas Kuhn's book came out was I able to appreciate how unexpected discoveries change the dimensions of a medical concept and foment revolutions in science. At this juncture I considered writing a book that would present this epiphany but, as an accomplished procrastinator, I found all the excuses needed to shelve the project. However, when, in recent years, the government and scientific community were unreasonably attacked for not delivering cures-on-demand for AIDS, cancer and other incurable conditions, I felt compelled to haul out my collection of stories about the medical discoveries that currently shape our lives and put them into a book for a general readership.

At most, they might squelch the rhetoric of special-interest groups, bandwagon celebrities and politicians, who promise quick cures in return for contributions to their particular causes; at least, they might dispel the notion that mass attacks on diseases are scientifically feasible and economically sound. Hopefully, the lessons in this book will prove to legislators and those responsible for planning future health priorities that a certain amount of basic research and way-paving must be in place before that rare combination of scientific reasoning, intuition, fortuities and good luck can force the final grand triumph. Or, as Dr. Comroe often said: "Some scientists make things happen; others make it possible for things to happen." We just have to wait for some cures until some scientists 'make it possible for them to happen.'

"Intuition in science," Victor Weisskopf, the internationally acclaimed theoretical physicist, once said, "consists of half-conscious knowledge...a certain feeling of how things work even if no exact information is available...and being right depends on a certain amount of luck." In deference to this assertion, the book initially started out as a compendium of lucky or accidental discoveries. When it began to appear that anarchy prevailed in the laboratory and that chance and accidents always led to unexpected rewards, I introduced episodes where these same factors led to failures or missed opportunities. Additionally, I brought in extraneous factors like wars, depressions, economics, politics and the changing patterns of diseases, to show that factors outside the laboratory also can have profound effects on the course of science.

As a science and medical writer for more than twenty-five years, editorial constraints of journals and magazines kept me from describing the lives and personalities of investigators and the basic research that paved the way for their discoveries. This book has permitted me to compensate for those omissions. First, I stress the valuable contributions of women scientists. Second, I detail discoveries of young people, who frequently change the dimensions of medicine since they do not have to unlearn misinformation in the guise of dogma. Third, in the text and personae sections, I present personality sketches of the main figures.

To fully comprehend a factual discovery, a certain amount of theory and basic research must be understood. Deciding on how much theory should be covered, and how detailed it should be, proved to be a thorny problem. Werner Von Braun, the aerospace genius, once said: "Basic research is what I am doing when I don't know what I'm doing." Conveying this idea proved elusive so I included just enough theory to understand the "big picture." Further details can be found in the sources and references for each chapter.

Despite the assistance of luck and good fortune, the principal players in this book never could have planned their experiments, collected and interpreted the right data, and figured out how to get the most out of their observations unless they possessed a good general education, substantial scientific training, experience, inquisitive minds and, above all, a willingness to dedicate most of their lives to hard work. In essence, this book combines Louis Pasteur's famous dictum, "Chance favors the prepared mind," with Stephen Leacock's pronouncement: "I am a great believer in luck and I find the harder I work the more I have of it."

Now an apology. I frequently use the word *serendipity*, although it really does not connote my true meaning. Horace Walpole invented the term (from the land of Serendip, now called Sri Lanka), which suggests an accidental discovery while looking for something else. Serendipity fits nicely for finding the missing mate to a pair of socks while looking for skis in the hall closet but does not apply to unexpectedly making a scientific discovery. *Serensagacity*, a word coined by Dr. Comroe, would have served better since it implies that there is a trained and disciplined mind prepared to capitalize on some accidental observation. But since that word never gained popular usage, I acquiesced to serendipity.

ACKNOWLEDGMENTS

Consultants who helped with individual chapters are acknowledged in the sources for those chapters.

The following individuals, librarians, archivists and institutions supplied material for the book as a whole: Barbara Houghton and Karen Moody, medical librarians, Sequoia Hospital, Redwood City, California; the staffs of the Lane Medical Library of Stanford University, the University of California at San Francisco medical library, and the corporate library of Roche Biosciences; the Centers for Disease Control and the Food and Drug Administration, Public Health Service, Department of Health and Human Services in Rockville, Maryland.

I am especially indebted to Robert Alway, MD, erstwhile dean, Stanford University School of Medicine, who provided personal and historic vignettes about the university and access to its archives and resources; to Gerald B. Kara, MD, former chairman of the Department of Ophthalmology at the New York Eye and Ear Infirmary, for his archival memory about the history of medicine; to Drs. Jerome Morse and Gerald H. Wagman, advisers in physics and chemistry; to Dr. Pierre F. Salgado, who inspired much of the philosophy contained herein; to Dr. Nicholas M. Guild, Ken Englade and Clell Bryant, who coached my writing; and, of course, my indefatigable editor Everett Smethurst, who acted as catalyst between thought and the printed page.

Yet, all my labors would have amounted to naught were it not for the support and contributions of my wife Kathleen—typist, editor, critic and muse.

chapter 1
DYES AND DESTINY

On the first day of July, 1924, President Calvin Coolidge's youngest son, 16-year-old Calvin Coolidge, Jr., developed a friction blister on his toe while playing tennis. When it became infected, doctors poured bichloride of mercury, an antiseptic, over the wound, applied hot packs to his foot, and elevated his leg. Still, red streaks inched along his skin from ankle to thigh. There was no doubt that blood poisoning was setting in. More heroic measures were needed.

The doctors now heated pus from his toe, to kill the germs in it, then injected the pus into another person who would build up antibodies and immunity and then donate blood to the patient. After that failed, young Calvin was given antitoxin made from horse serum—even though shock and death from allergy often went hand-in-hand with the procedure. Next, his doctors injected Mercurochrome, a popular but deadly skin antiseptic that contained mercury, into the hapless youth. Finally, the boy was transferred from the White House to Walter Reed Hospital where surgeons opened and drained the abscess. But nothing stayed the inexorable course of the poison. The President's youngest and favorite son died from intractable infection and blood

2 DYES AND DESTINY

poisoning. President Coolidge became despondent and never regained his composure.

The lad, in the words of Shakespeare, was 'a wretched soul bruised with adversity', because, as we shall see, five years before his untimely death, sulfanilamide, the drug that could have saved him, was being studied by Dr. Michael Heidelberger, at Rockefeller Institute in New York City, just a few hundred miles from the White House.

This tragedy illustrates how terribly vulnerable we were before the advent of antibacterial agents like sulfanilamide. In those perilous days a person with a serious infection, even when given a reasonable environment and good nursing care, stood no better chance of recovery than in the days of ancient Greece. A few chemical compounds were in use as medicines for malaria, syphilis, and sleeping sickness, and for the symptomatic relief of pain and diarrhea, but most of them were formulated into liniments, ointments, antiseptics and disinfectants—things to be used on the surface of the body. Generally they were as destructive to living tissues as they were to bacteria. Sulfanilamide changed that. It proved beyond all doubt that chemicals could kill germs inside the body without injuring the person taking them.

The ground-rules for fighting infectious diseases were initially laid down by Louis Pasteur and his co-workers and then carefully modified and expanded by the bacteriologists and vaccine makers who followed. They stipulated that invading germs could only be eliminated by the body's natural defenses, by harsh chemical antiseptics, and by surgical intervention—cauterization with boiling liquids and hot irons, incising abscesses and draining the pus and, as a court of last resort, amputations.

This paradigm was first disturbed by Paul Ehrlich (see Figure 1.1.) in 1890 when he proved that chemicals injected into animals could cure some infections. The rift widened, a decade later, when Roehl, one of Ehrlich's pupils, reported that he could eliminate the organism responsible for trypano-

Figure 1.1 Paul Ehrlich. (From the archives of the Deutsches Museum, Munchen.)

somasis, sleeping sickness, with a chemical compound called *Antrypol*. But the paradigm never really shifted until 1935 when Gerhard Domagk (see Figure 1.2.) discovered *Sulfanilamide*, our first effective chemotherapeutic agent.

To appreciate how rigidly a paradigm controls thinking and directs practices, we have only to look at the way President Coolidge's son was treated—which was the *best* treatment in the world before the paradigm for treating infections actually shifted.

There is an old saying, "Defeat is an orphan but victory has a hundred fathers." And this surely applies to the discovery of Sulfanilamide. You might say that the quest actually began with the Austrian scientist Wagner-Jauregg who set the stage for antibiotic discoveries when he laid down the ground rules: "The goal of chemotherapy," he said, "is the cure or prevention of infectious disease with chemical agents of natural or synthetic origin and the essential task is to attack pathogenic para-

Figure 1.2 Gerhard Domagk. (From the archives of the Deutsches Museum, Munchen.)

sites in a specific way without doing unbearable damage to the diseased organism." Earlier, however, toward the end of the 19th century, Paul Ehrlich introduced the word chemotherapy when he advised researchers to look for the "silver-bullet", a chemical that would target harmful germs and leave normal body cells alone. "The main task of chemotherapy", he said, "is finding, in a systematic manner, drugs which

Figure 1.3 Robert Koch. (From the archives of the Deutsches Museum, Munchen.)

prove to be directed specifically against the organism causing a disease." In this sense, then, the quest for sulfanilamide actually began in Paul Ehrlich's fecund laboratory at Berlin's Institute for Experimental Therapy—where he worked alongside the already famous bacteriologist Robert Koch (see Figure 1.3.),

6 DYES AND DESTINY

and upon whose recommendation he was appointed director of the Institute.

Certainly Domagk's discovery depended heavily on the way-paving by the pioneers of science, from the time of Louis Pasteur up to the 1930's, but in the final analysis, the big push in the right direction came from the turbulent times in Germany just after the first world war plus the government's unique relationships with industry and the academic world.

Gerhard Johannes Paul Domagk was born in the small hamlet of Lagow, Bandenburg, Germany, October 30, 1895. His father, a teacher, enrolled the boy in the scientifically orientated grammar school in Liegnitz and the affinity for science remained fixed for the rest of his life. The young Domagk had hoped to become a physician, but World War I broke out while he was in his first term at Kiel University. He gave up his studies to enlist in a Grenadier Regiment. After being wounded he was transferred to the German army medical corps where he received his first lessons in medicine. After the war he returned to Kiel University as a medical student and earned his MD degree in 1921. He married Gertrud Strube in 1925 and had four children, one of whom, his only daughter, Hildegarde, became the star player in a personal drama that was simultaneously a father's nightmare and a scientist's dream.

Professor Domagk was bright and had superb training under some of the best pathologists and physiologists of his time. He hobnobbed with the elite of Berlin, and was a popular figure with his sparkling blue eyes, high forehead with receding blond hair, trim 176-pound frame and jaunty gait. His earliest studies on the effects of X-rays on cancer and kidney disease attested to his proficiency as a polished investigator. In 1924 he published a widely acclaimed scientific paper showing how the human lymphatic system defends the body against infections. Four years later, in 1928, he was appointed extraordinary professor of general pathology and pathological anatomy at the University of Muenster and went on to become an editor of the German Journal of Cancer Research. On the strength of these attainments he was hired

DYES AND DESTINY 7

Figure 1.4 1929 Advertisement for chemicals and dyes – I.G. Farben Industrie (From the archives of the Deutsches Museum, Munchen.)

by Germany's giant dye and chemical monopoly, I.G. Farben Industrie. (see Figure 1.4.) That was the turning-point in his career.

8 DYES AND DESTINY

His field of study in industry, the relationship between dyes and medicine, was not new; indeed, it had preceded Domagk by three centuries. It began in 1628 when quinine was first introduced as a treatment for malaria. Though not a dye itself, quinine attracted chemists and pharmacologist who, through design and accident while experimenting with it, found many new dyes and drugs.

After the Spaniards conquered South America, King Philip handed the governorship to Don Luis Geranium Fernandez Cabrera Bobadilla y Mendoz, hereditary Alcalde of Segovia, Count of Chincohn, and Lord of 18 villages in the Kingdom of Toledo. The Don established himself in Lima, the capital of Peru and the center of Spanish power in the new world. His wife, who had malaria, was treated with medication from the bark of the Peruvian Quinquina tree, and in her honor botanists changed the name of the bark to cinchona.

The first written record of a malaria cure with cinchona bark appeared in 1630 when Don Juan Lopez de Canizares, the Spanish governor of Loxa, Peru, wrote that he was relieved of fever by drinking an infusion of cinchona bark. In 1811, a Portuguese naval surgeon, B. A. Gomez, came close to extracting the active principle, quinine, but failed because he used a gray bark which contains no quinine instead of the quinine-rich yellow bark. In 1820 two young French chemists, Pierre-Joseph Pelletier, age 32, and Joseph Caventou, age 27, extracted gray cinchona bark with alcohol, added a little water, a bit of alkali, and precipitated an essentially pure substance. But this, like Gomez's preparation, was completely ineffective in treating malaria. Later, when they repeated this process with yellow cinchona bark they ended up with what we now recognize as pure quinine.

Generations of chemists looking for quinine subsequently discovered, among other things, a treasure trove of pharmaceuticals: caffeine in coffee, atropine in belladonna or deadly night-shade, codeine and papaverine in opium, ephedrine in the Chinese herb Ma Huang, scopolamine in the scopola plant, and theophylline in tea. William Henry Perkins, an

English chemist, tried to make synthetic quinine and instead discovered, by the sheerest accident, synthetic mauve, Tyrian Purple, the first coal-tar or aniline dye.

The Germans, upon learning of Perkin's discovery, reproduced it, then went on to synthesize Alizarin Red, Congo Red, the fluorescent Rhodamines and Phenolphthalein. Besides acting as a dye, Phenolphthalein found another niche in medicine. When its red color was used to adulterate cheap wines it was, embarrassingly, found to be a potent laxative. It is the active ingredient in ExLax® and similar laxatives found on drugstore shelves. Indigo, another important synthetic dye, was discovered by Adolph von Bayer in 1897, earning him the Nobel Prize. The aniline dyes became the mainstays of both science and industry until other synthetic dyes came on the scene in the 1920s.

In another quinine-related discovery, Schulemann, a chemist with the German Dye Trust, tried to find if either one of quinine's components, quinoline or meroquine, would cure malaria. He failed, pigeon-holed his notes and started on another tack. Meanwhile, someone, somewhere, in the huge German Dye Trust, reported that the synthetic dye methylene blue could kill malaria germs. It was tested on people but proved unreliable; it also turned them blue. Chemists started overhauling methylene blue, tearing, adding, subtracting, and rearranging its molecules, but it always remained a dye. Then Schulemann's notes were accidentally found and chemists began tinkering with combinations of methylene blue and quinoline. This resulted in the discovery of plasmoquine, an effective treatment for malaria. About ten years later two other workers in the same laboratory found atabrine, another effective anti-malarial drug.

Even today, new uses for the anti-malarials quinoline, plasmoquine and their derivatives crop up unexpectedly and in the most serendipitous manner. A case in point, cited in the June 20, 1990 issue of the Journal of the American Medical Association, tells about a medical student who suffered from inflammatory bowel disease. When he went to

10 DYES AND DESTINY

Africa as part of his postgraduate training, he was given choloroquine to prevent malaria and, quite unexpectedly, his bowel disease went into remission. Tests showed that this was due to plaquenil a break-down product of chloroquine. A number of research projects still in progress have indicated that these drugs can affect our immune responses and interrupt the progress of inflammatory disease.

The paradigm that guided 20th century antibiotic research was initiated by Paul Ehrlich when he began looking for the chemical-bullet that would target harmful germs for destruction and leave normal body cells alone. He experimented with everything from inorganic and organic chemical compounds to exotic plant extracts and dyes. It was with dyes that he achieved his initial successes and, in the process, lit the beacon that revealed the path Domagk later followed.

Ehrlich was introduced to the aniline dyes by his cousin, Karl Weigert, the pioneer pathologist who devised methods of staining blood cells and tissues so they could be differentiated under the microscope. Ehrlich perfected these techniques and extended them to stain living tissues. In a series of experiments, advanced for his time, he injected the stain methylene blue into living rats. When he subsequently dissected them he found that all the nerve endings in the animals were stained blue.

"Why does the color hit only the nerve endings?" he asked himself. "There must be an affinity between this dye and the nerve cells. There must also be other special attractions between cells and dyes. If we could find dyes that possess germ killing properties they might fix themselves to bacteria and kill them without exerting any harmful effects on adjacent normal tissues," he reasoned.

Ehrlich pursued this idea by infecting rats with sleeping-sickness trypanosomes—tiny, unicellular, snake-like, parasitic, protozoans—and discovered that the rats could be saved from certain death by treating them with a recently developed dye, trypan red. Following-up on this spark of hope, Ehrlich then experimented with another germ closely related to trypanosomes, the one that caused syphilis, and found that he could kill it in test tubes, not with a dye, but with Atoxyl, an arsenic-containing substance that had been developed by two English chemists. But there was still no breakthrough; the arsenic caused blindness. Chemists under Ehrlich's intuitive guidance synthesized hundreds of variations of Atoxyl but only two seemed to have any chance of being effective, the 418th and the 606th variant.

When Ehrlich learned that a Japanese bacteriologist, Sahachiro Hata, had discovered a way to infect rabbits with *treponema pallidum*, the motile, spiral, microorganism known as the spirochete of syphilis, he invited Hata to his Institute and assigned him to test Atoxyl derivatives numbers 418 and 606. On August 31, 1909, Ehrlich watched Hata inject compound 606 into a large rabbit that had been infected with syphilis. Two days later no live germs could be found and shortly thereafter all signs of syphilis had disappeared. Compound 606, named Salvarsan, became the first drug to halt the ravages of one of our oldest sexually transmitted disease.

Even though there was a possibility that Salvarsan might cause blindness, it was nonetheless tested on two young laboratory assistants at Ehrlich's Institute. Fortunately they escaped unharmed. The first large-scale trials were made in St. Petersburg in 1909 by Dr. Julius Iverson who had used Salvarsan earlier during an epidemic of relapsing fever, a disease caused by a spirochetal organism similar to that of syphilis. On April 19, 1920, at the Congress for International Medicine at Wiesbaden, Germany, Ehrlich and Hata proudly announced that 51 out of 55 patients had been cured of syphilis with Salvarsan.

But chemotherapy proved to be a rogue in disguise, curing and killing with the same indifference. Ehrlich at first manufactured the yellow crystalline Salvarsan powder at his own laboratory. The powder had to be dissolved in water at the bedside and then injected into a vein. If the vein was missed and Salvarsan was injected into the surrounding tissues, it often necessitated amputation of the patient's arm. The arsenic also caused permanent deafness in some patients.

12 DYES AND DESTINY

As the cases of amputation and deafness built up, so did the criticism against Ehrlich. His health declined and, after a stroke, he died on August 20, 1915, at age 61. Just before his death he grumbled, "You say I accomplished a great work of the mind, a wonderful scientific achievement. My dear colleague, for seven years of misfortune, I had one moment of good luck."

In sum then, Ehrlich proved that the chemical structure of drugs determines how they behave in the body, that chemical compounds can be designed to attack specific targets, that dyes have affinities for certain cells, and dyes and chemicals can eradicate harmful bacteria. This wisdom triggered a mass screening of dyes through Germany to see if any had curative properties. It also gave Dr. Domagk a considerable boost when he began screening dyes for biological effects.

Samuel Taylor Coleridge, in *The Friend*, wrote, "the dwarf sees farther than the giant, when he has the giant's shoulders to mount on." In the same sense Domagk stood on the shoulders of Paul Ehrlich and the elite corps of German chemists who arose from the cataclysm of defeat in the first world war.

The country was physically and emotionally exhausted and virtually bankrupt, stripped of its colonies—and thereby its textile empire—by the victorious Allies. Lacking the colonial sources of fibers and the money to pay for imported wool and cotton, German textile factories would soon wither away and unemployment would skyrocket. The fear that its archrival, England, would soon take over the world textile markets goaded Germany into allocating huge resources toward the development of synthetic fibers.

Trained, talented, disciplined minds were recruited and pressed into service, spurred on by government subsidies and incentives. They soon contrived an astounding array of innovative products. Germany's technological resurgence after the first world war was no less phenomenal than that of Japan after World War II.

Within a decade after the first world war, Germany emerged as the world's leading manufacturer of textiles—

especially synthetic fibers. Rayon staple fiber replaced natural fibers and spun rayon textiles spewed from the looms. But when it came time to dye the textiles made with these synthetic viscose and acetate fibers, the results were less than satisfactory. Streakiness, uneven and inconsistent colors bedeviled the industry. This prodded researchers all over Germany to look for dyes that could color the new synthetics rapidly, cheaply and dependably. In the process, the Germans became the foremost dye producers as well as the world's most advanced chemists.

The huge chemical and pharmaceutical complex, I.G. Farben Industrie, invested heavily in time, money and effort to develop synthetic dyes. Much of its effort went into studying a particular red dye, para-amino-benzene-sulfonamide, called Prontosil, since it could color leather and was extremely color-fast. Its unique chemical structure made it an azo dye as distinguished from the older aniline dyes.

Heinrich Hoerlein, the brilliant and imaginative chemist that I.G. Farben Industrie lured away from Jena University, became fascinated with the potential moneymaker. He had studied sulfanilamide in 1909 just a year after its discovery by Paul Gelmo, a young Viennese scientist working on his doctoral thesis. The discovery had attracted attention because the compound was the basis for waterproof and permanent dyes. When Hoerlein joined I.G. Farben Industrie the first thing he did was to assemble a crack team of chemists who, in short order, synthesized a long list of dyes built around para-amino-benzene-sulfonamide.

The big corporations in the 1920s, like those of today, always kept a sharp lookout for ways to spin off their technology in order to develop new products, capture wider markets, and boost profits. The dream of unifying all the chemists in Germany into a mighty brain-trust came true when Bayer and other industrial giants integrated themselves into the vast chemical and pharmaceutical monopoly, I.G. Farben Industrie. Its medical division was handed over to Heinrich Hoerlein. New chemicals, drugs, and dyes came out of the laboratories in rapid profusion and Hoerlein soon realized that he needed additional experts to help evaluate them. Most pressing was the need for someone who could work with drugs. A search was started and among those in the stack of applicants was an obscure pathologist with a brilliant series of professional papers to his credit—Gerhard Domagk. He was chosen to direct I.G. Farben Industrie's Institute of Experimental Pathology.

Hoerlein personally introduced Domagk to the intricacies of the industrial-pharmaceutical complex and aimed him in the direction of patentable drugs and chemicals. But Domagk resisted change. From 1927, when he first went to work at the I.G. Farben Industrie, until 1930, when he became interested in biological effects of dyes, Domagk vacillated between studies on liver disease and cancer. By randomized trial and error and hit or miss procedures, he experimented with a wide and varied assortment of things—gold, tin, antimony, and arsenic, as well as chemical compounds synthesized by his colleagues—to see what, if any, effects they had on living tissues. Some worked in the test tube and not on mice. Some worked in the test tube and on mice but had too many harmful effects when tried on people. Although many colleagues and fellow scientists insinuated that he was jousting with windmills, Domagk persisted in screening chemicals for their biological effects on animals.

The visit to Dr. Domagk's laboratory at Elberfeld by Britain's renowned bacteriologist-immunologist Sir Almroth Wright gave a revealing glimpse of Domagk's dogged determination. Wright tried to convince his host that time was being wasted in the search for chemicals that could cure disease. "Bacteria and humans are made of the same stuff," he asserted. "Chemicals that kill one kill the other. Vaccines! Only vaccines can tell the difference!" Wright, whose pomposity exceeded his stature, proceeded to recite the litany of past failures with chemicals and his own wonderful accomplishments with the development of a vaccine against typhoid fever. There is no record of what Domagk said in rebuttal, but whatever it was, it made no impression on Sir Almroth. The Englishman would never espouse chemotherapy. Worse yet, he discouraged it. He strenuously barred all chemists from St. Mary's Hospital, London's prestigious research institution which he directed and inspired. "The doctor of the future," he said repeatedly, "will be a vaccinator." When Alexander Fleming came across the antibacterial action of a curious penicillium mold in 1928, his boss, Sir Almroth Wright, nick-named Sir Almost Wright, threw cold water on any attempt to pursue this lead. Fleming also experimented with sulfanilamide and some of its derivatives but put them aside when he failed to receive support from his colleagues, his employers or his government. He had to wait for a lucky accident to catapult him to fame with penicillin, an antibiotic more powerful than sulfanilamide

Erlich's concepts about germ-killing dyes were well known throughout Germany. Scientists reasoned that if azo-dyes form permanent bonds with proteins in textile fibers they might possibly attach themselves to proteins of germs and in so doing kill them. Therefore, it was quite natural for the German researcher Eisenberg to report in 1913 that bacteria in a test tube could be killed by the addition of the dye chrysoidine. The following year Tchichibabin and Zeide synthesized the red dye pyridium from chrysoidine, which also killed germs. Pyridium is still used to treat urinary tract disorders.

In 1930 Hoerlein visited Domagk, who was now director of the Pathological Laboratory in Elberfeld, in the Ruhr near Dusseldorf, and informed him that two of I.G.Farben's chemists, Fritz Mietszch and Josef Klarer had just come up with some new azo dyes. "You might want to look into them," Hoerlein suggested. Dr. Domagk subsequently set up meetings with the chemists where he learnt that Prontosil, a red azobenzene dye that had sulfonamide attached to its molecule, was being used to dye leather. It was generally believed that the dye worked so remarkably well because it attached itself to the protein fractions of fibers to form stable,

16 DYES AND DESTINY

permanent complexes. That being the case, it was natural to assume that the dye might combine with proteins in the protoplasm of bacteria and possibly form complexes that would prove incompatible with life—very much like carbon monoxide attaches itself to hemoglobin in red blood corpuscles and prevents them from carrying oxygen, causing death.

After two years of testing, Domagk found that Mietszch and Klarer's red dye could save mice after streptococci had been injected into their abdominal cavities. But he had to be sure, considering the criticisms that often stem from professional jealousies and rivalries. Domagk injected streptococci directly into the blood stream of laboratory animals and proceeded to save them with Prontosil. In that instant he knew he had made the greatest discovery of the century. The next step was to arrange for trials on humans.

As it happened, fate now gave Gerhard Domagk an unexpected shove—a scientists dream but a father's nightmare. His small daughter Hildegarde injured her finger with a knitting needle. The ugly wound got redder, bigger, and filled up with germs and pus. Then her entire limb became swollen, unmistakable evidence that bacteria were seeping into the child's blood stream and lymph system.

"Dare I try Prontosil?" the distraught Dr. Domagk pondered. Although the red dye had been proven to be safe for animals there was no assurance that it was safe for humans. But the child was as good as dead unless something could stop the progressive infection. Agonized, Domagk gave his daughter the new drug. Over the next several days doubt and fear gripped both parents. No one knew if this largely untested chemical would cure or kill. But finally the infection receded; Hildegarde survived.

Now, Domagk presented his experimental finding and data to his boss, Hoerlein. "I took 26 mice and infected them with lethal doses of streptococci. To 12 of them I gave Prontosil. The 14 animals that did not get Prontosil died, all of them. The 12 that did get the drug lived. Studies where the drug was given to infected and non-infected animals showed that there were no harmful side-effects to the drug."

The absolute success rate would have made most people suspicious. Had he forged his results? Was something wrong with the experiments? Hoerlein, familiar with Domagk's skill and integrity knew that he was above this. Top level executives met at once and patent attorneys were briefed on the potentials of the new substance. On Christmas day, in 1932, Mietzsch and Klarer patented Prontosil as an antibiotic.

Heinrich Hoerlein held a professorship at the nearby medial school in Dusseldorf and it was there that he snagged Dr. Hans Schreus and his young assistant, Dr. Richard Forester, and asked them to test the new Prontosil on humans. They did not need to look far. In the hospital at that time there was a ten-month-old infant dying of a pus-producing staphylococcal skin infection.

But using this child for a trial presented a quandary. Dr. Domagk had proven that Prontosil worked with streptococci, the chain-like bacteria that typically causes sore throats and rheumatic fever, but there was no telling what it would do with staphylococci, the abscess and pus producers. If this new drug was tried on a patient with staphylococcus and he died, there would be no way of knowing whether the drug or disease killed him. It would give Prontosil a bad name. Would it not be better to wait for a case of streptococcal infection?

As the night wore on, the baby's temperature shot upward and his labored breathing, separated by long death-like pauses, indicated that life was ebbing away. Unable to bear standing by and doing nothing, the two physicians decided to take the necessary risks. Dr. Schreus pulled a small vial with little brick-red tablets out of his pocket and told the nurse to give the patient half a tablet now and another half four hours later.

The hours ticked away interminably until a frantic nurse came running to the doctors protesting that her little patient had turned bright red. The doctors assured her that Prontosil was essentially a dye and it made people's skin red. The next morning the child received another half tablet of medicine. The following day the infant's temperature started to come down and he began breathing normally again. After having received eight half-tablets of the new red dye, the child was better. He was the first patient ever cured of staphylococcal blood poisoning, which up to then had invariably been fatal.

Throughout the next two years, 1933 and 1934, a few select doctors throughout Germany tested the new drug and all came forth with glowing reports. They were exhorted not to reveal their findings to the press or other scientists since time was needed to document the efficacy of Prontosil, and more importantly, to establish world-wide iron-clad patents.

Finally, in January 1935, Dr. Gerhard Domagk was permitted to publish the results of his experiments. The world learned about Prontosil when his paper appeared in print, February 15, 1935, three years after he started his experiments and two years after he tested Prontosil on humans. Dr. Domagk undoubtedly knew, long before his official pronouncement, that his new drug would revolutionize medicine and save countless lives, but he had to wait until his employers were ready to release the drug. He also must have remembered some of the horrible debacles of the past, such as the vaccine that was supposed to prevent tuberculosis but instead infected and killed healthy children, and he chose to delay his announcement until he was absolutely certain of his facts and data. His paper *Ein Beitrag zur chemotherapie der bakteriellen Infektionene* is a classic of strict experimentation supported by statistics, as well as unbiased critical evaluation of a new drug.

But Dr. Domagk made one big mistake. He failed to break Prontosil into its two components, the dye part and the sulfanilamide part. He did not perform any tests to see if either part worked by itself. This can be compared to a person taking a capsule of aspirin for a headache and attributing the cure to the combined effects of the gelatin in the capsule plus its contents, rather than to the aspirin alone. In effect, he had, as T.S. Eliot phrased it in MURDER IN THE CATHEDRAL, done "the right thing for the wrong reason."

In his defense, it should be emphasized that Dr. Domagk discovered the antibacterial properties of Prontosil while working for dye manufacturers. Prontosil itself was essentially a derivative of a red dye attached to a sulfanilamide molecule. It was only natural for him to believe that the colored part of the Prontosil dye somehow did the killing. When people asked, "How does it work?" Dr. Domagk stuck to his original erroneous concept, and replied, "The dye combines with the proteins in the bacteria and that's what kills them."

"Prontosil doesn't work in the test tube," Domagk said. "It only works in infected animals." He then went on to explain how he had divided his animals into infected colonies and non-infected controls and how both sets received the drug. All of the animals that had been infected were saved and none of the animals, in either the infected or the control group, had any adverse effects from the drug itself.

While testing chemicals for germ killing properties Domagk observed that certain substances had antibacterial actions in experimentally infected animals but this could not be seen in the test tube. By 1932, when he was working with sulfanilamide, he was certainly familiar with this phenomenon but made no attempt to explain it.

Not everyone rejoiced at his discovery. Peers and colleagues were immediately skeptical of Domagk's report. One hundred percent perfect results are virtually non-existent in the world of science. It seemed contrived. And why did I.G. Farben Industrie wait from Christmas 1932 until January 1935 to announce these results?

When the foremost chemist in France, Ernest Fourneau, read Dr. Domagk's report, he too questioned its accuracy and asked Bayer for some Prontosil so that he might corroborate Domagk's findings. Hoerlein invited Fourneau to Germany to discuss this request. "You know," Hoerlein began, "because of the Versailles Treaty we cannot protect our discoveries and inventions in your country. Prontosil cannot be patented in France. We might let you have a little of the drug from time to time, or even all you need, not only for experimental purposes but to treat your whole population, if you grant us some special arrangements to protect our investment." Fourneau replied that he would look into the political climate at home and try to get some concessions. This proved impossible. Competition between French and German scientists had escalated to acrimony ever since the Franco-Prussian war. Furthermore, the guns of August still resounded in the French mind and the Boche were not to be trusted. No! No deal! Besides, the French could duplicate the drug. Why pay for something they could get for nothing?

Germany's arch-rival, Papa Ernest Fourneau, as he was affectionately called by his peers, headed the brain-trust at the Pasteur Institute in Paris where he notoriously appropriated German patents. When Paul Ehrlich discovered his treatment for syphilis, compound 606, Arsphenamine, Fourneau and his trusted Rumanian co-worker, Constantin Levaditi, replaced arsenic in the molecule with bismuth and came up with a drug that worked equally well. When Hoerlein's chemists brought out the sleeping-sickness drug, Bayer 205, it was Fourneau and his team who analyzed it and gave its secret formula to the world. An almost identical scenario unfolded with Prontosil.

Domagk presented the formula for Prontosil in his research paper and Fourneau, Levaditi, and the Trefouels, a husband and wife team, succeeded in making an exact copy of the drug in no time at all. Tests on animals proved identical to Domagk's. Small samples were released to doctors at the Claude Bernard Hospital and again spectacular results against streptococcal infections were reported.

The Trefouels had noticed that 50 percent of the prontosil molecule was dye-related and that tampering with that part did not change the effectiveness of the drug. Was it really necessary? In a brilliant series of experiments they cut the dye part out of Prontosil and, within nine months, came up with a perfectly colorless and relatively inexpensive powder, para-amino-benzene-sulfonamide, later shortened to Sulfanilamide. Then, in classic experiments with Fourneau, they synthesized this compound and proved that sulfanilamide killed germs as effectively as Prontosil. Dr. Domagk's reasoning about his red dye was wrong.

Within a year, the Trefouels, working with French chemists Nitti, Bovet, and Fourneau, explained what had happened in Dr. Domagk's laboratory. In living tissues, para-aminobenzene-sulfonamide, the active part of prontosil's molecule, was split off and that stopped germs from growing. This was more than just an analytical exercise; it cost far less to produce sulfanilamide than Prontosil.

"Sulfanilamide is the part that works," Fourneau exclaimed. "Can we patent it?"

The Trefouels shrugged their shoulders and acknowledged that sulfanilamide had actually been discovered in 1908 by Gelmo and was patented in 1909, as part of a dye, by none other than Heinrich Hoerlein of Germany's I.G. Farben Industrie. A Cheshire-cat grin spread across Fourneau's face as he realized that if sulfanilamide had been patented in 1909 then the patent expired 10 years ago; France could use the new antibiotic freely.

The French and Germans played hide-and-seek with the new drug, giving intermittent peeks at it, while tantalizing the world with suggestions that they had a miracle drug under their wraps. It remained for the British obstetrician Leonard Colebrook, at Queen Charlotte's Hospital, to give mankind the information needed to use this new antibiotic. In 1936 he tested sulfanilamide in his maternity clinic and at the International Congress of Microbiology recounted its ability to save women who otherwise would have died from childbirth-related infections.

Dr. Colebrook endured periods of painful soul-searching. Should he try the wonder drug on only half of his patients and use the other half for controls? No! Colebrook chose to administer Prontosil to all of his critically ill patients and although he violated the basic tenets of science he nonetheless proved the remarkable effectiveness of this new germ killer. Shortly thereafter Colebrook was called upon to use Prontosil at St. Mary's Hospital. Ronald Hare, a bacteriologist, cut his finger with glass and developed a severe streptococcal infection and blood poisoning. In those days this was tantamount to death. Prontosil saved his life. Dr. Colebrook told the press, "Patients whom we would have given up before, now recover easily and without the long drawn-out desperate illness that would have previously been their lot."

In the audience at the International Congress of Microbiology, when Dr. Colebrook presented his startling report, was an American physician from Baltimore's Johns Hopkins University, Dr. Perrin Long. As soon as he returned home he repeated the French and German experiments and together with Dr. Eleanor Bliss, tried sulfanilamide on the wards of Johns Hopkins Hospital. They cured scores of patients who had come to them from all parts of the United States with hitherto incurable infections. Long and Bliss not only proved that para- aminobenzene-sulfonamide was the useful portion of Prontosil but were the first to suggest that it acted by interfering with the metabolism of micro-organisms rather than by killing them.

Contemporaneously, Dr. Albert Fuller confirmed the observations of the Trefouels when he found sulfanilamide in the urine of patients who had been treated with Prontosil. What had been called para-aminobenzene-sulfonamide until now, received its current name, Sulfanilamide, from the Council on Pharmacy and Chemistry of the American Medical Association.

American doctors were at first cautious in accepting the new Sulfanilamide. In his memoirs, Dr. Mark M. Ravitch, the distinguished American surgeon, recalled the days when he was a young pediatric intern, in 1936, and was asked by the hospital chief to translate a German article "by a Dr. Domagk which described a drug called Prontosil." Dr. Ravitch translated the piece but added his own assessment—this discovery was tragic and would probably set medicine back many years. Why? Simply because the pharmaceutical company, Eli Lilly, was beginning to produce serums in rabbits that would lead to the development of vaccines for pneumococci and other organisms. These never panned out, but, as we shall see in Chapter 7, largely through the genius of doctors Michael Heidelberger and Robert Austrian, a vaccine against pneumococci did ultimately materialize.

The stature of Sulfanilamide seemed assured when the President of the United States allowed doctors to treat his son with the new drug. The month of December in 1936 was characteristic of most winters in Washington, DC. Freezing rains, slushy snows, and bone chilling winds debilitated almost everyone, including 22 year old Franklin Delano Roosevelt, Jr., who came down with a streptococcal infection. At first his throat and respiratory passages were sore and inflamed but soon the infection spread to the sinuses. White House physician Dr. George Loring Tobey advised transferring the patient to Phillips House of Massachusetts General Hospital for treatment with the new Sulfanilamide.

"What does this drug do? How does it work?" asked the President as he rubbed the bridge of his nose with his thumb and forefinger, a mannerism used whenever confronted by serious affairs of state. Dr. Tobey replied that it was some sort of a chemical compound developed in Germany that seemed to kill germs.

"How safe is it?" the President asked. His own attack of poliomyelitis with its crippling paralysis was a constant reminder of the ineffectiveness of standard medical treatments. Trying to reassure the President, Dr. Tobey went over all the studies that had been carried out in Europe that indicated Sulfanilamide was safe and effective. "But," he added, "we have no way of knowing if any bad effects will show up next week, next year . . . or even in the next generation."

Doctors at that time knew that streptococcal infections often produced rheumatic fever, a disease dreaded because it damaged hearts and joints, and there was no complete recovery. Tobey now placed this possibility before Franklin and Eleanor Roosevelt and the patient's fiancee, Ethel Dupont. Would they agree to let doctors try the new sulfa drug? They did. The President's son recovered, and Sulfanilamide gained unqualified acceptance in the United States. Years later, Eleanor Roosevelt recalled in her autobiographical book *This I Remember,* that her son "was a guinea pig for the use of the Sulfa drugs."

The fears and doubts of many scientists—as well as those of the Roosevelts when they had to decide whether to permit the new drug to be used on their son—were by no means unfounded. Shortly after young Franklin Delano Roosevelt Jr. was restored to health, Elixir of sulfanilamide killed over a hundred people, mostly children. The owner of a small factory in Tennessee that made pharmaceuticals for animals decided to produce a liquid form of Sulfanilamide, which, at that time, was only available as tablets. The drug would not dissolve in water or alcohol but would go into solution when mixed with diethylene glycol, a toxic solvent. The factory produced gallons of pretty-colored, nice-tasting, easy-to-swallow medicine, called Elixir of Sulfanilamide—but it was lethal. The known deaths totaled 80; unreported fatalities accounted for about 100 more. When confronted with this fiasco the factory owner said, "My chemists and I deeply regret the fatal results, but I do not feel there was any responsibility on our part." The chief chemist committed suicide. The following year, June 25, 1938, President Franklin Delano Roosevelt signed into law the Federal Food, Drug, and Cosmetic act, which among other things required drug manufacturers to provide scientific proof of safety before marketing new products.

Almost at the same time, the drug was changing history half way around the world, in China. While Mao Tse-Tung was on his "long march" General Chiang Kai-Sheck's army attacked and severely wounded Mao's two ablest generals, Cheng Zihua and Xu Haidong. Dr. Qian Xinzheng, then a 25-year-old graduate of the German Tongi University in Shanghai, who was also wounded in the same battle, summoned up enough strength to treat his commanders with Prontosil. He later stated that they would not have survived without it. Zhou Enlai subsequently placed these men in leadership positions in the new government of China.

After its extraordinary success in the treatment of the President's son, sulfanilamide zoomed to unprecedented popularity and was heralded as one of the greatest discoveries of the age. The American pioneer allergist, Dr. Bret Ratner, in his classic *Allergy, Anaphylaxis, and Immuno Therapy,* wrote: "The introduction of the Sulfonamide group of drugs as bacterial chemotherapeutic agents has reanimated the hope that in chemotherapy we may find the key to the conquest of disease. Their wide variety of usefulness, and the amazing manner in which they act in diseases otherwise largely fatal, places them in the forefront of the major advances of our generation. Not since the discovery of diphtheria antitoxin has any remedial agent been so universally and unequivocally accepted. The extent of the ramifications of their discovery is beyond our immediate comprehension."

Like many medical breakthroughs, sulfanilamide went through the time honored cycle of cautious acceptance then flagrant abuse. Surgeons grasped the new drug and sprinkled it into wounds and body cavities during surgery, like so much pepper on a salad. Sulfanilamide was absorbed from inside the body in almost fatal amounts and made patients worse rather than better. It also acted as an irritant and caused massive adhesions and blockages within the abdomen. It also failed to reduce the incidence of post-surgical infections when used as a topical dust. Nonetheless, every American soldier carried a packet of sulfanilamide powder in his first-aid kit during World War II.

The sulfa drugs work in an interestingly deceptive way. The earliest human trials with sulfanilamide were spectacular enough to imply that it was *bacteriocidal,* that it killed germs on contact. Subsequent experimentation proved instead that it was *bacteriostatic,* that it interfered with the vital functions of bacteria. It stopped their growth and multiplication, thereby interrupting their life cycles. Sulfa drugs bind, therefore remove, enzymes that bacteria need for life. Just as we humans need proteins in our diets, bacteria require para aminobenzoic acid in their diet. Their survival depends on it. Accordingly they manufacture enzymes to absorb and digest this vital substance from their surroundings. Sulfanilamide and the family of sulfa drugs are chemically analogous to para aminobenzoic acid; they attract and bind all the germs' available enzymes. Without enzymes, bacteria can no longer utilize para aminobenzoic acid; they stop growing, stop multiplying, and die out. We humans would suffer the same fate if all our enzymes were used up by eating indigestible grass rather than digestible wheat. This phenomenon, called **competitive inhibition**, had been observed and documented by bacteriologists as early as 1931.

When the paradigm for fighting infectious diseases was being put together it was difficult to grasp the concept of **competitive inhibition**. No one realized that preventing bacteria from growing and multiplying until the body could "call up the reserves," would work as well as killing them. As soon as scientists understood this phenomenon, it led to the discovery of PAS, para-amino-salicylic acid, a drug effective against one of our oldest killers, tuberculosis.

Tuberculosis probably established its hold over mankind about 7,000 years ago. Ancient engravings on stone depict wretched, wasted people coughing up blood. Mummified remains in predynastic Egyptian tombs show that many people died from it. And tuberculosis is mentioned in the Rig Vida, a 2500BC sacred text of India. The disease reached its peak in Western civilization during the 18th century, often snuffing out life at the height of musicians', authors' and visual artists' most productive years. Even the mighty Paul Ehrlich contracted tuberculosis, which he diagnosed by staining some of his own sputum. By 1850—and into the 20th century—the white plague, or consumption as tuberculosis was then called, was the leading cause of death.

When Robert Koch identified a rod shaped organism, *Mycobacterium tuberculosis*, he discovered the greatest killer in the world. It had a waxy coating that made it difficult to stain and therefore almost impossible to see, even with the best magnification—one of the reasons that the tubercle bacillus had eluded the microbe hunters for so long. He grew the germ in his laboratory and produced tuberculosis in animals by inoculating them with it.

On March 24, 1882, Koch presented his discovery of the tubercle bacillus to the Physiological Society in Berlin. All the eminent men of science were there, including Paul Ehrlich. Most of them left the meeting unconvinced because they could not see anything through the microscope. Koch gave Ehrlich several slides smeared with his newly discovered bacilli and asked the master if he could somehow find a stain that would reveal them. Ehrlich tried several approaches but did no better than Koch until, quite by accident, he made the culprit visible to the world.

Ehrlich typified the cartoon version of the disorganized, absent-minded professor. After one particularly frustrating attempt to stain tubercle bacilli he threw some poorly stained slides on the stove in his laboratory and proceeded to other matters. Somewhere along the line a charwoman lit a fire in the stove. When Ehrlich next looked at these slides the tubercle bacilli were beautiful stained. Heating apparently enabled the stain to penetrate the waxy coating of the bacilli.

Meanwhile, back in his own laboratory, Robert Koch happened to see some dark spots on a boiled potato he had inadvertently left on his workbench. Under the microscope these proved to be clusters of microbes—BUT ALL OF THE SAME KIND. Until this point bacteria were grown in soupy meat infusions, liquid broth, in test tubes, and all the inhabitants swam freely and co-mingled with the contents and each other. Getting a pure culture or exact sample of germs was virtually impossible. Koch quickly grasped the potential and implications of his off-hand potato observation and went on to grow bacteria in pure isolated colonies, first on potato slices then on jellied or solidified nutrient media, which we still use today. Koch's solid culture media and Ehrlich's stains subsequently became the tools that permitted Domagk, and others like Bernheim and Lehmann, whom we will meet further on, to outwit the fiendishly clever tubercle bacillus.

Koch's discovery of *Mycobacterium tuberculosis* triggered an intense and universal search for a cure, not unlike our attempt to find a cure for AIDS today. But over the next 60 years the wily tubercle bacillus evaded all attempts to confine or kill it. Sanitariums, by providing good nutrition, lots of rest, sleep and mild exercise, helped reduce the number of deaths, but, in the final analysis, it was through improved sanitation and aggressive public health measures that the spread of TB was stopped.

In 1890 Robert Koch proclaimed that he had succeeded in finding a cure for tuberculosis but it proved to be his biggest blunder. He made a vaccine by growing tubercle bacilli on a glycerin broth for several weeks, killed the germs with heat, and filtered off the nutritive medium—the same method used by Emile Roux to make diphtheria toxin and antitoxin. It seemed to work, experimentally, and in November 1890 Koch published a report stating that his tuberculin material could cure consumption. The Kaiser awarded him the Order of the Red Eagle for this breakthrough. But this proved premature. A few months later, in January 1891, the pathologist Rudolf Virchow published autopsy results on 21 people who had been treated with tuberculin yet died from tuberculosis.

That put an end to tuberculin as a vaccine but not as a valuable tool in medicine. Tuberculin is still in use today as a means of determining whether a person has been infected with tuberculosis—a prime example of how a seemingly ineffective substance can still prove useful under different circumstances. Herman Melville had it right when he wrote, "Mishaps are like knives that either serve or cut us as we grasp them by the blade or by the handle."

The turning point in the fight against tuberculosis came with the discovery of streptomycin in 1944 and PAS, para-aminosalicylic acid, two years later. Selman A. Waksman, a microbiologist at Rutgers University, first identified the antibiotic streptomycin in a soil microbe. It proved to be a remarkably effective weapon against tuberculosis but it sometimes affected the nervous system and caused deafness. This prodded investigators to intensify their search for a better chemotherapeutic agent. Such a search had in fact started shortly after bacteriologists figured out that competitive inhibition worked for sulfanilamide and would work with other compounds. PAS, then, may be considered a direct off-shoot of sulfanilamide.

In 1940 the English scientist Bernheim found that tubercle bacilli needed salicylates, a substance similar to ordinary aspirin, in their diets. Bearing in mind the competitive inhibition phenomenon seen with sulfanilamide, researchers felt that all they had to do was to find a salicylate-like compound that would fool the tubercle bacillus—a compound that resembled the salicylate it depended upon for growth and nutrition but once consumed would prove to be absolutely useless. Most persistent in this line of investigation was the renowned Swedish scientist, Jorgen Lehmann, of the Stahlgrenska Hospital in Göteborg. Leaning heavily on Bernheim's findings, Lehmann exposed his cultures of tubercle bacilli to myriad salicylate formulations, but over a period of six years nothing worked. His bacilli really did not care whether or not they had salicylates in their diets. Just short of the moment of despair, Lehmann realized that Bernheim had used a more virulent strain of bacilli than his. When he switched to a "pathogenic" strain, one that produced severe tuberculosis, his germs eagerly consumed their salicylates. Then, in one of the most exhaustive pursuits in science, he studied more than 50 derivatives of salicylates to find out which would fool the bacilli, which would replace the usable salicylate with an unusable form and thereby slow or stop bacterial growth. In 1946, PAS, para-amino salicylic acid, proved to be such an inhibitor. Trials on humans started in 1948, and the following year, 1949, it was approved for the treatment of tuberculosis.

We now come to one of the strangest, if not one of the luckiest coincidences in the annals of medicine. PAS was initially accepted with doubt and reservations. Many physicians felt that PAS lowered temperatures by virtue of its likeness to aspirin rather than by its ability to eradicate tubercle bacilli. Indeed, after extensive trials it became apparent that PAS alone was not a very strong agent. Streptomycin, which came on the scene almost simultaneously with PAS was also beginning to fall by the wayside. Tubercle bacilli developed such rapid resistance to the antibiotic as to render it virtually useless. It certainly would have been abandoned if not for what happened next. A study was designed to see whether PAS or streptomycin was the better anti-tuberculosis agent and under what circumstances either one might be used to advantage. When tested alongside controls who received no drugs, streptomycin proved to be the better of the two. But as soon as resistance appeared this advantage was lost. Somehow a number of patients received both PAS and streptomycin and results in that group exceeded all expectations. PAS PREVENTED THE DEVELOPMENT OF RESISTANCE TO STREPTOMYCIN. This coincidence saved streptomycin from the junk pile and gave medicine a truly effective combination against tuberculosis.

Dr. Domagk was also involved in the search for a tuberculosis cure and happened to be the mediator of another important accidental discovery. Working with colleagues Behnisch, Mietzsch and Schmidt, Domagk reported in 1946 that he was able to stop the growth of tubercle bacilli in the test tube with chemicals belonging to the thiosemicarbazone family. Because they were toxic and had so many harmful side effects when tried on humans, they were relegated to the role of "second-line drugs," to be used only when tubercle bacilli were resistant to PAS and Streptomycin.

But again providence intervened and a serendipitous observation showed how Domagk's thiosemicarbazone could be detoxified and even strengthened by linking it to a chemical related to the B-Complex vitamins. This gave us Isoniazid, which was our best anti-tuberculosis agent until the recent emergence of resistant strains.

In 1945, a year before Domagk reported the anti-tuberculosis effects of thiosemicarbazones, the intrepid biophysiologist Vital Chorine observed that part of the B-Complex vitamins, the nicotinamide fraction, protected rats from experimental leprosy. Since there is a close relationship between *Mycobacterium lepra* and *mycobacterium tuberculosis*, Dr. Chorine repeated those experiments, this time infecting guinea pigs with tubercle bacilli. Again nicotinamide prevented the disease. But this proved to be a hollow victory. What worked in the laboratory did not work with humans in the clinic. Somewhat later, when it became known that an active part of the nicotinamide fraction was isonicotinic acid, it was chemically combined with Domagk's thiosemicarbazone, giving us Isoniazid. It was first tested by Drs. Irving J. Selikoff and Edward H. Robitzek in New York's Sea View Hospital in 1952 when this author was a pediatrician there.

There is an old adage that says, "when an idea is ready to be born it finds many midwives." This was proven with the birth of Isoniazid. Two Czechoslovakian chemists, Hans Meyer and Josef Malley synthesized isonicotinic acid hydrazine, Isoniazid, in 1912, but could find no medical use for it. In the years 1950 and 1951 it was again simultaneously isolated and patented by Squibb and Hoffman-LaRoche in the United States and I.G. Farben Industrie in Germany.

Isoniazid also led us to the discovery of a most remarkable family of drugs, the antidepressants. When Isoniazid was administered to patients with tuberculosis, doctors noticed that they became happy, hyperactive, even euphoric. This was initially attributed to 'getting better'. But when the medicine was stopped, even when there was marked improvement in the patient's condition, they became sullen and depressed. It did not take scientists long to figure out that the medicine had some neuropsychiatric effects in addition to its antitubercular properties. Biochemists manipulated Isoniazid into *Iporoniazid* which became the fountainhead for antidepressant drugs known as monamine oxidase inhibitors, MAOIs. This family of drugs, after a brief decline in popularity, is now resurfacing and assuming its place as a major tool in psychiatry, especially in the treatment of depressions, panic anxiety, stress related problems, and other mental disorders.

By the 1980s, after the advent of streptomycin, PAS, and Isoniazid, tuberculosis all but disappeared. A death from TB stood out as a statistical curiosity. Prophets and soothsayers, citing the marvelous accomplishments of these drugs plus the steady flow of antibiotics, vaccines and genetically engineered products, predicted the demise of all infectious diseases. At that time we had on hand well over 25,000 different antibiotic products.

What the prophets failed to take into account was nature's capriciousness that makes germs resistant to discoveries almost as fast as they are made. By 1985, for instance, treatment with anti-tuberculosis drugs had lowered the death rate in the Western technologically advanced countries to less than 10 per 100,000 annually. However, in the last nine years, in the United States, there has been a 20% increase in the number of tuberculosis cases and a concomitant rise in deaths. Tuberculosis today causes more adult deaths in the world, about 8 million new cases each year, than any other single infectious agent. Phoenix-like, tuberculosis survived its own death. Beginning in 1987, acute rheumatic fever also began to reappear throughout America, particularly in Pennsylvania and Utah.

The spread of the immunodeficiency virus and the increase in the number of cases of Acquired Immunodeficiency Disease, AID, account for many of the new cases of tuberculosis. Patients with immunosuppression, of course, cannot generate antibodies and are completely powerless to combat all infections. But more importantly there has been an increase in opportunities to spread the disease. Inmates closely-quartered in prisons, homeless people packed into shelters, and the influx of immigrants harboring drug-resistant strains of tuberculosis, have added fuel to the fire. To a lesser extent, the resurgence of TB can be attributed to our cavalier attitude toward the disease and failure to enforce the isolation measures that proved so effective in the past. Hospitals caring for active cases of tuberculosis were often at fault. Many had rooms with positive air pressures relative to other parts of the hospital; active tubercle bacilli were spread from the patient's rooms to corridors and other parts of the building. Patients who had been assigned to isolation rooms were often found in hallways, lounges, gift shops and other common areas. Problems were further compounded when multi-drug-resistant tuberculosis was transmitted to health care workers.

There is a strong possibility that the emergence of drug-resistant bacilli could rekindle new epidemics. In the 1950s Marjorie Pyle, of the Mayo Clinic, described the appearance of bacteria resistant to streptomycin. Two years later, in 1952, Joshua and Esther Lederberg demonstrated how the genetic material in the chromosomes of bacteria changed so that they could adapt to changing environments; in essence they became resistant to antibacterial substances and then produced progeny that were equally resistant. In 1970 Hugo David, an investigator with the Center For Disease Control, showed that drug resistance in tubercle bacilli resulted from the spontaneous and random occurrence of mutations in bacterial chromosomes. Many wild strains of tubercle bacilli proved to be resistant to anti-tuberculosis drugs even before contact with them.

Tuberculosis is not as highly contagious as measles but it does infect about one-third of the people who come into contact with an active case. The most common route of tuberculosis transmission is through the air. Coughing creates bacteria-laden droplets five to ten microns in size which are easily inhaled. In these infected people, about 90% mount and sustain an immune response that prevents development of the disease. In the remaining 10%, where the disease becomes established, the patient generally harbors large numbers of normal non-resistant tubercle bacilli plus a small number of resistant mutants. While susceptible bacteria are killed off by drugs, the drug-resistant bacilli continue to grow and multiply, resulting in severe, overpowering, fatal disease. Also, when patients fail to take their medicines or take them sporadically, they just kill off the susceptible germs and allow the other more hardy bacteria to reproduce; these then do not respond to treatment when drugs are administered again in proper dosages.

There are several techniques for identifying drug- resistant strains of tubercle bacilli. The most common one entails seeding culture plates with bacteria, flooding them with anti-tuberculosis drugs, and observing if any colonies survive. An intriguing new method undergoing development should make resistant bacilli light up like a neon sign. Samples of tubercle bacilli are subjected to a battery of anti-tuberculosis drugs then exposed to a virus that infects them. The virus is genetically engineered to carry the gene for Luciferase, an enzyme that fires up Luciferin, the material that enables fireflies to emit light. When Luciferin is added to a questionable bacterial sample, resistant bacteria light up, since only live bacteria, those resistant to anti-tuberculosis agents, can pick up the gene for the needed enzyme. The absence of light indicates drug susceptibility.

Isoniazid is still one of our most effective weapons against tuberculosis. When the infecting strain of M. Tuberculosis is vulnerable to Isoniazid and the medication is taken as prescribed, the drug can prevent latent tuberculosis infection from progressing to clinically active disease. However, because of the constant threat of drug resistant strains, the initial phase of treatment of tuberculosis requires several drugs and antibiotics plus Isoniazid.

In addition to leading to the discovery of the anti-tuberculosis drugs, PAS, and Isoniazid, sulfanilamide also pointed the way to the development of drugs used in treating high blood pressure and heart disease. After sulfanilamide came into common usage, several reports called attention to the fact that it increased the excretion of salt in the urine. Dr. Karl H. Beyer, Jr. picked up on this oddity and directed his team of research chemists at Merck Sharp & Dohme pharmaceutical company to find out how and why this happened. The search for a drug that would rid the body of excess water had been going on for over a decade. Whereas other investigators tried to increase the output of urine, Dr. Beyer decided to focus his efforts on the excretion of salt, knowing that water would follow the salt. In 1955, by chemically altering sulfanilamide, he hit upon chlorthiazide, the drug that revolutionized the treatment of heart disease and high blood pressure. Chlorthiazide forces kidneys to eliminate salt. As salt leaves the body it carries water with it. This reduces the load on the heart and blood vessels, lowers blood pressure, and rids the body of water that causes swelling of the feet and puffiness in other parts of the body. After extensive clinical trials by more than 1,000 physicians, both here and abroad, it was introduced as Diuril(tm) in January 1958. Since then we have seen a whole group of drugs called Thiazides or Thiazide congeners, that have come to us through manipulations of sulfanilamide.

In his later years Dr. Domagk turned to the greatest challenge of all, the treatment of human cancer. Although success evaded him, he clearly understood the intricacies of cancer and time has borne out his admonition: "One should not have too great expectations of the future of cytostatic agents." Domagk also had a keen perception of the ills that still beset society. He wrote, "If I could start again I would perhaps become a psychiatrist and search for a causal therapy of mental disease which is the most terrifying problem of our times."

In October 1939, American, French and British scientists nominated Dr. Gerhard Domagk for the Nobel Prize in Medicine for his discovery of sulfanilamide. When the telegram announcing the award arrived in Berlin the Kultur

Ministerium shot back a curt note to the Swedish Foreign Office declining the honor. Hitler had forbidden the acceptance of Nobel prizes. The Nazis were still fuming over Carl von Ossietzky who had been awarded the 1936 Nobel Peace Prize but, because of his opposition to the ethics and morality of the Nazi regime, had been put away in a concentration camp. In language as subtle as an iron fist, Dr. Domagk was officially warned not to accept the prize if he knew what was good for him and his family.

The 44-year-old doctor aged overnight. Withdrawn to his darkened study, he wrestled with the alternatives: forfeiting the Nobel Prize meant giving up a place alongside the immortals of medical history and prize money worth $35,000; accepting it would put him and his family in danger of imprisonment, or worse.

Dr. Domagk's three sons, Gutz, Wolfgang and Jung felt helpless in their mournful house and retreated within themselves or stayed in their rooms. Hildegarde, his teen-age daughter, wept inconsolably as she relived those perilous days when "Papa" used his new and as yet untried discovery to save her life. By default the task of ameliorating the doctor's anguish fell into the hands of his wife, Gertrud, the woman who had married this ingenuous young doctor 14 years before and remained steadfast throughout the roller-coasting emotions that beset a scientist's life—ups with discoveries, promotions and honors, downs with corporate intrigues, academic infighting and failed experiments. She constantly reminded her husband that his greatness would in no way be diminished by not claiming his Nobel Prize. It was he who had handed mankind the Silver Bullet that had been postulated but never found by Paul Ehrlich, the era's foremost medical scientist.

Dr. Domagk wanted to believe his wife. He composed a letter to the rector of Caroline Institute, the administrator of Nobel Prizes, in which he acknowledged the honor bestowed upon him but regretted that he could not go to Sweden until such time as his government allowed him to do so. A copy was sent to the German Ministry for Foreign Affairs just to make sure that there were no misunderstandings or thoughts of subterfuge. Rather than assuaging the authorities it disturbed them. The doctor was promptly arrested at gun-point and jailed. After nerve-racking interrogations by the SS he was released only to be arrested again. When he formally and officially declined the Nobel Prize by signing a letter drafted by Nazi officials, the harassment stopped.

What worried Dr. Domagk most was not the loss of a medal and prize money but that his luck was running out, that providence was asking him to pay up for the favors and advantages so lavishly bestowed on him throughout the research that led to sulfanilamide. For indeed, fortune had smiled on him in more ways than one: The blind screening of compounds has seldom led to the discovery of a new biodynamic substance, yet that is what happened in the hands of Dr. Domagk. He achieved success even though he relied upon a false premise and an erroneous explanation of how and why his new drug worked. And his choice of germs for test purposes was a lucky 1-in-100 long shot.

How close Domagk and the world came to missing this key to the conquest of disease has only recently been appreciated. We now know that the number of bacteria resistant to sulfanilamide and related drugs far exceeds those that are susceptible. A short time ago, for instance, tabloids in England blazed headlines reading "KILLER BUG ATE MY FACE". They were referring to a devastating disease caused by a group of virulent streptococci, bacteria growing in long chains, that make a substance capable of disintegrating and liquefying the tissues that hold the body together. The infection spreads rapidly, sometimes too quickly for antibiotics to work. It was such a vicious germ that killed Muppet creator Jim Hensen in 1990. Had Dr. Domagk used this particular virulent strain or type B hemolytic streptococci, which are resistant to sulfa drugs, instead of the vulnerable type A, his experiments would have ended in dismal failure.

Dr. Domagk was also luckier than scientists before him who had failed to recognize antibiotics within their grasp. Ehrlich concentrated much of his research on the effects of organic chemical compounds on parasites and bacteria in test tubes or on plates with culture media rather than on infected animals and humans. He could not see the full potential of chemotherapy, that is, the means by which chemical or other agents stop germs from growing in living organisms. He abandoned this line of research short of a breakthrough. The brilliant American biochemist, Michael Heidelberger, suffered a similar fate.

In 1919, four years after Ehrlich's death, five years before Calvin Coolidge Jr. succumbed to blood poisoning, and sixteen years before Domagk reported his discovery of sulfanilamide, Michael Heidelberger, a young research assistant at New York's Rockefeller Institute, came within a hair's breadth of making the very discovery that catapulted Dr. Domagk to fame. Heidelberger had read about the 1908 discovery of sulfanilamide by Paul Gelmo and thought that this new compound might have some germ-killing potential. He synthesized sulfanilamide in his laboratory and went so far as to prove that it was well tolerated by warm-blooded animals. But his crucial tests failed. Relying upon the great microbe hunters, Louis Pasteur, Paul Ehrlich, Robert Koch, Emile Roux and Emile Von Behring, who insisted that bacteria must be killed in order to prevent and cure diseases, Dr. Heidelberger focused his sights only on bacteriocidal or lethal effects of sulfanilamide. On finding that the bacteria in his test tubes were not being killed when sulfanilamide was added to them, he concluded that the compound would be useless in fighting infections. Had he performed his experiments on live animals, as Domagk did, the course of history would have been different.

Gerhard Domagk died April 24, 1964, at the age of 69, in Burgberg, Germany. He lived long enough to see the advent of penicillin and watch it push his sulfanilamide aside, almost to abandonment. He undoubtedly suffered through the indignities that Michel de Montaigne wrote about in the 16th century: Whenever a new discovery is made people first say, "It's probably not true." Thereafter, when it has been proven to be true they say, "Yes, it may be true but it's not important." Finally, when sufficient time has elapsed to fully demonstrate its importance, they say, "Yes, surely it's important, but it is no longer new."

Yet in recent years, as bacteria have become smarter and more resistant to antibiotics, there has been a resurgence in the use of the sulfa drugs. They are still used for urinary tract infections, some types of pneumonia, dysentery, ear infections and skin disorders.

In 1947 Dr. Gerhard Domagk finally went to Sweden to claim the Nobel Prize that had been denied him by the Nazis, but all he received was a medal because, by then, the prize money had disappeared. To compensate for this loss providence has seen to it that the name Domagk will forever reign alongside the immortals of medicine. When the time comes for historians to record mankind's fight against infectious diseases they will logically divide the struggle into four epochs: The first will belong exclusively to Louis Pasteur who established the connection between germs and disease and developed the mechanics of putting laboratory acquired knowledge into practical use. The second will belong to Paul Ehrlich, Robert Koch and Gerhard Domagk for handing us the concept as well as the end-products of chemotherapy. The third will include the penicillin and antibiotic pioneers, Sir Alexander Fleming, Sir Howard W. Florey, Dr. Ernst B. Chain, Selman A. Waksman and all the other scientists who gave the world natural and synthetic antimicrobial and antiviral agents. The final epoch, still in progress, will be shared by Linus Pauling, Francis Crick and James Watson, who unraveled the structure of our nucleic acids, and the genetic engineers who apply this knowledge to gain an understanding about AIDS, cancer, viral infections, malaria and third world scourges.

PARADIGM POINTERS

A. Paradigm shifts, like volcanic eruptions, appear only after many years of silent pressure-build-ups. It took almost a half century for scientists to build up enough pressure to dislodge the old theories that relied on killing disease producing germs outright with new concepts based on **competitive inhibition**. Most of the antibiotics and antibacterials we enjoy today resulted from this shift.

B. Scientists have to be especially careful changing paradigms. A wrong interpretation of results, even in a flawless experiment, could be disastrous. Here is a situation not too different from Dr. Domagk's first trials with sulfanilamide, except that rabbits and staphylococci were used instead of mice and streptococci.

In 1908 Dr. Saltykow at the Kanton Hospital in San Gallen, Switzerland, injected staphylococci into rabbits and observed yellow spotted thickening on the valves of the heart. He concluded that staphylococcal infections produced heart valve damage and stated this, categorically, in his scientific paper. But he was wrong. He was really seeing fatty deposits in the heart resulting from cholesterol-rich foods that were being fed to his experimental animals—the very same changes that go hand-in-hand with cholesterol and high fat diets that doctors are warning people about today. When these changes appear in large arteries of the body they lead to atherosclerosis, or hardening of the arteries, a very old disease, unmistakably present in Egyptian mummies. Although hardening of the arteries was painstakingly described by the great anatomists of Padua, Italy, in the 1700s, and by pathologists throughout the 19th and 20th centuries, no one had any clues as to how or why this condition developed.

When the laboratory rat became commonly available, it was used in a wide variety of experiments to explain this phenomenon. The experiments always failed. Laboratory rats on high fat diets do not develop atherosclerotic changes. In one of those lucky breaks, where a discovery in one discipline unlocks the secrets in another, science found a way to get around this roadblock.

At the turn of the 20th century the Russian scientist, Ignatowski, working at the laboratory of Diagnostic Clinical Science in St. Petersburg, tested different feeds on animals. In a lecture presented in 1907 to the Military Medical Academy, titled "To The Question of the Influence of Animal Nutrition on the Organism of the Rabbit," he proved that feeding certain diets with a high fat content, which we now know contain large amounts of cholesterol but was unrecognized as such at the time, produced hardening of the large arteries in the rabbit. Later, Ignatowski said that proteins did the same thing. However, what he called protein was really a mixture of milk and egg yolks, essentially more cholesterol. Serendipitously, Ignatowski gave us the rabbit model which is still standard for heart and arterial disease research. If scientists had continued to utilize rats for experiments they would have missed the harmful effects of high fat intakes and we could still be enjoying unlimited butter, eggs, fat marbleized steaks, and second helpings of ice cream.

POST SCRIPT

1. Acute rheumatic fever has come back despite having been virtually stamped out. Beginning in 1987 it began to appear throughout America, particularly in Pennsylvania and Utah. There is still no explanation for this recurrence, but even more disconcerting, is the fact that this is happening even though we now have many more antibiotics than we

had when sulfanilamide was first tried on President Franklin Delano Roosevelt's son.
2. Statistics show that tuberculosis started its decline long before there were any antibiotics. This was due to better nutrition, sanitation, isolation of active cases, and good public health administration. The best example we have of how well public health measures can work is seen in the events surrounding the 1854 cholera epidemic in England. John Snow, a shy, isolated, celibate, and zealous crusader against the use of alcohol, noted that more than 500 fatal cases of the disease had been reported within a ten day period, most of them centered at the intersection of Cambridge and Broad Streets, Golden Square, London. Although the nature of bacterial diseases was unknown at that time, John Snow intuitively reasoned that the disease was being transmitted by some element related to the water-well at that site. He removed the handle of the Broad Street pump, and the epidemic ended.
3. There are numerous examples of scientists, long before Almroth Wright and his ilk, who allowed their personal biases to obscure their vision. Joseph Priestly discovered oxygen but refused to believe it. He later wrote:

"... it was very slowly, and with great hesitation, that I yielded to the evidence of my senses. When I... compare my last discoveries relating to the constitution of the atmosphere with the first, I see the closest and the easiest connection in the world between them... That this was not the case, I attribute to the force of prejudice, which unknown to ourselves, biases not only our judgments... but even the perceptions of our senses— and the more ingenious a man is, the more effectually he is entangled in his errors; his ingenuity only helping him to deceive himself, by evading the force of truth." Roughly a hundred years later, physiologist Claude Bernard said the same thing: "It is the things we do know that are the great hindrance to our learning the things that we do not." Restated: Accepted paradigms are often the greatest hindrance to our learning new things.

chapter 2

A LESSON FROM THE WIZARD OF OZ

While jogging at Camp David on May 4, 1991, the President of the United States, George Bush, suddenly developed shortness of breath and felt his heart racing madly out of control. His physicians were summoned and, fearing the worst, they rushed the almost 67 year old president to Bethesda Naval Hospital. To everyone's relief, an over-active thyroid gland, not heart-trouble, was found to be the cause. The official diagnosis was hyperthyroidism , Graves' disease, eponymous of Dr. Robert Graves who, in his 1835 Clinical Lectures described the classic picture of the condition: rapid heart rate with an irregular beat, fatigue, appetite disturbances, intolerance to heat, insomnia, agitation, apprehensiveness, and a tremor often severe enough to interfere with writing. Abnormally prominent or bulging eyes is one of its occasional, baffling, complications. Before there were drugs to control this condition, surgery was the only recourse. But it was always a tricky business. Removing too much thyroid gland led to hypothyroidism—a lethargic, zombie-like state where all body activities slow down—and a life-long dependence on thyroid suppliments. Removing too little did nothing to help the patient. Fortunately for President Bush and countless other victims of super-active, run-away thyroid glands, surgery became less of an imperative with the discovery of drugs that could counteract such conditions.

Around 1943, two groups of investigators, unknown to each other, almost simultaneously stumbled across chemical compounds that subsequently proved to be effective in treating hyperthyroidism. The success of one of the groups arose from Dr. Domagk's discovery of sulfanilamide and attempts to modify and improve it. Apparently, when sulfanilamide came on the market in January 1935, everyone understood that this 'miracle-drug' was taken by mouth, it was absorbed into the blood stream from the stomach, and was carried to diverse parts of the body where it eradicated infections. "If we could make sulfanilamide insoluble" scientists reasoned, "it would not be absorbed from the stomach but remain in the digestive tract where it could cure diarrhea and intestinal infections. By sterilizing the inside of the gut it would greatly reduce the chances of infection following stomach and intestinal surgery." In due course an analog of sulfanilamide called sulfaguanidine was developed. It was relatively insoluble, it stayed in the intestinal tract and effectively inhibited bacterial growth there, as predicted. But it still had to be tested before it could be approved for general use. The person entrusted to do the testing was Dr. Elmer Verner McCollum, America's foremost biochemist, the discoverer and co-discoverer of several of our most important vitamins. McCollum, together with his team, doctors J.B. and C.G. MacKenzie, at Baltimore's Johns Hopkins University School of Hygiene and Public Health, fed sulfaguanidine to laboratory animals and began looking for untoward effects. No grave toxicity was noted but the animals did develop unmistakably large neck swellings. These proved to be goiters, enlargements of the thyroid glands.

While Dr. McCollum's studies on sulfaguanidine were going on, another, independent group of researchers, also in Maryland, came across similar thyroid enlargements in animals while looking into the possibilities of using the chemical compound phenothiourea as a rat poison. In 1941 they reported that phenothiourea causes goiters in laboratory animals; in 1943 McCollum reported that sulfaguanidine did the same thing. Somewhere between these two reports, Dr. Jeantet, in France, observed that workmen in contact with the chemical aminothiazole developed enlarged thyroid glands. He too issued a scientific report, which, when combined with those originating in the United States, aroused the curiosity of investigators around the world. Was this sheer coincidence or was there some common factor that could explain why three different chemical substances caused goiter? It was soon proven that they all contained thio molecules, segments in a chemical compound where sulfur replaces oxygen, and this was responsible for the effects on the thyroid gland. Within the next two years, by 1945, E. B. Astwood was able to show that thio molecules interrupted the production of thyroid hormones by the thyroid gland and had the potential to control hyperthyroidism. In rapid succession, a number of thio-containing chemical substances were developed. They became the precursors of drugs that have, over the years, given us mastery over the thyroid. Propylthiouracil, PTU, emerged as the drug most commonly prescribed in the United States for Grave's Disease. We still have to be careful in using these drugs. Some block thyroid hormone production so completely that the gland, in a futile attempt to meet the body's demand for thyroid hormones, becomes abnormally large, duplicating the type of goiters produced by disease and iodine deficiencies.

The thyroid gland, just below the Adam's apple in the neck, is shaped liked a butterfly with one wing on each side of the windpipe; it usually weighs less than an ounce. It uses iodine to make thyroxin and triiodothyronine, two hormones that regulate the speed with which the body transforms nutrients into energy.

The human requirement for iodine, normally 100 to 250 micrograms a day, is taken into the body with food and water. It is absorbed into the blood from the gastrointestinal tract then selectively concentrated in the thyroid gland, where it is bound to hormones. The thyroid releases its hormones directly into the blood stream. The kidneys excrete iodine but constantly make adjustments in accordance with the amounts taken into the body with food. If for any reason a person's thyroid gland stopped working, he or she would probably not feel any different because the gland stores a one to three month's supply of hormones. Growth and activities of the thyroid gland are regulated by the thyroid-stimulating hormone, TSH , which is made by the anterior pituitary gland located at the base of the brain. When there is a shortage of iodine in the body the output of thyroid hormones falls off, less appears in the blood, and this signals the pituitary gland to release more TSH. In response to the stimulus of TSH the thyroid gland increases in size, often to that of a soft-ball; this is a goiter.

Besides disease and drugs, there is another strange cause of thyroid swelling which accidentally came to light about thirteen years before McCollum linked goiters to chemicals. Dr. A. M. Chesney and his co-workers at Johns Hopkins University noticed that their colony of rabbits, which were normal at the time of purchase, developed extremely large goiters for no apparent reason. After sifting through the possibilities of infection, cancer and vitamin deficiencies, the culprit was found. It turned out to be a diet of cabbage. In his 1928 report on this phenomenon, Dr. Chesney presented the long anecdotal history of similar observations—natives in many parts of the world said that certain foods caused neck swelling—but now there was proof-positive that such a relationship really existed.

Scientists all over the world, especially those in New Zealand, picked up this thread of information and soon succeeded in producing goiters in laboratory animals by feeding them cabbage, brussels sprouts, turnips and the seeds of cabbage, rape and mustard plants. These all belong to the mustard family, *Brassicace Cruciferae* therefore the term brassica-seed goiter is commonly used to describe this condition. Today, rape seed and rape oil are immensely popular in

China, India, Japan, Europe, Poland, Sweden and parts of Germany, and could play a role in goiter formation in people living in those countries.

Soy beans, rutabaga roots, ground nuts, cassava (manioc), sorghum, yams, maze, peas, beans, apricots, prunes, cherries, sugar cane, almonds, and bamboo shoots have also been found to induce goiters, but the effect is noticeable only when the diet is limited to these foods and when there is a deficiency of iodine in the body. Shortly after the introduction of soy containing baby-foods and soy-based formulas into the United States pediatricians began to see puffy thyroids in infants. These subsided when soy bean preparations were removed from the diet. Does this, then, mean that eating any of the foregoing foods will cause goiter? Not usually. The only people who are seriously affected are those who live in parts of the world where soil and water are iodine-deficient and their diet is limited to cassava (manioc), as in certain parts of Africa.

Iodine, the element so inextricably tied to thyroid size and function, was also discovered quite by accident. When Napoleon began his rampage through Europe the British navy blockaded France and the armies of Prussia and Austria cut off the supply of chemicals needed to make gunpowder. The Emperor retaliated by ordering his scientists to make saltpeter, potassium nitrate, from local sources. Traditionally, saltpeter came from wood ashes and decayed vegetable matter but French scientists soon found a better and cheaper source in sea weed that washed up in abundance on the Atlantic coasts of Normandy and Brittany. At special installations called *nitre plantations* seaweed was burned in large metal vats and nitrates for gunpowder were extracted from the ashes. In 1811, Bernard Courtois, who had studied chemistry before taking over his father's nitre plantation, happened to come upon the scene while the vats were being cleaned with hot acid and watched as intense violet colored vapors arose from the muck and deposited lustrous metal-like crystals on the cooler parts of the vessel. Courtois performed a few elementary studies on these crystals and found that they combined with hydrogen and phosphorous but not with oxygen. They formed an explosive compound when mixed with ammonia. Lacking proper facilities and expertise, Courtois asked his friends, professional chemists Charles Bernard Desormes and Nicholas Clement to carry on the studies, to see if they could shed some light on the nature of these crystals. They accepted the challenge and within a year were prepared to announced that they had discovered a new element. The brilliant French chemist Joseph Gay-Lussac immediately corroborated this finding and christened the new element *iode*, a Greek word meaning violet in color. Mysteriously, some of the crystals found their way into the hands of the enemy. In 1813, England's Sir Humphry Davy devised ingenious experiments to confirm that the substance was indeed a new element and proceeded to anglicize its name to iodine, conforming with its chemical cousin, chlorine.

Hawkers and hucksters instantly and shamelessly exploited this marvelous new discovery. No part of the body was spared from iodine. It was inhaled as a vapor. It was put into pills, candy, cigarettes, soaps and hair tonics. It was advocated as a contraceptive. People were so impressed with what seemed to be a divine blessing that they strung little bottles of iodine around their necks, like amulets, to ward off evils and disease. In Geneva the iodine droppingbottle replaced perfume and jewelry in the evening salons. In France, it was common practice for over two centuries, to keep iodized salt and plain pure iodine in open bottles at schools and in bedrooms. In hospitals, strips of gauze saturated with an iodine derivative, iodoform, hung from the ceiling like flypaper.

But iodine was not accepted into medicine with similar fervor. At first it was warmly embraced, then used improperly and finally discarded. Eight years after Courtois discovered iodine, Andrew Fyfe, a lecturer in chemistry at the University of Edinburgh, reported that common sea sponges also contained large quantities of this new element. Jean-

Francois Coindet, a prominent physician in Geneva, Switzerland, who treated many goitrous patients with the centuries-old remedies of burnt sponge or seaweed, put two and two together, and correctly guessed that the violet crystals were the active ingredient in his crude remedies. In 1820 he reported great success in treating thyroid conditions with pure iodine instead of burnt sponges but cautioned against using it indiscriminately. He stressed that when given in excessive doses in the treatment of goiter, iodine produced annoying and harmful symptoms, like fast pulse, palpitations, increased appetite, emaciation, insomnia and muscular weakness. We now know that people can become hyperthyroid and show the exact same symptoms described by Coindet when supplements of iodine are added to the plentiful amounts already in food and water. Unfortunately, iodine came into prominence long before the physiology of the thyroid gland was understood. It was often prescribed excessively. The hyperthyroidism and skin rashes it produced soon led to its abandonment as a remedy for thyroid disorders.

Throughout the 19th century scientists grappled with goiters, often clumsily, often with great insight. On the clumsy side we see bacteriologists , at the time when the germ theory of infectious disease became generally accepted, desperately trying to find the organisms responsible for thyroid disease. Of course, they failed. On the insightful side we see the multitalented mining engineer, Jean-Baptiste Boussingault, who, as early as 1831, reported that the natives in Colombia, where he was working at the time, had a variety of neck swellings but those who used salt from certain natural deposits with a high iodine content did not show such disfigurements. He related how he tested different sources of salt and proved that the iodine content was inversely proportional to the number of goiters encountered. The affliction was seldom seen where iodine levels were high but often seen where levels were low. Nonetheless, this knowledge remained unrecognized and unused.

The fate of French chemist, Gaspard Adolphe Chatin, who, in the mid-19th century, proved that goiter was related to iodine deficiencies, was no better than that of Boussingault. First he carefully and conscientiously analyzed the iodine content of the water, soil and vegetables in many parts of Europe. Then he compared the number of cases of goiter with the levels of iodine in these areas. Wherever his measurements showed deficiencies in iodine he also encountered more than average number of cases of goiter. But when Chatin declared "Iodine deficiency is the principle cause of goiter," his contemporaries rejected this as nonsense. The French Academy of Sciences concluded that there was no correlation between goiter and deficient intake of iodine. And the great and famous founder of pathology, Rudolph Virchow, as late as 1863, rejected Chatin's claim. The pioneering surgeon and 1909 Nobel Prize winner, Emil Theodor Kocher, who was instrumental in designing the original operations for thyroid disease, also discouraged the use of iodine.

Interestingly, Chatin's data and evidence were re-examined in the 1960's and his values almost exactly matched those obtained with modern apparatus and advanced techniques. Both studies placed the concentration of iodine in sea water at about 50 to 60 micrograms per liter, the same as in human serum—a further clue that we humans emerged from the sea.

Chatin was finally vindicated by the German pharmaceutical chemist Eugen Baumann, who, in 1895, proved that iodine was invariably hooked up to a protein in the thyroid gland and was essential in its normal physiology and hormone production. He also demonstrated that thyroid extracts could sustain patients whose thyroids had been surgically removed. Edward Kendall, on Christmas Day in 1914, at the Mayo Foundation, purified, isolated, and discovered the active principle of the thyroid gland, the hormone thyroxin. C.R. Harington later synthesized it.

Iodine, as a remedy for goiters, went through periods of acclaim, rejection, and ultimate recognition, but its use as an antiseptic never suffered this opprobrium. Tincture of iodine, the antiseptic that everyone remembers from childhood, gained favor early and kept its preeminent position for more than a century. The English surgeon John Davies used iodine in alcohol to disinfect wounds as early as 1839. It remained a popular and effective antiseptic for battle injuries during the American Civil War and again in World War I but was supplanted by newer and less irritating antibacterials in World War II.

The most innovative use of iodine came as a consequence of work on the atomic bomb. In 1934, Madam Curie's daughter, Irène Joliet-Curie and her co-worker husband, Frédéric Joliet, created elements that were radioactive. That same year Enrico Fermi embarked on his systematic production of radioisotopes, including radioactive iodine. In 1946, an iodine radioisotope, I^{131}, became available from the nuclear reactor at Oak Ridge, Tennessee.

Normally the thyroid gland accumulates most of an administered dose of I^{131} within 24 hours. Abnormally high uptakes are seen in people with iodine deficiencies or hyperactive thyroid glands; low uptakes are seen in people taking antithyroid drugs and those with hypothyroidism. Sensitive instruments, when placed in front of the neck, over the thyroid gland, can record radioactivity and demarcate the exact location, size and function of the gland. This technique has enabled physicians and researchers to quantify iodinated compounds which are present in amounts too small to be detected by biochemical means and has proven invaluable in tracing hormone distribution. In 1952 Dr. Rosalind Pitt-Rivers employed iodine radioisotopes to discover triiodothyronine, the thyroid's other hormone besides thyroxin.

In select cases of thyroid over-activity, large doses of radioactive iodine may be used to destroy part or all of the gland. It is the treatment of choice in patients who cannot tolerate antithyroid drugs or surgery but is not administered to children since it may cause cancer of the thyroid in later life. Because of its relatively long half-life, I^{131} has largely been replaced in medical practice by I^{125}. In essence, radioactive isotopes of iodine serve as both test and treatment.

Goiter, as mentioned earlier, refers to a puffy swelling in the lower part of the neck due to thyroid gland enlargement. It may be caused by drugs, diet, disease, or lack of iodine. When it appears in a large segment of the population where there is a known deficiency of iodine in the soil, water and diet, it is called endemic goiter. This type is still found in the Alps, the Himalayas, the Pyrenees, the Carpathians, the entire Andean chain, and the mountains of New Zealand. Himalayan Goiter, though nominally an endemic goiter, has acquired its own name in deference to the unusual severity of iodine deficiency in that part of the world.

The effects of iodine deficiencies are generally reversible in adults but in the fetus they are catastrophic. Babies in the womb, when deprived of iodine, commonly end up as cretins, mentally retarded, deaf-mute, dwarfs. These infants do not feed properly, they have a large protruding tongue, umbilical hernia, thick dry skin, and diminished or absent reflexes. Later they develop a hoarse cry, the eyes appear to be set wide apart in a round face, they become constipated and their teeth erupt later than normal.

Studies have shown that the mother's thyroid hormones do not cross the placenta. Fetal thyroid function is completely autonomous and thyroid hormones in the fetus are not influenced by hormones of maternal origin. The fetus, however, does need iodine to make its own thyroid hormone and its only source is from the mother's contribution to the placental circulation. In endemic areas, where the supply of iodine is meager, the mother's thyroid captures the little iodine that enters her body so that none gets to the fetus. Since the growth of the skeleton and the central nervous system in the developing baby are dependent on thyroid hormones, they stop or are drastically affected. These disturbances are permanent and irreversible and explain the abnormalities found in cretinism.

Endemic cretinism, like endemic goiter, applies to widespread cretinism in a population living in a area of known iodine deficiency. The Pan American Health organization (PAHO) in 1963 defined a cretin as: an individual with irreversible changes in mental development born in an endemic goiter area and exhibiting a combination of neuromuscular disorders, speech and hearing disorders associated with deaf-mutism, impairment of somatic (muscle and skeletal) development and hypothyroidism. In areas with severe iodine deficiency, as Central Africa, the incidence of congenital hypothyroidism and cretinism is about 500 times higher than that in the rest of the world. Oddly, cretins from endemic areas often show normal thyroid function by radioactive iodine uptake tests. It seems that these tests do not necessarily reflect the condition which existed in fetal or prenatal life when the irreversible damage was inflicted. Also, we cannot account for the fact that only some and not all babies deprived of iodine in the fetal stages become cretins.

The term cretin came into usage long before its cause was even suspected. The Runer Musterbuch of 1215 depicts a figure with a large goiter, stupid expression and brandishing a fool's staff. The Encyclopedia of Jacques de Vitry, published in 1220, describes people who have goiters and cannot hear or speak. Diderot's Encyclopedia, in 1754, contained this definition: an imbecile who is deaf, dumb and has a goiter hanging down to the waist. Some lexicographers maintain that the name comes from Pauvre Chretien or Bon Chretien, a harmless or innocent waif.

As early as the middle of the nineteenth century authorities in Europe recognized the deplorable economic and social ravages of goiter and cretinism. In France alone, in 1874, the goiter commission estimated there were 500,000 goitrous people and 120,000 cretins and cretinoid idiots within its borders. Kocher reported that 80% to 90% of school children in Berne, Switzerland, had goiters, more than in the rest of Europe. As a result of this finding Swiss physicians and scientists took an early lead in the study of thyroid abnormalities and were among the first to join the United States in its advocacy of iodized salt. The Swiss also made hearing tests compulsory in schools and dramatically proved that hearing losses in the population decreased concomitantly with the nation-wide introduction of iodized salt. Physicians have come to regard bilateral loss of hearing with intact ear drums as a reliable marker for cretinism.

Geologists agree that intense glaciation over long periods of time is the underlying cause of the depletion of iodine in mountainous countries. Apparently, iodine, like chlorine and bromine, was present all over the earth during its primordial development. Glacial flow, rain and melting snow tend to wash it away with topsoil. Rivers carry it to the ocean. It returns to earth with dust and rainfall, only to be lost again when deluges remove iodine rich surface soils. In the Antarctic's Shackelton Glacier, for example, where permafrost prevents the loss of surface soils, tests reveal that the surface has much more iodine than the deeper layers.

In Norway and Sweden the incidence of goiter coincides with the direction of glacial flow. Fig 2.1 The topography of these countries is tent-like. On the steep-side where the flow is highest, iodine is washed away and goiters are seen with great frequency. On the other flatter-side of the 'tent' where glacier flow is minimal, goiters are rarely encountered.

All mountainous areas of the world where torrents of water wash iodine out of the soil have extremely high rates of hypothyroidism. A World Health Organization (WHO) survey documented that the overall prevalence of hypothyroidism in Nepal ranged from 6% in some parts to as high as 94% in other parts. Mental and physical retardation went hand-in-hand with these figures. In the highlands of Papua, New Guinea, cretinism was extremely common and goiters were seen in 80% of the adults in some enclaves. South America, especially the Andes chain, has always had a very high incidence of cretinism with deaf-mutism. In the mountainous region of Ecuador 0.4% to 8.2% of the population show signs of cretinism. A survey in 1940, in Argentina,

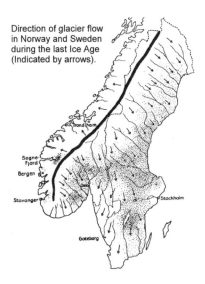

Figure 2.1 Dotted areas show distribution and density of goiters.

revealed goiter, hypothyroidism or cretinism in almost half of the school children and 12% to 18% in army recruits.

Endemic goiters and cretinism also persist in flat areas such as Central Africa, China, Japan, Brazil, Finland, Holland, Ireland, Algeria, Tunisia, Morocco, Sudan and Egypt, Ivory Coast, Zaire, Cameroon, and Kenya. In North Zaire, hypothyroidism with goiters can be seen in 1% to 7% of the entire population.

The United States, long ago, had endemic patches around the Great Lakes, the Dakotas, Montana, Colorado, Idaho, Utah, Minnesota, Ohio, the Appalachian Range, and the Pacific Northwest. In 1800, the noted American physician Dr. Benjamin Barton, published a monograph that depicted goiters among Indians living along the shores of Lakes, Ontario and Erie. Early farmers and ranchers in Michigan continually decried the loss of hogs, cattle, horses, chickens, and sheep from goiters and deformed offspring. When salt deposits were discovered adjacent to the Great Lakes, toward the end of the nineteenth century, it was added to the diets of animals, and magically, goiters subsided and the animals flourished.

"Salt prevents goiters! Salt makes animals healthy!" became the slogans of the time. We now know that these salt deposits contain appreciable quantities of both bromide and iodine, which, incidentally, are still extracted on a commercial basis from those sites. As we shall see later on, such speculations about the iodine content of salt and goiters helped Dr. David Marine when he tried to convince people that iodized salt could prevent goiters. Today, thanks to iodized salt, the incidence of goiter in the United States ranges from a low of zero in some states to a high of 7.3% on the southern shores of the Great Lakes; 5% is the national average.

Throughout this book we will learn how medical discoveries, made by chance or design, were eagerly snatched up and used to combat or prevent the afflictions of people. Here, however, we come across an inglorious episode in world history WHERE A PREVENTATIVE AND CURE HAS BEEN DISCOVERED, PROVEN TO BE EFFECTIVE, BUT NOT UNIVERSALLY IMPLEMENTED.

In 1960 the World Health Organization, WHO, published a comprehensive monograph that dramatically called attention to the physical retardation, mental impairment, deaf-mutism, and high infant mortalities that resulted from the appalling persistence of endemic goiter and cretinism. It cited studies that conclusively proved that iodine deficiency was the basic cause underlying these conditions. Dietary components, especially cassava and contaminated water were also listed as possible contributing factors. The most condemnatory part of this report stated that experiments, trials, and prevention programs carried out by WHO in the previous decade proved unequivocally that these blights could be eliminated through relatively cheap iodine supplements such as iodized salt—there was no excuse for not utilizing this knowledge. Yet now, more than 30 years after that clarion call for positive action, endemic goiter and endemic cretinism continue to impair the quality of life in Latin

America, Southeast Asia, the Himalayan region, Central Africa, and parts of Southern Europe. Mexico and seventeen states of South America also have moderately high rates of endemic disease. Five to ten percent of the population in Central Africa still display the most obvious features of cretinism.

In L. Frank Baum's classic, FROM THE WIZARD OF OZ, the wizard tells the scarecrow, who has just asked for a brain, "I'll stuff your head with brains. I cannot tell you how to use them, however. You must find that out for yourself." In much the same way, science, well over three-quarters of a century ago, handed society the tools capable of virtually wiping out the havocs brought on by iodine deficiency but sadly, they have not been universally adopted. Perhaps, as the Wizard cautioned, world leaders have brains but must learn to use them.

From what has been said so far, it is obvious that iodized salt figures prominently in the crusade against endemic goiter and cretinism. This then is a good place to pay homage to the quiet, unassuming, American genius, David Marine, who solved the riddles of goiter, shifted the paradigm of goiter, then showed how iodized salt could eradicate this menace.

The year of 1905 was an exciting time to be alive. American railroads advertised 18 hour trips between New York and Chicago in the world's fastest streamlined, long-distance trains. Eugene V. Debs and the Industrial Workers of the World actively promoted socialism and unionism. Suffragettes agitated for women's rights. And Grover Cleveland, retorted in the Ladies' Home Journal: "Sensible and responsible women do not want to vote. The relative positions to be assumed by man and women in the working out of our civilization were assigned long ago by a higher intelligence than ours."

On a bright sunny morning in July of that year, 25 year-old David Marine, having just graduated from Johns Hopkins Medical School, stepped off the train in Cleveland, Ohio, to further his medical training. He had been born on September 20, 1880 and spent most of his young life on the family farm in eastern Maryland. Despite having had only local schooling without formal high school training, he entered Western Maryland College at age 16 and successfully completed the classical course, the only one offered. He then moved on to Johns Hopkins University , where he pursued his first love, zoology. After several months of graduate work he decided to switch to medicine where the challenges and rewards were greater. By plunging into the study of German and chemistry he fulfilled the entrance requirements for medical school, breezed through the curriculum, and now, here he was in Cleveland, ready to take on post-graduate work.

The newcomer presented himself to Dr. W. P. Howard, Professor of Pathology at Western Reserve University School of Medicine, who in turn put the young man at ease by telling him about his friendships with Johns Hopkins' dean, Dr. Welch, and luminaries like Sir William Osler. Then, getting to the business at hand, Howard asked Marine what line of research would interest him.

"Goiter, sir," Marine shot back. At Johns Hopkins he had observed the work of the eminent surgeon, Dr. William Halsted, who partially removed the thyroid gland in dogs, either before or in early pregnancy and observed that they delivered puppies with enlarged thyroid glands. By adding iodine to the diet of these surgically altered dogs, Halsted enabled them to deliver puppies with normal glands. Marine was determined to find out how and why this happened.

"You know," Dr. Howard said, "goiters stumped even the great Virchow in the late 1860's. What do you have in mind that might be different or better than your predecessors?"

"For one thing," Marine replied, "they didn't know much about iodine in those days and Virchow studied only human thyroids. I intend to look at goiters in many species." Though nominally a doctor of medicine, David Marine knew a great deal about fish and mammals from his early education in zoology. He expounded on his observations on fish that suffer from goiters and related stories from ancient peoples who believed that the cause was to be found in water. Finally, Marine clinched his case by reminding his host that Cleveland was at the center of America's goiter belt; what better place to study goiter?

Dr. Howard was so impressed by his new research fellow, that, as the two men shook hands before departure, Marine was invited to work at the University's H. K. Cushing Laboratory of Experimental Medicine, which housed some of the world's best research equipment. David Marine literally moved into the laboratory. He collected and examined countless normal and diseased thyroid glands from humans as well as from fish, dogs, sheep, hogs and cattle, the latter, grown locally and as far away as Texas, a non-goiter area. By correlating the iodine content in normal and abnormal thyroid glands Marine concluded that iodine was an important factor in all classes of animals that have goiters.

Initially this endeavor was entirely a one-man affair. Marine performed all dissections and autopsies, preserved and cataloged tissue samples, prepared and stained slides for microscopic examinations, and meticulously recorded everything he saw or did. Sundays were spent doing chemistry and assaying the iodine content of his tissue collections. On his way to and from the laboratory he would frequently stop to examine the many dogs with swollen necks that roamed the streets to prove to himself that they indeed had enlarged thyroid glands. To no one's surprise he became recognized as one of the University's most accomplished comparative embryologists and pathologists. As his work mounted and experiments broached wider areas, Dr. Marine enlisted the help of a young surgeon, Dr. Carl H. Lenhart, to help with surgical procedures on dogs and sheep, and Dr. W. W. Williams, a pharmacologist, to test for iodine.

In one of their earliest collaborations Marine and Lenhart studied the thyroid glands in fresh and salt water fish from several areas in the United States. In their 1910 report, now a classic, they proved that Lake Erie's bass, pike and trout, suffered from a higher than normal incidence of goiters. When these goitrous fish were put into troughs filled with running Lake Erie water that had been fortified with iodine, the goiters disappeared. Dr. Marine soon advanced the hypothesis, which is still valid today, that the lack of iodine causes the thyroid gland to swell and increase in size. This increase in size is not just due to puffiness but a real increase in the number of cells in the gland. To feed these extra cells there is a proliferation of blood vessels throughout the gland. When the intake of iodine returns to normal the gland shrinks, but never to its original size. Such cycles of swelling and shrinkage are commonly seen in areas where goiter is endemic.

This paradigm, that an insufficiency of a dietary component could cause disease or malfunction in the body, was really advanced for its time and laid the foundations for our current understanding of vitamin and mineral deficiencies. It was not until two years later that Casimir Funk shook the world with his proof that certain vital amines, later shortened to vitamins, could affect health, normal growth and the many functions of the body.

In 1909, in an event that later proved to be pivotal in his understanding of the thyroid gland, Dr. David Marine was approached by the Blooming Grove Hunting and Fishing Club, an exclusive enclave in the mountains of eastern Pennsylvania, and asked to find a cure for what was believed to be cancers of fish in their hatchery. We will never know for sure why Dr. Marine was selected for this job since all records from the Club were lost in a disastrous fire. This seems like a fortuitous meeting because young Dr. Marine was still unknown outside of scientific circles at that time.

Marine and Lenhart traveled to eastern Pennsylvania and in no time at all announced that the fish in the hatchery had goiters, not cancer. The condition became noticeable in the fry as soon as artificial feeding was started. The thyroid gland became slightly enlarged at four months of age and grossly visible by the second year. After noticing that native wild trout in nearby streams did not develop goiters, Marine and

Lenhart surmised that those fish were getting iodine from their diet of natural foods. To prove this the two men placed fish from the hatchery with goiters into adjacent free-flowing, native streams and watched the goiters disappear.

The problem had to be in the fish hatchery.

"What do you feed your fish?" Dr. Marine asked one of the hatchery workers.

"Fish food," was the snide reply.

"But what is in the fish food?"

Looking at the sack of fish-food-pellets nearby, the man read off its contents, "Ground beef, hog liver and heart."

Marine knew these were deficient in iodine. Then, in one of the most convincing trials ever mounted, Marine and Lenhart placed twenty-one healthy, tumor-free fish in a trough filled with the same water available to all fish in the hatchery and watched for the development of goiters. As soon as there was evidence of thyroid swelling, iodine was added to the water, morning and evening. One fish was taken every two days for complete anatomical and chemical studies. After 8 to 10 days of living in iodine enriched water there was evidence of renewed hormone production. By the twentieth to the twenty-second day the entire thyroid gland appeared normal. Moreover, the fish looked clean and robust and showed remarkable changes in weight and color. Whether or not the trout from the study were cooked for dinner was never noted.

Before departing the scientists recommended a diet of ground ocean fish, a good source of iodine. When Marine returned to the Club in 1913, two years after this diet had been in effect, he quickly noted that that there no goiters, the fish were stronger, showed more girth, and did not die the moment they were removed from water.

By now Dr. Marine was convinced that most goiters were the result of iodine deficiency and felt it was time to put this knowledge into practical use. He administered iodine to a small number of children at Cleveland's Lakeside Hospital Clinic, and saw no ill effects, but he needed a large treatment program to prove to the world that goiter could really be prevented.

When a proposal for a city-wide trial was put before officials of the City of Cleveland, the Health Commissioner exploded, "No! Not in this town will you thrust your iodine on the public."

"But sir," argued the young doctor, "I've got proof that small amounts of iodine will prevent goiter. Cleveland is right in the middle of America's goiter belt. It has one of the highest goiter rates in the nation. Don't you feel morally obligated to try something that could end this blight?"

"My dear Dr. Marine," the commissioner replied, "Right now I'm trying to get people to accept small pox vaccinations. You know I've been fighting an uphill battle ever since vaccinations became compulsory in this state. I don't need any more friction. Try the school board. They may allow you to experiment in their classrooms."

Dr. Marine followed this advice and approached the Cleveland school board with all the facts and figures he had accumulated to support his proposed study. The chairman of the school board was a physician, so Marine had high hopes for success. Instead, the chairman told his board, "This man wants to poison our children." Another failure.

Here the matter rested until 1916. While lecturing to the second-year medical class in pathology, Marine stated that any doctor who did not prescribe iodine for children with thyroid enlargement was guilty of negligence. When the lecture was over, a student came up and asked him why he had not done more to push iodine as a public health measure. After Marine recited his litany of rebuffs, slights, and insults, the young man told him he had taught school in Akron, Ohio, knew all the local politicians, and felt sure he could arrange a clinical trial in the public school system in that city. We have no record of what transpired but the Akron city fathers did gave Marine the green light to go ahead with his experiment. The deciding factor, it seems, was that *A likely impossibility is always preferable to an unconvincing possibility* as Aristotle once said.

The Akron study was simple and direct. First, 3,872 adolescent girls were surveyed to determine the natural prevalence of goiters. More than half, 56.41% were found to have enlarged thyroids, 43.95% were normal. Then the normal girls were divided into two groups. With written consent from parents, one group was given a sodium iodide solution ten consecutive school days in autumn and spring; the other group, the controls, received no medication. After a year of observation, this is what the researchers found:

	Iodine	No Iodine
Thyroids Remaining Normal	283 - 100%	637 - 74%
Normal to slight goiter	0 - 0%	103 - 11.8%
Small goiters (unaltered)	2 - 0.5%	103 - 11.8%
Small goiters (disappeared)	141 - 33.5	10 - 1.2%
Large goiters (unaltered)	34 - 66.7%	106 - 95.5%
Large goiters (decreased)	17 - 33.3%	5 - 4.5%

Dr. Marine proved the correctness of his paradigm. Iodine prevents and helps to cure goiter.

On April 6, 1917, the United States declared war on Germany and on May 18th, President Wilson, following Secretary of War Baker's advice, issued a formal proclamation introducing universal conscription. Dr. Marine knew that sooner or later he would be called up for medical corps duty, so he took the bright second year medical student, Oliver Perry Kimball under his wing, and prepared him to carry on the Akron studies. Shortly thereafter Marine entered the army and served in Ohio, Michigan, and France.

Between 1917 and 1920 doctors Marine and Kimball conclusively proved that iodine deficiency was the cause of endemic goiter and that small supplements could prevent the disease. Michigan, introduced the iodination of salt in 1924, at which time 38.6% of its inhabitants had goiters. By 1951 the incidence had dropped to only 1.4%.

After Marine had proven to the world that iodized salt was inexpensive and effective in preventing goiters, it was adopted throughout the United States and Switzerland. In the United States salt is iodized to a concentration of 1 part in 100,000; in Switzerland the salt is iodized to 1 part in 200,000. The World Health Organization recommends 1 part in

100,000.. Unfortunately world-wide acceptance was interrupted by the second world war.

In the United States the use of iodized salt is voluntary. There is no legislation to enforce it. Non-iodized salt is still available on grocery shelves. In most other countries of the world iodination is regulated by legislation. Indeed, legislation mandating iodized salt has been credited with the drastic reduction in the incidence of goiter, cretinism and deaf-mutism in Central American and Caribbean countries. In 1976 iodized salt was being used in 43 countries and was available on a voluntary basis in 9 others.

The legendary detective, Sherlock Holmes once said, "The more featureless and commonplace a crime is, the more difficult it is to bring it home." This truism became painfully apparent as national leaders in many parts of the world failed to recognize the 'featureless, commonplace, crime' of persistent endemic goiters and cretinism. As far back as 1960, the World Health Organization urged all nations to adopt health programs that could easily and inexpensively blotout these scourges, yet today, a quarter of a century later, pockets of endemic goiter and endemic cretinism persist.

This crusade actually had its beginning in 1948 when the Third World Health Assembly asked WHO to initiate and support surveys and studies on endemic goiter and then follow up with programs for prevention. WHO, United Nations Children's Emergency Fund, UNICEF, and the Food and Agricultural Organization of the United States, FAO, organized and launched the International Cooperative Study of Goiter. By 1951 Massachusetts General Hospital, Harvard Medical School, and the Central Hospital University of Cuyo, Mendoza, Argentina, set up pilot programs in South America, Africa, India, Indonesia and Europe. Subsequently, many other universities and scientific foundations joined in.

In 1960 there were about 200 million people in the world suffering from goiter. Since then iodine supplements have been started in a few countries and nonregional iodine-rich foods have penetrated some isolated areas. Rising economies have made iodine sources more accessible. While these measures may have narrowed the geographic distribution of goitrous people, they probably have not affected the total number because of the increase in world population.

The World Health Organization has targeted the year 2000 for the elimination of endemic goiter and cretinism. This seems feasible inasmuch as iodination programs have virtually eliminated these conditions in Europe, North America, Asia, New Zealand and Australia. Tasmania currently uses iodized bread to prevent goiter. Iodination of salt is easy, efficient, and so cheap that it should not raise the price of the salt. In areas where iodized salt was difficult to distribute or not acceptable, as in Ecuador, Papua, Peru, Zaire, the Himalayas and Andes, a small dose of iodized oil, injected as infrequently as every three to five years, has proven to be effective in correcting iodine deficiencies, improving the physical development and IQ of children, and greatly reducing the number of new cases of cretinism.

The discovery that iodized oil could be used for the prevention of iodine deficiencies and the control of endemic goiter deserves special attention because it illustrates the role of serendipity in the discovery process and shows how a commonly used substance often finds an equal or better use in another distant or unrelated field. When iodinated compounds were found to be radio-opaque to x-rays, around 1921, iodized oil was utilized by radiologists to show the contours of hollow organs in the body. This technique was eagerly adopted by pulmonologists who instilled iodized oil into the bronchial tubes of patients then used x-rays to produce detailed pictures of all the airways to the lungs. On one occasion, Dr. D. Jamieson, a chest physician in New Guinea had instilled iodized oil into the bronchial tubes of his patient John Gunther and they both remarked that the oil was innocuous, it persisted for quite a long time and was only gradually removed. They reported their observations in a scientific journal. Two doctors, Clark and McCullagh, happened to learn of this report and got the idea of injecting iodized oil into people living in iodine deficient areas. Beginning in 1957, they carried out their scheme in New Guinea and by 1960 collected enough data to prove, unequivocally, that a single injection of iodized oil could reduce the incidence of goiters and cretinism.

Studies have since shown that a single injection of a teaspoon full of iodized oil corrected severe iodine deficiency. It was effective for four and a half years and, if administered before pregnancy, prevented cretinism. On Idjwi Island, in central Africa, only one case of cretinism occurred in two districts containing 1,181 inhabitants who had received iodized oil—a woman with a cretin child had been absent from the village at the time of injections. To date, iodized oil has been successfully employed in Argentina, Brazil, Ecuador, Peru, Papua, New Guinea, Nepal, Zaire and some sections of Ubangi.

The cost of injected iodized oil is between 4 to 174 times that of iodized salt, since trained personnel and expensive materials and implements are needed. Nonetheless, the cost of treating an individual amounts to about 35 cents. This figures out to seven cents per person for a year of protection. Iodized oil can be given less expensively by mouth but it tastes awful. Some countries are trying to hide iodide in candy and hand it out once a week or several times a year. No legislation is needed to implement these programs since they are extensions of public health services.

Iodized oil is efficient, long-lasting, effective and slightly more expensive on a per capita per year basis than iodized salt. Nonetheless it is a remarkably small item in a national budget when measured against the economic benefits. More importantly it insures ready acceptance and bypasses political, social, or cultural objections.

Then what is holding up the universal adoption of iodized salt and iodized oil in areas where goiter and cretinism are still endemic?

The most formidable obstacles are internecine frictions, political instability, poverty, and overpopulation. Most of the third world lacks experience in methods of improving public health. And even when public health measures are instituted the most urgent priorities are directed toward combating infectious diseases, wide-spread parasitic infections and malnutrition. It is difficult to distribute large amounts of coarse rock sea salt to populations in isolated and remote areas of South America, Himalayas, New Guinea, Indonesia, and Asia.

Both iodized salt and oil injection programs require a sound political system, good administration, and personnel to take care of financial, administrative, and technological problems. It requires cooperation between the producers and distributors of salt and iodine and acceptance by the public. Although the world stands ready to offer a helping hand, governments must contribute to their own national life by eliminating endemic goiter once and for all. There is no disease which impedes national development and the quality of life so much and which can be eliminated for so little.

In 1924, Michigan's Ann Arbor News proclaimed that that severe smallpox had broken out in a nearby town, authorities were preparing for the poliomyelitis season and iodized salt was now available in stores. By 1978, the World Health Organization offered a prize for any report of a case of smallpox. Poliomyelitis was, by now, a rare disease. But the promise that iodized salt would eliminate the scourges of iodine deficiency—endemic goiter and cretinism—has yet to be realized.

PARADIGM POINTERS

A. Paradigms tend to be stubbornly persistent and often misleading. For example, throughout the 19th century when large numbers of microbes were being discovered, almost every disease or abnormality was attributed to them. This type of paradigm worked very well for Cholera but when applied to goiter became a protracted stumbling block. The persistent search for a germ as the cause of goiter seriously delayed the discovery of its true cause. Although Jean-Baptiste Boussingault and Gaspard Adolphe Chatin both produced evidence that iodine deficiency, not infection, was the cause of goiter, it took another century before the goiter paradigm would change.

B. Often, as Salman Rushdie expressed it, "Facts are hard to establish, and capable of being given many meanings. Reality is built on our prejudices, misconceptions, and ignorance, as well as on our perceptiveness and knowledge."

Here is a good example of this. When farmers in areas around America's Great Lakes found that adding salt to the diets of livestock improved animal health, salt became a ritualistic additive to all feeds. Only later, when Dr. David Marine proved that it was the iodine in the salt, not the salt itself, that produced these wonderful results, did the absolute faith in salt diminish.

POST SCRIPT

The following three items are related to iodine and thyroid gland function but are placed here because they have implications of their own. The first alerts us to the possibility of getting too much iodine. The second tells us something about a new aspect of thyroid disorders and illustrates how some breakthroughs must bide their time until they can be appreciated. The third alerts us to a relatively new and disturbing cause of goiters in fish.

1. There is some concern among health authorities and environmentalists in the United States that we may be adding too much iodine to foods and salt. Increased usage of iodinated compounds in food technology have generated iodine loads much greater than those which people encountered at the time when iodized salt was first introduced. Iodine containing compounds are now used as

dough conditioners and for sanitizing bakery equipment. Bread and bakery products are consistent contributors of iodine. When dairy cows are fed soy bean protein they also receive large doses of iodine to offset the effects of this goiter inducing diet; iodine ultimately spills over into milk. Iodine containing compounds employed to clean teats and udders and in sanitizing milking machines and industrial dairy equipment, complicate matters further. Also, iodine-containing medications that prevent foot-rot and soft lumpy-jaw in cattle, have the potential to add iodide to milk and milk products.

Unnatural quantities of iodine spill over into the environment:
- from manufacturing and chemical industries - from processed sewage that is used as a fertilizer
- from swimming pools that use iodine as germ inhibitors because it is cheaper than chlorine.

Besides these incursions, iodides are used for coloring beverages, ice cream, fruits, cereals, multivitamins, antacids, as well as therapeutic and diagnostic drugs. For instance, one antacid tablet and a bowl of pink cereal can supply 800 micrograms of iodine, a significant overdose when you realize that the recommended daily allowance of iodine is less than 300 micrograms. Nonetheless, the U.S. Food and Drug Administration has repeatedly stated that "there is no evidence that dietary iodine intakes in the USA in recent years have lead to any untoward effects."

2. It is now generally accepted that people with certain thyroid diseases have antibodies in their serum that attack their own thyroid glands. But this knowledge came to us through a long, circuitous route.

In a series of ground-breaking experiments in 1956, doctors Witebsky and Rose proved that rabbits could be induced to make antibodies that would attack their own thyroid gland. This was heralded as a great advance in medicine at that time but actually, the same observations had been made in 1911 by A. Papazolu in Paris, France. He reported that serum from patients with certain thyroid disorders attacked extracts from their own thyroid glands. Papazolu's report was entirely rejected, then neglected, because at that time it seemed inconceivable that any mammal could form antibodies against constituents of its own tissues.

When the paradigm for allergy was being constructed it stipulated that only things alien to the body could evoke allergy. When Papazolu challenged this he was ignored.

Papazolus' and Chatin's experiences were almost identical in that they could not be understood or accepted by contemporaries: SOME PARADIGM SHIFTS MUST WAIT UNTIL THE TIME IS RIPE FOR THEM.

3. In 1906, sports-fishermen rejoiced when Coho Salmon were introduced into Lake Michigan and Lake Erie tributary streams. The success of the program exceeded expectations but in 1972 thyroid nodules and goiters were found in 44% of mature fall spawning Coho Salmon from these waters. Thinking back to Dr. David Marine's work, authorities shrugged this off as a sign of iodine deficiency. The affliction, however, kept spreading. Over the next four years, from 1972 to 1976, the goiter prevalence in Coho salmon in Lakes Michigan, Erie and Ontario had increased from one to six percent. There had been no change or decrease in the iodine content of these waters during that time frame. What could it be? Studies then confirmed that PCBs, polychlorinated biphenyls, in these highly polluted waters was affecting the fish. Coho salmon now act as sentinels for the detection of environmental contaminants that cause goiters. The next chapter examines this problem more fully.

chapter 3

ELEGY FOR A PLANET

If, as we have just read, fish will serve as sentinels of water pollution, what will warn us about pollution elsewhere? We could wait until birds stop singing and frogs no longer cavort in marshlands but that might be too late. Maybe science will hand us the tools to keep us from annihilating all living things on this planet. But after what happened in Michigan just two decades ago, it makes you wonder if we can rely entirely on this less-than-perfect technology.

From high above, in the early 1970's, Michigan's lower peninsula looked like an orderly patch-quilt of brown squares, the fallow fields, separated by the yellows, greens, and blues of crops, vegetable gardens and pastures. At ground level the landscape resembled a scale-model miniature railroad. Here and there pastel ginger-bread houses with white-washed fences, stainless-steel silos, and red barns dotted the gently rolling, velvet green, countryside. Bumper crops and healthy livestock promised prosperity and happiness to everyone.

Toward the end of 1973, not unlike the biblical visitation of plagues upon Egypt, disasters swept over the land. Cows began losing weight and milk production dropped off. Cattle developed rheumy eyes, abnormal hoofs, aborted or gave birth to sickly calves; many just became lethargic and dropped dead. Farmers naturally assumed that a transient illness had attacked their herds and it would soon disappear. As losses continued to mount, consternation and apprehension rose proportionately and pretty soon fear and anxiety gripped the people. Panic set in when chickens, pigs and other livestock keeled over and died by the hundreds. Economic ruin seemed imminent. Whatever this mysterious killer was, it threatened to wipe out Michigan's agricultural belt, the source for 85% of the meat and almost all of the eggs, chickens, and milk consumed in the state.

Farmers whose herds were attacked during the early months of this cataclysm had no way of knowing that their troubles were part of a state wide pattern of devastation. Local agency inspectors responded to pleas for help and in due course they were assisted by state health departments, veterinary agencies, and Grange groups. These initial investigations, however, were limited and restrained. The trouble in Michigan did not get any media coverage or qualify for federal intervention—it was just considered to be a brush fire.

Eight months later, in April 1974, the cause of the holocaust was found. Bags of flame retardant that contained PBB, polybrominated biphenyls, somehow became mixed up with bags of magnesium oxide, a feed-supplement for livestock. When the flame retardant was mixed together with fodder and fed to cattle, large-scale poisonings followed. Even after PBB was fingered as the trouble maker, it took another year and a half to localize PBB contaminated areas. The scariest part of this entire story is that if not for a remarkable series of fortunate coincidences and lucky accidents there is no telling how long these gruesome events would have gone unchecked.

Biphenyls are highly toxic organic compounds. A six-sided group of carbon atoms, called a phenyl ring, is chemically hooked to another phenyl ring, making it a biphenyl. Variable amounts of bromine or chlorine can then be added to this biphenyl making it either polybrominated biphenyl, PBB or polychlorinated biphenyl, PCB. Four to ten bromine or chlorine atoms are generally used for commercial products. Manufacturers frequently mix chlorine and bromine together in the same product. In the field, investigators seldom come across clear cut cases of PCB or PBB contamination; more often than not, mixtures are found. Then, too, slightly different compounds may exhibit widely different degradation patterns or may shift from one form to another. And to complicate matters further, poor quality control will vary the toxicity of different end products.

PCB and PBB are not new compounds. Their heat exchanging properties were recognized in the 1930's and manufacturers in the western industrial world had been using them in transformers, paints, varnishes, synthetic waxes, insulators, capacitors, hydraulic fluids, and in caulking compounds for over 40 years. They were also used as plasticizers. Until recently PBBs and PCBs were used to flameproof pajamas, curtains, and household fabrics, but were discontinued when found to be poisonous. Bromine replacement compounds for flameproofing children's clothes have also been banned because they tend to produce cancer in animals.

Despite their known toxicity, PBBs and PCBs are still found in many parts of the world as well as throughout the United States. The San Francisco Bay Area Rapid Transit system used them to cool its train transformers and was under investigation by the Occupational Safety and Health Administration because of potential hazards to employees. Manufacturers of transformers and capacitors still use both PBBs and PCBs. In Japan, PCB production increased more than 50 fold between 1954 and 1968, and it is widely used as a solvent in carbonless duplicating papers, in heat transfer equipment, and as an ingredient in plastics.

Chlorinated biphenyls are better heat exchangers than biphenyls containing bromine. But PBBs, though less active are more dangerous than their chlorine counterparts. Once they get into humans and animals, PBB's seek out fatty tissues, and if undisturbed, stay there forever. Since they are almost never eliminated, they act like land-mines just waiting to explode.

Sometime during the summer of 1973, at the Michigan Chemical Corporation plant at St. Louis, Michigan, about thirty 50-pound bags of *Firemaster*, a flame-retardant containing PBB, were inadvertently included in a truckload of *Nutrimaster*, a magnesium oxide additive for cattle-feed. Cognizant of the dangers inherent in a possible mix-up of these two products, *Firemaster* was regularly sold in red bags and *Nutrimaster* in brown bags. When the company temporarily ran out of red bags it put *Firemaster* in brown bags. Although the brown bags were correctly labeled as *Firemaster* they nonetheless ended up at the central farm bureau of Michigan where they were mixed with fodder and distributed throughout the state. Unknowingly, farmers then fed this to their livestock.

The reason *Nutrimaster* was introduced into cattle feed in the first place was because cattle suffer from a delicate calcium and magnesium balance at the time of calving. Cows as well as calves often develop a spastic condition, tetany, due to low calcium values in their blood. Veterinarians learned long ago that the addition of magnesium to the cattle diet prevented this troublesome complication, so magnesium oxide became the standard supplement for cattle.

Late in the summer of 1973, Frederic Halbert, a dairy operator near Battlecreek, Michigan, unfortunately and fortunately, happened to be the farmer who received the most heavily PBB contaminated consignment of feed. Some of his cows ate as much as a half a pound of pure PBB. This was unfortunate. The fortunate part was that Halbert was no ordinary dairyman. He had earned a masters degree in chemical engineering and worked three years for Dow Chemical Company before returning to the family farm in 1971. Except for these unusual circumstances and for the fact that Halbert was exceptionally resourceful and persevering there is no telling how long it would have taken to find out what was destroying animal life on Michigan farms.

When cows stopped eating, lost weight, and no longer produced milk, neither Halbert nor veterinarians had any notions as to where to start looking. Blood samples from sick cows were analyzed and carcasses of several cows were autopsied but no cause of death could be identified. Halbert did everything possible to get authorities, both at the Michigan Department of Agriculture and the U.S. Department of Agriculture's National Animal Disease Center in Ames, Iowa, to look into the matter. Over a period of time, local, state and federal agencies did pay attention to his pleas for help and threw their expertise into the fray. Secretions, excretions, hair, horn and tissues were collected from living and dead animals. Everything was examined and cultured for parasites, bacteria and viruses. Soil was analyzed. Statistical surveys were set into motion and data was logged on huge charts, like war maps, which soon pinpointed the ravaged areas. But there were no recognizable patterns. Much to the consternation of biostatisticians and health authorities the charts failed to duplicate the known patterns of bacterial and viral epidemics. Apparently this was more than an isolated incident and more devious than epidemics in the past.

Disease causing bacteria were first eliminated as part of the problem. They just could not be detected from any source. Despite the use of selective culture media and sophisticated growing techniques, with and without oxygen, no infectious germs showed up in any of the countless test tubes and petrie dishes housed in laboratories across the state. Viruses proved more troublesome and required additional time for their identification; but they too were eliminated as the cause of trouble. Poisons, toxins and toxic wastes were also suspect, but since this part of Michigan was primarily an agricultural area, far removed from industrial centers where such products might be found, this possibility seemed remote. In essence, there were no clues as to the nature of the killer or killers. Where and how to start searching was a problem in itself. Air? Water? Soil? People? Animals?

Epidemiologists, public health specialists, veterinarians, bacteriologists, virologists, pathologists, soil analysts, biostatisticians and teams of experts and technicians all failed to come up with logical answers. Like blind men touching an elephant, where each thought it to be a tree, a snake, or large leaf, depending on the part they happened to feel, so it was that all specialists saw only the part of the picture that fell within their purview. Veterinarians saw only animals; epidemiologists saw only people; bacteriologists saw only germs; and biostatisticians saw only numbers; no one saw the 'whole the elephant'. But they were not uninformed. Everyone was aware of Rachel Carson's warnings about the dangers of DDT and they also knew that PCB, polychlorinated biphenyl, had, in the past, contaminated animal-feed. It was only when their tests for these two substances, plus countless other toxins, came back negative, that they threw up their hands in despair. It must be remembered that Halbert's feed contained PBB, which is relatively stable and hard to detect.

The break in the case came late in 1974 at Ames where, by 'pure dumb luck', a peculiar reading showed up on the gas chromatograph analysis of Halbert's feed. Some technicians forgot to turn off the chromatograph during their lunch hour and by the time they returned a wildly unfamiliar reading had appeared on their read-out. "Our machine is turning out the Rocky Mountain range" one of them remarked. Yet no one knew what the reading meant or even whether it meant anything at all. Environmental pollutants like DDT and PCB tend to breakdown into their elements and usually show up as early emerging peaks on a chromatograph, but PBB, because of its exceptional stability, is revealed by a very late emerging peak. In this instance, when it was observed, its meaning was not immediately understood.

Just about this time word came to Ames from Washington Headquarters of the Animal and Plant Health Inspection Service that they would have to call off their studies because the Center's funds were being earmarked for other purposes. This too turned out to be both unfortunate and fortunate. Unfortunate because it frustrated Halbert and threatened to put an end to the investigation just as people were learning that something was wrong. But is was also fortunate in that it forced Halbert to turn elsewhere for help. It led him to perhaps the only person who could resolve his problem, George Fries, a scientist in the Pesticide Degradation Laboratory at the USDA's Agriculture Research Center at Beltsville, Maryland. A few years earlier Fries had obtained some PBB for experimental work and had mastered the methods for its detection. He happened to be one of the few scientists in the world who knew how to recognize PBB on a gas chromatograph.

Halbert sent Fries samples of his feed and also some of the data that had so far been collected.

"Do you suppose those odd late emerging peaks have any significance? or are they artifacts?" Halbert asked Fries during one of their telephone conversations.

Instead of replying, Fries shot back another question. "Does Michigan chemical Corporation make PBB?"

Why was he asking that, Halbert wondered? "Yes. They make *Firemaster* and they also make *Nutrimaster*."

With one quick intuitive thrust Halbert had drawn the critical linkage between *Firemaster* and *Nutrimaster*. And, sure enough, by April 29, 1974, Fries, using gas chromatography and mass spectrometry proved that Halbert's feed sample contained PBB.

Gas chromatography and mass spectrometry, the sophisticated technologies employed to sniff the air, soil, and water in the troubled areas of Michigan, evolved from the spectroscope, which also came to us through accident. In 1859, the German chemist Robert Wilhelm Bunsen, the inventor of the Bunsen burner, that nice little gas-fed flame found in every class-room chemistry laboratory, passed the light emitted from heated chemicals through a prism expecting to find a smooth continuous spectrum going from red to violet.

Instead, unexpectedly, he saw a series of sharp lines. He soon discovered that every inorganic substance had its own characteristic brightness and location of lines, its specific fingerprint, and used this to identify elements and compounds by intense heat rather than by cumbersome chemical analysis. Gustav Kirchoff and Bunsen built their first spectrograph from an old telescope and a cigar box. Gas chromatography and mass spectrometry have extended this basic concept to organic materials.

When a substance is vaporized in very intense heat, say a spark or electric discharge, it dissociates into its constituent ions or molecules which give off their own characteristic spectral or mass responses. Electron detection coupled with computers then give readouts that identify and quantify the emissions. Spectrometric analyses are usually made photographically or electronically because most of the lines for analytical purposes lie in the ultraviolet rather than in the visible ranges. These proceedures are currently used by science and industry to identify chemical substances qualitatively and quantitatively.

With 20/20 hindsight we can now understand why no single epicenter of trouble could be found, why the source of the killer could not be located, and why it spread in such a mysterious way. It took a little more than a year to 'finger' PBB, and during that time, people throughout the state of Michigan inadvertently carried the deadly poison away from its place of origin.

As people in towns and cities became sick they created new statistics which drew attention away from the main culprit on the farm. Along with most of the population of the Michigan peninsula almost every Detroit housewife picked up some PBB by way of the food chain. Like cigar smoke in a florist shop, the damage drifted beyond the animals, beyond the land, and beyond the areas of rural isolation. Random sample studies by the state health department and University of Michigan showed that about eight million of Michigan's 9.1 million residents exhibited detectable levels of polybrominated biphenyls. Fully 96 percent of the samples of breast milk taken from nursing mothers contained PBBs, even in the far removed and less densely populated northern part of the state. These findings were subsequently confirmed by Dr. Irving Selikoff and his researchers from the Mount Sinai School of Medicine in New York, who did the quintessential survey of Michigan's tragedy.

In May 1974, after surveying the damage and examining mounds of carcasses, Michigan health authorities set an arbitrary level of PBB in fatty tissues above which live stock would have to be killed. Thirty thousand head of cattle, 1.6 million chickens, 5 million eggs, and thousands of sheep and hogs were destroyed and buried in a remote mining area—very much like radioactive wastes. Farms were quarantined if a single sample of meat or milk was found to contain PBB in levels above minimum permissible limits.

This however proved frustrating because much of the PBB that was not absorbed by animals went back into the soil by way of their manure. PBB is not degradable. Since it will not go away or disintegrate, large areas had to be quarantined against further use. No longer could land be used for animals or vegetables. The nation's finest farmlands were now irretrievably contaminated, gone.

In November 1976, Dr. Selikoff and his 35 member team set up shop in Grand Rapids' Kent Community Hospital and launched one of the most intense and exhaustive environmental/medical studies of this century. They examined all the residents of quarantined farms, people who ate food from quarantined areas, residents and consumers of products from non-quarantined farms, and workers from the PBB manufacturing plant. The investigators spent four to six hours with each subject. They took detailed dietary, occupational, medical, and family histories; they performed physical examinations with special attention to skin, eye, and nervous system problems; they took chest x-rays and performed a battery of laboratory tests; and bits of fat were snipped out for analyses.

Similar examinations and tests were performed on 238 residents of dairy farms in Marshfield, Wisconsin who had no PBB exposure; they acted as controls. The results showed a significant increase in the prevalence of skin, muscle, bone, and neurological symptoms in the Michigan residents, particularly those aged 16 to 55. Men who had been accustomed to working vigorously on their farms from sunup to sundown complained of weakness and the need for 14 or more hours of sleep a day. Some said that they took their lunch in the fields so that they could "sneak" a nap. Men in their early 30's were found to have swollen, deformed, painful joints and various skin sores, including halogen acne, also known as chloracne. This is quite similar to the acne commonly seen in teen-agers but the pimples turn into abscesses and these then become oozing and scabby sores. Analyses of the fat samples from 896 residents of six Michigan cities revealed that 101 contained PBB; 90.2 % of the serum samples also proved positive for PBB.

When children were examined by the University of Michigan's Dr. Mason Barr they too showed many more symptoms and abnormalities than the Wisconsin controls. However, there were no differences in the youngsters from quarantined and non-quarantined Michigan farms. This was strange.

"How come quarantined and non-quarantined children had identical problems?" was the inevitable question whenever Dr. Barr presented his data.

"It appears," he replied, "that whenever a farm was quarantined the family became cautious and carefully selected food from safe sources. People from non-quarantined farms assumed that they were safe and probably continued to eat food containing PBB."

To find out if exposure to PBB had any 'time-bomb' effects the Michigan Department of Public Health in collaboration with the Center For Disease Control, the Food and Drug Administration, the National Institutes of Health, and the Environmental Protection Agency undertook a study of exposed individuals and healthy controls. The highest serum

PBB levels were naturally found in people working in chemical plants and factories. Next highest were the members of quarantined farm families. Then, in decreasing order, came the farm product recipients, residents of non-quarantined farms, and healthy volunteers. Levels were significantly higher in males than in females.

No positive associations were found between serum concentrations of PBB and the number of symptoms that were reported. As a matter of fact the people with no detectable PBB in their serum complained of more illnesses than those with measurable quantities. This was no fluke because in two previous studies in Michigan there was no relationship between the frequency of symptoms and illnesses and the amounts of PBB in the patient.

"Why is it that people with little or no PBB in their bodies were sicker than those with high levels?" investigators were asked. "Contamination isn't synonymous with disease," Dr. Selikoff explained. "Stress appears to mobilize the chemical in the body; this leads to a wide variety of patterns of expression." Other investigators added, "factors other than PBB absorption may be responsible for the production of symptoms. Of course, psycho-emotional and psycho-somatic factors can provoke symptoms which are exactly the same as those produced by toxins."

Euripides had his finger on the truth when he said, "The errors of parents the gods turn to the undoing of their children." Nowhere is this better proven than in the manner in which PBB affects fetuses and babies. All babies born to contaminated mothers and nursed for six months had a large body burden of PBBs. Half came through the placenta and half from the breast milk. A decade ago PBB was found in the breast milk of women living mainly in the Michigan peninsula but now, however, it can be detected in breast milk of women throughout the United States. Except for sterilization or abortion there is no way of preventing PBB's from crossing the placenta from mother to unborn child. Nursing, however, is something we can control, but dare we outlaw that mighty maternal symbol, breast milk? And would it do any good?

These are really moot questions because with or without breast milk our progeny is doomed to coexist with PBBs. These chemical compounds are fat soluble and become imprisoned in fat. Since it is impossible to get rid of all fatty tissues in the body, they will always serve as a reservoir for fat soluble toxins. The only way to get rid of them is through the loss of fat, but, humans as well as animals lose fat sparingly. The few ways that fat does leave the body, unfortunately, is through the placenta into a fetus and in breast milk. Most of our fat stays with us almost forever. Dieting, exercising and losing weight is not the answer. The overweight person with a body load of PBB who decides to shed a few pounds faces a terrible enigma: with any weight loss there would be a concomitant disturbance of fat cells which would release PBB into the circulation. This in turn might cause more havoc than having it bound down in fat. No one is quite sure what damage these substances produce during their transit in the blood. Ultimately, PBB is redeposited in parts of the body where the fat content is stable.

PBB and PCB have become firmly fixed into the food chain and the environment. Plants, then animals, and eventually humans inhale or ingest these substances, store them in fat, and transmit them to their offspring. Or, when they die, the toxins go into the earth, then into plants, then into animals, and the cycle repeats itself.

These compounds are not degradable. They are destined to remain in the environment for inordinately long periods of time. The only known method of destruction is to incinerate them at very high temperatures. But this would only transfer toxic pollutants into the air.

The greatest threat to society comes from our general indifference to the menace posed by that these chemicals. As a case in point, as far back as the 1960's health authorities were aware of the fact that carbonless copying paper containing PCB was commonly recycled into cereal boxes and that it inevitably migrated from cereal boxes to the cereal people ate. But little concern was shown and little action taken. This complacency was interrupted in 1966, when a report from Sweden found PCBs everywhere in our environment and alerted the world to the presence of a time-bomb in its midst.

Industry, too, must shoulder its share of the blame for disseminating potentially harmful chemicals. For example, the General Electric capacitor plant near Albany, at Hudson Falls, New York, had been leaking PCB into the Hudson River for 30 years, *with a state permit to do so*. Other manufacturers also dumped their waste materials into adjacent watersheds. Inevitably, PCB and PBB found their way into the waterways of the United States, especially the Hudson River and Great Lakes. Although these rivers and lakes do not figure prominently in the drinking water supplies of major cities, the chemicals in them were picked up by micro-organisms, later by larger organisms, and in due time by fish. When people ate these fish they added to their stores of PBB acquired from contaminated meat, eggs, milk, and milk products. People adjacent to Lake Ontario, with freezers full of such fish, especially fat salmon, found themselves sitting on a powder keg. Fishing in the Hudson River was, for a time, banned because of the high PCB content in fish.

There is a time honored axiom among environmental epidemiologist that says, if a cataclysm is destined to appear anywhere on earth it will surely surface first in Japan. Being overcrowded, highly industrialized, small in size with no place to dispose its industrial wastes, conditions are just right for catastrophe. Even Japan's Food and Agricultural Organization admits that the nation uses ten times as much pesticides per acre under cultivation than the United States and, as expected, the pesticide content of human tissues, body fat and mother's milk, is also ten times higher.

Japan and its people serve as ideal modules for environmental pollution studies. A decade before Michigan suffered its PCB debacle, Japan went through an almost similar series of events. Sadly, the United States benefited little from that experience, proving once more, as George Santayana so aptly put it, that those who would ignore the lessons of history are doomed to repeat them. Had authorities paid the least bit of attention to the report by M. Kuratsune, T. Yoshimura, J. Matsuzaka, and A. Yamaguchi, in Environmental Health Perspective entitled: *Epidemiologic Study In Yosho, A Poisoning Caused By Ingestion Of Rice Oil Contaminated With A Commercial Brand Of Polychlorinated Biphenyls* much of the pain and suffering in Michigan might have been avoided.

In 1968, five years before the Michigan debacle, a particular Japanese cooking oil, Kanemi rice oil, was inadvertently contaminated with PCB when leaky pipes permitted the two products to mix. Of course no one ever intended to allow PCB to flow in or near pipes conveying cooking oil, but it did happen, and some 1,600 people consumed the heavily laced oil before it could be recalled. Almost immediately skin and liver diseases ravaged the victims. Dark pigments and chloracne blotched the skin, hair fell out, eyes swelled to puffy slits, jaundice, nausea, and the torments of hepatitis took hold. This was labeled *Yosho*, rice oil, disease. Women who happened to be pregnant when they consumed contaiminated oil had defective fetuses or malformed babies. The women who were not pregnant at the time but conceived a year or two later, also delivered PCB to their fetuses and gave birth to babies with marked discoloration of the skin. These PCB babies were called Cola-colored babies because of their peculiar brown cast. Some of the newborns had chloracne, many were weak and feeble, and most of them did not grow or develop as well as their uncontaminated siblings. Kumamoto University's Dr. M. Harada, proved that fetuses and babies became tainted with PCB when it crossed the placenta or appeared in breast milk.

In a mass effort to find out just how and why these chemicals cause such devastating damage, scientists both here and abroad moved to their laboratories and methodically tested PCB, PBB, and their related breakdown products. Working at the University of Wisconsin, J. R. Allen used the rhesus monkey as a model for studying the immediate and long range effects of PCB. Reporting to the National Conference on Polychlorinated Biphenyls sponsored by the Environmental Protection Agency, Washington, DC, in 1976, he described a tremendous fetal wastage in pregnant monkeys fed PCB contaminated food. The number of live births was reduced because fetuses did not develop normally, aborted spontaneously, or were otherwise rejected by the uterus. Fetuses that did mature to term had levels of PCB comparable to those found in American women. Newborn monkeys tended to be on the small side and many of them had chloracne. And when these newborns nursed at the breasts of their contaminated mothers they received additional dosages of PCB, which, when added to that delivered across the placenta, proved fatal. If breast feedings were interrupted early, infant monkeys tended to survive. PCB effects seemed to be reversible.

Experiments with animals suggest that PCB's can cause liver tumors and other malfunctions. Unfortunately, there is a maddening variability in the response of experimental animals to chlorinated and brominated hydrocarbons as well as to their many breakdown products. To date there is no absolute proof that PCB and PBB cause cancer.

The Toxic Substances Control Act was passed in 1976 and then, in 1977 the Environmental Protection Agency developed regulations that banned further manufacture of equipment and devices that contained PBB or PCB. But here is the kicker. Whereas the United States and other wealthy and technologically advanced nations are in a position to clean up their acts poorer sovereign nations can not afford to do likewise. The world had best find an answer because as late as 1979 there was another repetition of the Japanese *Yosho* type of poisoning, this time in central Taiwan. About 2,000 people consumed PCB contaminated cooking oil for a period of over 9 months and came down with *Yu-Cheng, oil disease*. The symptoms and long-range debilities were the same as in the American and Japanese tragedies, but here, possibly because of the extended follow-up, both victims and babies exposed to PCBs while still in the uterus were found to be mentally retarded or slow learners.

Despite all good intentions the situation has gotten out of hand. PCB is now found in wild animals as well as in remote parts of the world. Drs. Marcus and Dora Wasserman of the Hebrew University Hadassah Medical School, Jerusalem, have proven that the levels of PCB are rising in plankton, fish, and birds. In the United States eagles have 14 parts per thousand in their fat, peregrine falcons 2 parts, and herons 0.9 parts per thousand. Scandinavians hold 0.85 parts per million in their fat, Austrians up to 3.5, Germans, in Cologne, 6.8, in Munich 10 parts per million.

At present there is no practical way of rolling back this tide of global contamination. Hopefully, by accident or intent, scientists will discover ways to do this; if not, then, as Roger S. Jones put it, in his book *Physics for the Rest of Us*, "should we be so foolish as to annihilate ourselves in an atomic holocaust or through the strangling pollution of the earth, it will not make the least difference in the scheme of things. The planets, stars, and galaxies, will continue their cosmic schedules, completely oblivious to our passing. So much for human significance."

PARADIGM POINTERS

CHANGE THE PARADIGM AND THE WORLD CHANGES WITH IT. The paradigm that foretold the fate of man-made refuse such as chemicals, mine-tailings and by-products of manufacturing came into being during the infancy of European and American industrialization. It assured us that there was enough land, water and air to safely dilute or destroy anything that we humans made and cast aside. This theory began to fall apart when the hazards of radioactivity were recognized and it fell into disrepute when Rachel Carson, in 1962, called the world's attention to the dangers of chemical pesticides. The current paradigm calls for regulated control over the disposal of hazardous wastes and adds incentives to prevent pollution.

POST SCRIPT

The PBBs and PCBs are certainly toxic, as we have seen, but the herbicide Dioxin is worse. This, the most toxic man-made substance known, was used in Agent Orange to defoliate jungles in Vietnam during the war with that country, and many U.S. servicemen were exposed to it. About a decade after the war ended Maude de Victor, a claims representative in the Veterans Administration Chicago office, noticed that an unusually high number of Vietnam veterans were sick and dying. By February 1978 Ms. de Victor had assembled a list of roughly 30 veterans who had symptoms suggestive of Dioxin poisoning and brought the matter to the attention of federal agencies. The Veterans Administration was reluctant to accept responsibility for something that happened over ten years earlier. "Why didn't these people become sick before now?" was the unfailing refrain.

The answer lies in this chapter. Dioxin, like PBBs and PCBs is taken up by a person's fat; it stays there without causing any obvious symptoms. When these veterans lost weight, due to aging or illness, their fat broke down and released enough Dioxin to produce poisoning.

Strangely, Yosho, PBB and PCB, Yu-Cheng, and Dioxin poisonings, all took place within a few years of each other—the four horsemen of the apocalypse.

chapter 4
AFFLICTIONS OF AFFLUENCE

When we introduce new technologies, ostensibly designed to improve our health and happiness, we often, simultaneously, add insult and injury to our environment and ourselves. Air-conditioning, perhaps, is the best example of such dubious progress—especially in light of what happened at the American Legion Convention in Philadelphia, in 1976.

That saga actually began about a decade earlier when, in the midst of the hot, humid, month of July, in 1965, the staff and officials of St. Elizabeth's Psychiatric Hospital Washington DC came face to face with an explosive epidemic in their institution. Suddenly, without rhyme, reason, or forewarning, 62 mental patients came down with pneumonia. By the next month, August, the disease had killed off fourteen of them. Everyone was perplexed. Pneumonia was usually a winter problem but these were summer attacks. Also, there were too many cases and too many deaths to pass off as normally anticipated or routine infections. Fearing that this brush-fire of cases might rage out of control, calls for help went out to neighboring health departments.

Public health experts and technicians scoured the hospital and examined hundreds of blood and tissue samples looking for the cause of the outbreak. When nothing was found, specimens from the sick and dead were sent the Centers for Disease Control in Atlanta, Georgia, with the hope that the Federal Government's experts with special, advanced, technologies, might shed some light on this mysterious outbreak. But here too all tests proved negative. In due course the blood and tissue samples were consigned to a deep freezer—remaining there for eleven years until they were found to be a vital link to another weird outbreak, in another city, in another group of people—that of the 1976 epidemic of the American Legion Convention in Philadelphia.

Inasmuch as the Declaration of Independence had been signed in Philadelphia and the Constitution of the Unites States of America was forged there, the Pennsylvania Department of the American Legion chose this city both to celebrate the bicentennial of those events and also to stage, from July 21 to July 24, its 58th annual convention. Thirteen candidates for high office were up for election. Some 4,400 Legion delegates and members with families, relatives and friends, from all parts of the country, swarmed into the city and commandeered four Philadelphia hotels. A frenetic, mad-cap ambiance seized the crowd as they regaled at banquets, marched in star-spangled parades, danced in the streets, and attended meetings at the headquarters hotel, the Bellevue-Stratford. On the second night after the assembly convened two Legionnaires developed fevers, felt a tightness in their chests, coughed incessantly, and complained about aching muscles. These were older individuals so it was presumed that they had a cold or flu and they received little attention. As others began to feel queasy they headed for home. This dispersed the sick people so it was impossible to appreciate the extent of the epidemic. Records later indicated that between July 22nd and August 3rd, 149 conventioneers had been struck.

A week after the convention the Pennsylvania Department of Health received an avalanche of reports concerning a pneumonia-like illness and deaths among the people who had visited Philadelphia hotels. One hundred eighty-two cases with twenty nine deaths were reported; seventy eight per cent were men. But the oddest thing about these reports was that 82% of the cases appeared in American Legionnaires and most of them visited or stayed at the Bellevue-Stratford Hotel.

The deaths in Philadelphia triggered a media blitz. Daily tallies of the sick and dying appeared in the press, television and radio. Thomas Payne, a Legionnaire, faced television cameras and described his joint pains, troubled breathing and feeling awful. "Even though I received large doses of penicillin," he said, "my temperature peaked at over 106 degrees." Jim Kelley, another Legionnaire said "I had headaches, was sick in the stomach, was sore in the kidneys, sore in the chest, and couldn't stand; wished I was dead." Headlines across the country splashed provocatives: EXPLOSIVE OUTBREAK, KILLER PNEUMONIA, MYSTERIOUS AND TERRIFYING DISEASE, LEGIONNAIRE KILLER. On August 2, the press officially dubbed the menace LEGIONNAIRES DISEASE.

City and state health departments were called in to explain the mystery and to squelch the bad-press. But the culprit remained elusive; no recognizable cause could be found. Dr. William E. Parkin, Chief Epidemiologist of the Pennsylvania State Health Department said, "It may be one year, five years, or one hundred years before our technology becomes efficient enough to cope with it." On August 2, 1976 the Pennsylvania Health Department sent an urgent appeal for help to the U.S. Public Health Service's Center For Disease Control in Atlanta. Universally acclaimed for expertise in epidemiological intelligence, CDC identifies itself with two symbols: a shoe with a hole punched out of the sole and a miniature barrel of Watney's Ale. The ale is served at the John Snow Pub in London, a shrine for epidemiologists since that is where the city's 1854 cholera epidemic was squelched—by removing the handle of the Broad Street pump.

Dr. David W. Fraser, chief of the Special Pathogens Branch of the Bacterial Diseases Division in the Center's

Bureau of Epidemiology, and Dr. David Heymann were dispatched to the scene to spearhead the investigations. Dr. Fraser had received his MD at Harvard Medical School in 1969 and after an internship and residency at the Hospital of the University of Pennsylvania served as Special Projects Associate in epidemiology at the Mayo Graduate School of Medicine in Rochester, Minnesota. He returned to the University of Pennsylvania to teach at the medical school then moved on to CDC, where he earned the rubric *disease detective*. Dr. Heymann, was young, lean, shy, and fresh out of his postgraduate training in clinical medicine. With them went 31 field workers. One group concentrated on hotels and defined the characteristics of the disease. A second group, in an effort to determine who was getting the disease and where they got it, studied patients and their families. The third group tried to find out what was different between the sick and the unaffected people in the same area. Twenty-three epidemic intelligence officers were assigned to tracking down individuals who had gone home or wandered off to other areas. The search for the cause of the epidemic was launched.

Normally, when CDC personnel are handed a complex problem they first draw an algorithm, a flow-chart in which the pieces of evidence, as they surface, are placed into proper relationship to each other. The investigators then throw their might in the direction of the flow, the direction of the preponderant evidence. But this would not work here since nobody knew whether the cause was a bacteria, a virus, a fungus, or a parasite. Maybe it was not even an infectious agent; the cause could be a chemical, a poison or toxin. No one could explain why some people got sick and not others. Influenza was the prime suspect, sabotage came next.

CDC, with valuable assistance from public health workers of the Pennsylvania and Philadelphia Health Departments, questioned over 4,400 Legionnaires and their families. Then nurses and staff physicians at hospitals where Legionnaires had been treated were questioned and patient records were examined. The four Philadelphia hotels that had hosted the

AFFLICTIONS OF AFFLUENCE 93

Legionnaire's convention underwent special scrutiny. Experts and technologists collected air, water, soil, dirt and vacuum materials from every room that had been occupied by the Legionnaires. Ultimately they spent 90,000 man hours and $4 million testing and examining dust, pigeon droppings, pesticides, plastics and resins, and souvenirs. None contained incriminating evidence of microbes or toxic chemicals. Some of CDC's personnel even ensconced themselves at the Bellevue-Stratford hotel itself for the entire length of the investigation, which took more than a year.

The next tack was to look at the environment. The prime suspect was *paraquat*, a deadly, nonselective herbicide. Now, older and wiser after their experiences in Michigan with PCBs and PBBs (see CHAPTER 3, ELEGY FOR A PLANET) the environmental team performed over 300 gas chromatography and mass spectral analyses by the end of August. If anything abnormal was present in the air or in the soil tell-tale spikes would show up on recording graph paper. There were no spikes.

By now CDC knew this much: Seventy-two people who contracted Legionaries disease were not Legionaries but had stayed at or visited the Bellevue-Stratford Hotel. The disease only struck people who had stayed in Philadelphia. Food and drink were ruled out as a cause because many of the people who remained healthy ate at the same restaurants and at the same hotels as the people who became ill. The infection hit the lungs the hardest. Fevers rose to 104°F and 105°F and victims suffered with chills, headache, dry cough, diarrhea, confusion, kidney and liver complications. The acute illness lasted seven to ten days and one out of six died. But the most comforting evidence was that this thing was not contagious. When sick people returned home their immediate family members did not come down with the disease.

Early on, CDC began to suspect that the culprit was airborne. Data showed that disease victims spent about 60% more time in the lobby of the Bellevue-Stratford Hotel than those who did not become sick. Working around the clock, medical and toxicology specialists examined blood, tissues

Figure 4.1 Joseph E. McDade, M.D. (Courtesy of *Centers for Disease Control and Prevention*, U.S. Public Health Services.)

and samples of excretions of those people who were definitely infected. When autopsies were performed on patients who had died everything was carefully tested and specimens were preserved by deep freezing for later studies. Bacteriologists tried to identify the culprit by culturing secretions, excretions, blood and tissue extracts from affected individuals. Nothing would grow despite trials with over 1,400 different types of nutritive media and special techniques. Using traditional stains and standard light microscopes, investigators saw nothing. The electron microscope proved no better. Next came antibody tests. If a victim's blood, when mixed with antibodies against known microbes, caused clumping reactions in the test tube, then the cause of a particular disease could be identified. Standard and florescent antibodies were employed to test for, among other things, influenza, Q-fever, mumps, measles, adinoviruses, and unconventional organisms called chlamydia and mycoplasma. When none of the antibodies worked, investigator were forced to conclude that the causative agent was either new, something against which there were no antibodies, or something old that had changed its characteristics so that antibodies no longer recognized it.

Batteries of tests were also directed at diseases transmitted by ticks and mites and those caused by various yeasts, molds, fungi and viruses; again failure. Microbiologists tried to isolate viruses by inoculating chicken eggs, monkey cells, human cells, guinea pigs and mice with victims' blood, tissues and excretions; nothing showed up. Lung, liver and kidney samples from Legionnaires who had died were now tested for heavy metal poisoning. The most likely suspects were mercury, arsenic, thallium, nickel and cobalt. A total of 23 potentially toxic metals were tested; all proved negative.

Not only did CDC have to fight a ruthless, invisible assailant but it also had to contend with the hubris of personalities seeking media attention. Everyone, from Capitol Hill down to small town gossip mongers accused the investigators of dragging their feet, or worse still, holding back vital information. When the Pittsburgh Post-Gazette broke the news that two elderly people died shortly after getting swine-flu shots, during that $135 million ill-fated campaign to vaccinate the nation, accusing fingers were pointed at the vaccine as well as at swine flu itself. Personnel at CDC were pulled off the Legionnaires puzzle in order to look into these allegations. They were subsequently disproved but, again, after a nettlesome distraction and at no little cost.

Such reprehensible attacks, as the record proves, had no basis in fact. CDC chased down every real and spurious clue even when logic suggested that they were on a 'wild goose chase'. For example, before the Legionnaires convened in Philadelphia, hundreds of magicians held their annual convention at the Bellevue-Stratford. The question arose did they bring any unusual chemicals with them for their tricks or illusions that might have triggered this epidemic? After many hours of work and sleepless nights CDC found that the magicians used nothing at the hotel that could in any way be incriminated.

In the same vein, Dr. William Sunderman, Jr. of the University of Connecticut School of Medicine, after studying some samples of Legionnaires Disease, concluded that unusually high levels of the metal nickel were present. This set up a search for nickel carbonyl. It was later proven that the nickel had come off surgical instruments used in the original autopsies.

The first positive clue about the nature of this weird epidemic surfaced in October 1976, roughly two months after the convention, when a pathologist who had been in contact with victims of Legionaries Disease came down with the disease. Scientists now had convincing proof that an infectious virus or bacterium was the trouble-maker and when it got into humans it preferentially invaded the lungs. Since 'this thing' would not grow in the laboratory in standard test tubes or petri dishes with the ordinary bacterial promoting growing media, investigators believed that they were dealing with a virus, so they switched to special methods designed to study viruses. This, unfortunately, proved self-defeating because antibacterials are added to viral-type culture media—wiping out the bacteria encourages viruses to grow. Without knowing it, microbiologists destroyed the very thing they were looking for.

By using statistics that clustered cases around sources of infection, precisely as Semmelweis had done in the previous century to link child-bed fever with dirty hands, CDC was forced to conclude that Legionaries Disease was caused by bacteria. Since standard microbiological and antibody tests proved entirely negative, and nothing could be grown in the laboratory, investigators concluded that they were looking for a hitherto unknown organism.

Dr. Joseph McDade (see Figure 4.1.) cracked the case almost inadvertently, by not using bacteria-finding tools. Unlike the experts around him he eschewed petri dishes, test tubes, biocultures, and experiments with mice. Dr. McDade was basically a research microbiologist in the Leprosy and Rickettsia Branch of the center's Virology Division and felt more comfortable with rickettsial-hunting technologies that leaned heavily on culturing organisms in living tissues.

Rickettsia, the class of organisms so familiar to McDade, were named after Dr. H.T.Ricketts who pioneered the study of Rocky Mountain Spotted Fever, and, while investigating an epidemic of typhus fever in 1910, contracted the disease and died. These peculiar organisms possess many of characteristics of both bacteria and viruses: they assume many shapes, are visible with ordinary light-microscopes, they multiply within certain cells of susceptible animals and they thrive in various insects in nature. Rickettsial diseases are transmitted by the human body louse, rat flea, ticks and mites, therefore they flare up whenever man and rodents become close neighbors. They provoke severe and acute illnesses like typhus, rocky mountain spotted fever, scrub typhus and Q fever, all of which are accompanied by high fevers and a skin rash.

Joe McDade epitomized the ideal scientist. His sharp blue eyes missed nothing as they peered through thick heavy glasses. An intense mien accompanied his total dedication to work. He was a meticulous researcher and everything in his laboratory was neat and orderly. After obtaining his Ph.D. in microbiology from the University of Delaware, in 1967, he studied rickettsial infections at the U.S. Army Biological Center at Fort Detrick, Maryland, for two years, then became Director of the cell-production department of Microbiological Associates, Inc. for another two years. Just before joining CDC, from 1971 to 1975, he served as a research associate in microbiology at the University of Maryland School of Medicine, spending most of his time in Cairo and Ethiopia studying the suspected extrahuman cycles of epidemic typhus.

This background put Dr. McDade in an ideal position to run down the germ responsible for Legionnaires disease. His discovery was fortuitous but not a blind hit or miss foray. Since victims of the Legionnaire malady all developed pneumonia it was only logical to look for the rickettsial organism that causes Q fever since pneumonia is a common component of that disease. McDade, aware that Rickettsia grew only in living tissues, inoculated guinea pigs with extracts from lungs obtained from deceased legionnaires. He then homogenized the guinea pigs' spleens where the germs would be concentrated and injected portions into the yolk sacks of embryonated chick eggs. After an incubation period of seven to ten days the yolk sacks were examined microscopically. Although McDade saw no Rickettsia, he did observe some unidentified rod-shaped bacteria.

There was a small element of luck in McDade's switch to living tissue as a culturing medium and in the selection of the guinea pig for test purposes. Researchers at CDC customarily use mice for microbiological testing because the small rodents are relatively inexpensive, easy to handle, and there is an extensive library on their anatomy, physiology and genetics. Indeed, throughout 1976, CDC ran countless tests on mice but failed to find the cause of Legionnaires disease. When Dr.

McDade switched to guinea pigs for test purposes, a species known to foster the growth of Rickettsia, he serendipitously discovered a new bacillus.

There was no way of telling, at that time, that this was indeed a new bacillus. It refused to take up bacterial stains therefore could not be readily seen. By sheer determination and endless hours of peering through the microscope Dr. McDade determined that these strange objects were rod-shaped. But nothing more could be ascertained.

McDade's progress was abruptly interrupted when he was assigned to another pressing problem elsewhere. But thoughts about the funny-looking rods in his yolk-sac extracts never left his mind. Two days after Christmas, still 1976, Joe McDade decided to take another look at the slides he had made four months earlier. There, under his powerful microscope, he again found the almost-invisible tiny rods. The following month, in January 1977, in collaboration with CDC's Charles C. Shepard, the net began to close in on the culprit.

On the assumption that survivors of Legionnaires Disease had manufactured antibodies against the 'thing' that had attacked them, blood samples were taken from thirty-three former, recovered patients and antibodies were extracted from the serum. Sure enough, the antibodies reacted positively with germs growing in McDade's infected chick yolk sacks. This proved that the yolk sacks held the germ that caused Legionnaires Disease. To make sure that they had targeted the right organism, extracts from the yolk sacks were mixed with serum samples from people who had never been exposed to Legionnaires Disease and, as might be expected, there were no positive reactions.

Now that the prime suspect and identifying antibodies were on hand, Dr. McDade retrieved the blood and tissue samples that had been placed into a CDC deep-freeze eleven years earlier from the pneumonia epidemic in St. Elizabeth's Psychiatric Hospital, in Washington, DC. Amazingly, they too reacted positively with the yolk sac organism—proving that they contained antibodies identical to those found in

Legionnaires survivors. On Tuesday, January 18, 1977, at a press conference attended by the Surgeon General and CDC staffers, the announcement was made that the St. Elizabeth's epidemic and Legionaries disease were caused by the same 'bug'.

"If this thing will grow in chick embryos, in guinea pigs, and in the serum of human beings, why the heck can't we see it under the microscope? It's there. It's just not showing up," scientists reasoned. What they did not know was that the organism was tiny, translucent and therefore impossible to see with the ordinary microscope. This impasse was removed when Francis W. Chandler and his colleagues at CDC tried to stain the newly discovered 'unknown' organism and found normal bacterial stains did not work but that silver impregnation techniques which had been developed 50 years earlier for the detection of spirochetes, like that of syphilis, made the rod-shaped interlopers distinctly visible. "But," as McDade was quick to point out, "we couldn't be sure whether we were seeing Rickettsia or small bacteria."

Robert E. Weaver at CDC then found that he could grow the new germ on the type of bacterial nutrient media used for the culture of the gonorrhea germ. This media contains 1% hemoglobin, from red blood cells, and therefore contains iron. Iron, as it turned out, proved to be a prerequisite for the growth of the new bacterium. Weaver subsequently inoculated this medium with a heavy suspension of yolk sack material and was able to grow and isolate, for the first time, the agent responsible for Legionnaires Disease. This conclusively proved that it was a bacterium and not a rickettsial organism, since the latter requires living tissues for survival. It was officially labeled *Legionella pneumophila* in deference to its 'love for the lungs'. It took another decade—the time needed to developed DNA hybridization techniques—for Donald J. Brenner and Arnold G. Steigerwalt, at CDC, to prove that this was not a hybrid of a known bacterium but an absolutely new species. James C. Feeley and his associates at CDC then brought to light the fact that *Legionella pneumo-*

philia must have cysteine, an amino acid, in addition to iron, to grow. That explained why previous growing attempts with ordinary media failed.

Now, armed with a way to grow and stain their quarry, McDade and Shepard peered at it through their microscopes to learn more about it. "The presence of a few organisms in a guinea pig's spleen is far from establishing that they had caused the patient's deaths," Shepard said. It was time to move on to the crucial tests; time to establish a new paradigm for Legionnaires' disease.

By now the researchers had powerful new tools to work with: specific antibodies from the Legionnaires' disease victims, florescent antibody techniques, ultraviolet-light microscopes, and a way to grow the strange rod-shaped germs. The case was clinched when rod-infected yolk sac material was injected onto the foot-pads of guinea pigs and they developed symptoms similar to Legionnaires' disease. When the same germs were recovered from these animals, Koch's postulates were fulfilled; effect and cause were positively linked.

At this juncture the disease detectives determined that Legionaries' illness was not transmitted from person to person but that germs had to be inhaled in order to establish an infection. But they still had no inkling about its hiding places and the routes of spread. This is not too surprising because Legionella normally and characteristically lives in pond scum, the gooey detritus that collects around stagnant water, especially that of cooling tanks used with air-conditioners. The organism prefers dark, nutrient rich, oxygen poor, environments, and sometimes lives inside other cellular organisms. Investigators missed it for a long time because they looked down instead of up. They kept looking at soil, housedust, dust mites in rugs, yeasts and fungi in damp places instead of looking up at the water cooling towers on the roofs of hotels. For a while CDC even suspected that employees might be silent carriers of the germ, innocently spreading it to hotel guests. Ultimately the germ was found hiding in the Bellevue-Stratford Hotel's water-cooling tower where evaporating water cools the air for air conditioning. Clinging to the biofilm or scums along the edges of the cooling tower, Legionella were actively pumped into the hotel's lobby and hospitality suites during the hot month of July.

The Centers for Disease Control once estimated that about 2,000 to 6,000 people die every year from Legionnaires Disease and this has probably been going on for decades before it was recognized. Legionnaires' disease is not a new disease; it is actually a recently recognized old disease. Case reports from Scotland and Spain of an unknown type of pneumonia appeared in the medical literature for many years before the discovery of *L. pneumophilia*. After science learnt about the bacillus and its behavior, it was generally accepted that those cases were indeed Legionnaires' disease.

While still tracking down the Philadelphia epidemic, CDC spotted cases in eleven different states. By September 1977, a year later, it found three hospital outbreaks in Ohio, Vermont, and Tennessee. At the Wadworth Medical Center, a veterans' hospital in Los Angeles, where an outbreak claimed sixteen lives, many of the staff members and three per cent of the patients were found to be infected. It even appeared in a brand new hospital that had not even been occupied. *Legionella* then cropped up in Nottingham, England and elsewhere around the world.

In the few years since its discovery the disease has been reported in almost every part of United States as well as Europe, Australia, Canada, and the other continents. Legionella is not restricted to urban settings. In Spain, recently, L. pneumophilia, hiding in shower heads and toilet tanks, caused havoc in a sparsely inhabited rural hamlet.

By 1978 Legionella bacteria were discovered in soil, ponds, cooling towers, water driven condensers, slow flowing creeks, mud, polluted and silty water, construction sites, and in steam turbines. Upon closer scrutiny germs were detected in shower heads, grocery store vegetable misters, hot tubs, fountains and humidifiers and water aerosolizers. Paradoxically, Legionella

AFFLICTIONS OF AFFLUENCE 103

thrived in the very sources that we associate with cleanliness—toilets, showers, and tap water pipes and faucets. Apparently, Legionella can survive over a year inside pipe biofilms and can withstand water temperatures from freezing to 110 to 115 degrees Fahrenheit, the temperature of hot water normally found in homes, office buildings and hospitals. It can be killed by raising the temperature to 170 degrees but this would scald users. Even distilled water occasionally contains small numbers of Legionella organisms. The ordinary amounts of chlorine used in water purification, 0.2 parts per million, may not be adequate to kill high concentrations of bacteria.

In the United States alone we see three million cases of pneumonia a year, and in one-third of them no causative organism is identified. About 500 to 1,000 cases of Legionnaires Disease are reported to CDC annually but since not all cases are proven, the incidence is probably higher, more likely 25,000 to 50,000 cases a year. Air-conditioning was originally blamed for the steady increase in the number of cases. Laws now mandate frequent and thorough cleaning of all cooling towers and large air conditioning systems.

Roughly 30 years before the Philadelphia epidemic, in 1947, while working with the U.S. Food and Drug Administration, F. Marilyn Bozeman came across four types of Rickettsia-like organisms that just did not fit the descriptions and classification of known Rickettsia. Clinical and autopsy specimens from patients with respiratory illness and high fever had been sent to her for diagnosis. She inoculated guinea pigs with suspect infected tissues then inoculated embryonated eggs; thus she stumbled upon these strange germs. Since they could not be identified at the time, they were stored in a deep freezer. In 1978 they were retrieved and tested with the now readily available *L. pneumophilia* antibodies. Three of the four Rickettsia-like agents were not related to Legionnaires' disease but the fourth was virtually identical with it.

The most enigmatic event of the entire Legionella saga took place eight years before Legionnaires' disease even appeared in

Philadelphia. On July 2nd, 1968, during a heat wave, 95 of the 100 people who worked in a single building of the Oakland County Health Department in Pontiac, Michigan, developed high fever, headache and muscle aches. Some had diarrhea, vomiting and chest pain but there was no pneumonia. The entity, labeled *Pontiac Fever*, lasted only three or four days and all patients recovered. No cause for the outbreak was ever found. Interestingly, everyone in the building who had been there when the air conditioning was in operation became ill; none of the people in the building when the air-conditioning was off were stricken. Guinea pigs were exposed to an aerosol of water from the air conditioning system but they showed no ill effects. Serum specimens were collected from the Pontiac Fever patients and were preserved at CDC. When tested for antibodies to the newly isolated Legionella, in 1977, they reacted positively. Apparently Pontiac fever had been caused by the same agent that caused Legionaries disease. But rather than clearing things up the waters became muddier. Pontiac fever patients did not have pneumonia; 95% of the people exposed to germs came down with the disease; the incubation period was one to two days. By contrast, Legionaries disease almost always produced pneumonia; it struck only 5% of the people exposed to it; the incubation period was two to ten days. In other words the same organism appears to have different behavior in different epidemics. The paradigm at present allows for two forms of Legionellosis, the severe life-threatening pneumonic type seen in Legionaries disease and the mild, self-limiting, non-pneumonic form, Pontiac Fever.

Even after solving the riddles of Legionnaires Disease and uncovering the defects in the Bellevue-Stratford Hotel's air conditioning system, CDC was continually challenged by the unpredictable nature of **L. pneumophilia**. Going back to the 1965 outbreak at St. Elizabeth's Hospital scientists have concluded that it was probably not triggered by air conditioning. During that summer, several sites on the hospital grounds had been excavated for the installation of new lawn-sprinkler systems. Stephen B. Thacker and John V. Bennett of CDC found that patients whose beds were closest to the excavations were the ones who came down with the disease. It is believed that dust carrying **L. pneumophilia**, raised in the process of disrupting the soil, spread through the air and infected the patients.

In 1978 nine people at Indiana University, Bloomington, Indiana, came down with Legionnella. CDC traced the bacillus to a cooling tower but they also found it in a natural stream that flowed along the campus and in the earth bordering the stream.

In 1983 five cases of Legionnaires disease at the University of Chicago Hospital were traced to a portable room humidifier and to the inhalation of aerosolized tap water from jet nebulizers.

In 1988, at Stanford University Medical Center, in California seven patients who had undergone surgery for the replacement of heart valves, developed Legionnaires disease. There were no new cases after the hospital switched to sterile distilled water for bathing post-operative patients.

In 1989 CDC investigated cases of Legionellosis that sprung up in a Seattle Hospital between 1982 and 1988. Fourteen patients developed the disease, half of them in 1988. Seventy-nine per cent of them became ill in October through March when cooling towers were shut down. This outbreak was traced to contaminated respiratory equipment, nebulizers and oxygen humidifier bottles.

In 1990, in conjunction with the Louisiana Office of Public Health, CDC confirmed 34 cases of Legionaries disease, including two fatalities. Investigations proved that a mist system used for spraying vegetables at a Winn-Dixie supermarket in Bogalusa, Louisiana, had been contaminated with **L. pneumophilia** and passed the bacteria on to shoppers.

On July 15, 1994 CDC was notified by the New Jersey State Department of Health that six people on board the cruise ship Horizon, on its way to Bermuda, had contracted pneumonia. By August 10th the count had risen to fourteen patients. Tests proved conclusively that they had Legionaries disease and it was probably acquired in contaminated whirlpool baths.

The disease is commonly found in cigarette smokers but not in cigar or pipe smokers. It is also more common in heavy drinkers. Gregory A. Storch and William B. Baine of CDC found cases more frequently among travelers, construction workers, and people living near sites of excavation or construction. In another study, employees who worked outside of a building had much higher levels of antibodies to Legionella than those that worked inside. Since the germ lives in the soil this is not too surprising. People who have just had surgery and immunosuppressed individuals, either with AIDS or undergoing anti-cancer therapy, are particularly prone to getting Legionnaires Disease.

Since its discovery, over fifty strains of **L. pneumophilia**, falling into four serological groups, have been identified. Most of the 1976 Philadelphia cases belong to group one. Culturing for Legionella infection is still the most sensitive and specific means of diagnosing the disease but other methods, like antibody tests on respiratory secretions and radioimmunoassay for the detection of antigens in the urine, are faster and less expensive.

L. pneumophilia is resistant to a wide spectrum of antibiotics. In the Philadelphia epidemic most patients were treated with tetracycline after doctors found that penicillin did not work. Currently erythromycin and rifampin are favored for treatment.

Still unanswered: Why is this organism so persistent since it is so hard to grow in the laboratory and has such strange nutritional requirements? Why does it succeed so well in its weird scum-like environment? Why does L. pneumophilia cause two different diseases, Legionaries Disease and Pontiac Fever?

PARADIGM POINTERS

Robert Hudson, medical historian at the University of Kansas stated, in 1978, "The Philadelphia event remains unsettling

because it shows the very real limitations of our tools for investigating an apparently new microbial disease."
While this premise is still basically true it fails to take into account the fact that providence has a way of interfering with the normal course of events—that unexpected discoveries, like that of L. *pneumophilia*, will shove the unknown into the realm of the known and trigger revolutionary paradigm shifts.

POST SCRIPT

1. The air conditioner itself came into being quite by accident. William Carrier, in 1904, devised an air cleaner that moved air through a spray of water. When he noticed how much cooler he felt under the sprayed air he engineered a device that has since become the standard air conditioner. Individual room air conditioners were designed by H.H. Schutz and J.Q. Sherman in the early 1930's.
2. The media fanned the flames of fear when Legionnaires Disease first erupted. Notables seeking votes, contributions, or publicity, found it expedient to clamor for immediate answers and miracles on demand. Congressman John M. Murphy, a particularly venomous Democrat from Staten Island, NY, without any scientific background, accused CDC of being lax, even derelict in carrying out its duties and investigations. While CDC was 'in, over their heads' sifting, sorting and tracking all the clues that had been collected, the Congressional Subcommittee for Consumer Protection, part of the House Interstate and Foreign Commerce Committee, released a report that condemned CDC's efforts in Philadelphia and also accused the agency of sabotaging the inquiry. "It appears to be the consensus of opinion" the report stated, "that the failure to save, take, and keep free from contamination the tissues of the victims of the epidemic is clearly the reason that ultimate resolu-

tion of the cause of Legionnaires Disease may never be found."

Read the paper, listen to the radio, watch the television news report, or browse the internet, and you will hear and see the same vituperation fired at the research arms of government, scientists, academia and pharmaceutical companies for not curing all of our current ills. Times may change but not human nature.

chapter 5
TOO MUCH OF A GOOD THING

The previous chapter dealt with an affliction brought on by new technology and the one before that implicated toxins. Here we meet problems where both technology plus toxin are to blame. But the toxin in question in not one that you would ordinarily think of as a dangerous poison; it is oxygen. And the technology is part of a seemingly innocuous object, the incubator for newborns that allows extremely small premature babies to survive.

Without food, humans survive for about thirty days; without water, three days, and without oxygen, three minutes. With the exception of certain unusual people, such as pearl and sponge divers, this three minute threshold for oxygen determines our precarious grip on life.

Dependency on oxygen is a relatively new adaptation. When our planet was first formed it did not contain much oxygen so early development of life must have occurred essentially under absent or low oxygen conditions. Things changed when blue-green algae evolved and began to use water for energy, which, in conjunction with photosynthesis, liberated free oxygen into the air. Ultimately our atmosphere became infused with 21% oxygen, its present state. This all-pervasive atmospheric pollution with oxygen certainly altered the pace and direction of evolution, so much so, that all respiring organisms eventually became enmeshed in a cruel bind. They needed oxygen for life but were injured by it unless it was strictly controlled.

Too little and too much oxygen are equally bad for humans. We huff and puff as we ascend to higher altitudes and conk out two miles up where the air thins out and oxygen becomes scarce. On the other hand, oxygen pressures higher than those of our atmosphere are toxic to life. Professor Norman Ashton, who we will meet presently, put it this way: "all living matter is vulnerable to the injurious effects of oxygen and man is certainly no exception. It is strange indeed that the element upon which our very existence depends, should at the same time be so alien, and withstood only by the antioxidant defense mechanisms our cells have elaborated. We are born into and live in a poisonous atmosphere .... we all age and die because, it is thought, we lose (our defenses.)"

Nice pink cheeks and a rosy complexion, the salubrious effects of breathing oxygen, were observed shortly after its discovery and records going back to 1780 show that it was administered to babies for that purpose. But oxygen never really caught on as a reliable medical tool until British scientists, appalled by the suffering of soldiers exposed to poison gas during World War One, began to look into the physiology of respiration. The first reward from their studies was the development of oxygen rooms and oxygen tents for adult patients. The second was preliminary designs for oxygen-enhanced infant incubators. By the 1920's Western physicians began using oxygen with less trepidation, especially for influenza and pneumonia and for babies who were blue at birth. In the 1940's and 1950's, permanent installations in hospitals and clinics piped oxygen directly to the patient's bedside while portable tanks brought it into the home. Oxygen became as important as the newly discovered antibiotics.

Full term babies who have enjoyed 39 to 40 weeks of development in their mother's womb weigh five and a half pounds or more at birth and are equipped to thrive in the earth's harsh environment. Infants born prematurely, however, are either deficient or totally lacking in enzymes that protect the body from the onslaughts of oxygen. The critical size for a baby's survival is somewhere between one to three pounds, a stage reached around the fifth to sixth month of pregnancy. Newborns weighing less than this are ill-equipped for independent existence outside of the mother. Only 40% of babies born between 23 and 25 weeks of gestation survive and half of them will need special education because of blindness, cerebral palsy, or learning disabilities. Medical technology cannot as yet reproduce the mother's intrauterine environment but neonatologists in special intensive care nurseries, by using oxygen judiciously, can salvage many of these extremely 'unripe', tiny, prematures.

In the two decades after World War I, American pediatricians focused their attention on problems related to newborns generally and prematures, particularly. Besides working out the details for proper feedings and maintaining body temperature they designed practical incubators that would simulate conditions inside a mother. This led to the development of a glass incubator that maintained a constant temperature and high humidity yet permitted unobstructed observation of the unclothed baby within it. When oxygen had proven its value, it too was delivered into the incubator.

Physicians in those early years felt obligated to super-oxygenate babies. Premature babies regularly become blue and stay that way for inordinate periods of time: their lungs remain collapsed or underdeveloped; softness of the ribcage and abdominal distention hampers breathing; poor synchronization with swallowing often clogs the windpipe and lungs with food; brain hemorrhages devastate their immature central nervous systems; they have poor circulations and develop pneumonia easily. At first oxygen was supplied through tubes and an inverted funnel placed over a baby's face. In

1923, the pioneer American pediatrician Harry Bakwin, MD, proved that the funnel method of delivering oxygen was inefficient and that poor oxygenation was the main cause of infant deaths. He subsequently devised fine rubber tubes that could be introduced into the baby's nose or back of the throat for better oxygen delivery.

Up to this point there was little risk of too much oxygen because the methods of administration were less than perfect and levels of oxygen seldom exceeded 30%. In 1938, however, Dr. Chappell developed an air-tight incubator that could deliver and maintain high concentrations of oxygen. When commercially manufactured, in the early 1940s, it was universally adopted.

Back in the 1940's and early 1950's, a number of pediatric physiologists pointed out that premature babies often breathe irregularly and, though not visibly blue, they may actually be suffering from an oxygen deficiency. When placed in an atmosphere of 70% to 90% oxygen their breathing usually became regular and no harmful effects could be detected. This sanctioned the use of high concentrations of oxygen for the routine treatment of all premature babies.

The association between blue babies, lack of oxygen, and brain damage was always suspected but hard to prove. Nonetheless, animal experiments during these two decades conclusively proved that oxygen starvation caused severe, irreversible damage to the brain and central nervous system, and was responsible for cerebral palsy. It then became incumbent upon doctors to prescribe oxygen for all threatened newborns. Medical text books and health authorities recommended that all prematures be maintained in 40% to 60% oxygen regardless of the presence or absence of respiratory symptoms. The U.S. Children's Bureau stated that an oxygen concentration of 38% to 42% would be good for prematures and added that an oxygen concentration of 46% "is relatively safe for an indefinite period whereas 100% oxygen may be given for 24 hours." It was also believed that the newborn infant was more resistant to higher pressures of oxygen than the adult. By the end of 1948 faith in the value of oxygen was explicit and universal. Pediatricians faced with neurologic damage and high mortality rates in prematures felt justified in using as much oxygen as necessary to protect their tiny patients. New and better incubators delivered more oxygen to more babies in more premature centers and intensive care nurseries. No one ever suspected that this new technology might be causing trouble. In those early days, bubbles in water and air-flow gauges indicated the rate of oxygen delivery to a tent or incubator. By the end of World War II these were supplanted by oxygen analyzers that measured the per-cent of oxygen in air samples. Actual assessments of the amount of oxygen in a person's blood required large blood withdrawals and laborious chemical tests. Precise micromethods for determining the amount of oxygen in very small blood samples were perfected in the 1960's and 1970's, and these became the mainstays in most U.S. hospitals. Although fast and efficient this technology nonetheless left large blocks of time, especially in tiny prematures with poor blood supplies, where there was no way of knowing exactly how much oxygen was in their tissues. Currently, with pulse oximetry, where red and infra-red light placed on one side of a blood vessel passes through the skin to a photodetector on the other side of that vessel, we can evaluate oxygenation continuously, reliably, and without the need for blood samples.

Between 1940 and 1942 a disturbingly high number of babies with a strange eye disease and blindness began showing up in hospitals and clinics across the country. Theodore L. Terry, MD, an ophthalmologist at Harvard University, perplexed by the sudden, unexplained, frequent appearance of this new condition elected to look into it. He first examined statistics and found that this odd eye problem appeared almost exclusively in premature babies especially those with low birthweights. At the Boston Lying-In Hospital approximately 12% of the small premature babies had suffered from eye damage. He then meticulously examined affected eyes and discovered that an abnormal string of fibers somehow developed between the lens and the retina, the light-sensing tissue at the back of the eye. In due time the retina was pulled loose from its source of nourishment and this caused blindness. Dr. Terry surmised that there were two possible explanations for this series of events. Normally, blood vessels appear in the fluid that fills the space inside the developing eye but they disappear later on. If these vessels failed to disappear, they could conceivably form an attachment to the retina and later pull it loose from the back of the eye. Another possibility presumed that there was an overgrowth of connective tissue fibers from the membrane adherent to the back of the lens. This cord-like attachment to the retina later dislodged it and caused blindness. In his preliminary report made in 1942 Dr. Terry coined the term *retrolental fibroplasia*, RLF, as a label for this new eye disease. It means excessive fibrous tissue behind the lens.

Over the next decade, from 1942 to 1952, over 100 articles were published in Western medical journals calling attention to the steady rise in the number of cases of RLF. By 1950 at least 8,000 infants were totally blind from the disease and by 1953 the figure rose to 10,000. During this same period in Great Britain, RLF struck as many as 40% of very small premature babies. Unfortunately these revelations received comparatively little notice because they came to light at the same time that the devastating malformations in babies due to *Thalidomide* captured public attention.

The first clues as to what was going on were discovered by ophthalmologists William Councilman Owens, MD and his wife, Ella Uhler Owens MD. Upon examining the records of the Harriet Lane Home, the children's hospital of Johns Hopkins University, they found that in 128 children born prematurely between 1935 and 1944, before high doses of oxygen became routine, none of them had RLF. They also confirmed Dr. Terry's statistics; 4% of recent newborns had RLF but in prematures the incidence was as high as 12%. In studies carried out as early as 1941 these intrepid researchers were able to show that RLF developed two to six weeks after birth. But after this brilliant start they chose a detour rather than the main road—they presumed that a deficiency of Vitamin E was responsible for all the trouble. This was neither a capricious whim nor a wild guess. It was based on the well known fact that premature babies do not tolerate fats, they absorb them poorly, and unless the fat soluble vitamins are given by injection, prematures suffer from deficiencies. "Of the fat soluble vitamins A, D, K and E," they stated, "the vitamins A and D have been provided routinely and vitamin K is usually given early after birth. Vitamin E alone of the fat soluble vitamins has not routinely been included." Pointing to the central nervous system lesions produced in vitamin E deficient animals and the greater susceptibility to Vitamin E deficiency in immature animals, these researchers were not too far out in postulating that Vitamin E deficiency was at fault. They began a controlled study by giving Vitamin E to alternate prematures in the nursery of the Harriet Lane Home and collected enough data to show that vitamin E supplements played a protective roll in reducing the incidence and severity of RLF. When this was reported to the world at large, many people felt that it was inhumane to withhold such an important therapy and the investigators were advised to discontinue their experiments and give vitamin E to all infants. In the ensuing years however, there was no corroborative evidence that vitamin E was useful and it fell by the wayside. By 1953 the administration of vitamin E had been gradually discontinued in most nurseries throughout the world. DO NOT LOSE SIGHT OF THIS EVENT. IT WILL BECOME IMPORTANT LATER IN THIS CHAPTER.

Doctors Owens and Owens were not the only brains to be misled by 'red herrings' in the frenetic search for the true cause of this strange eye disorder. A noted researcher, W.A. Silverman, MD, listed some 70 factors that he speculated could be the possible cause. These included such things as too much light, too many vitamins, too few vitamins, too much iron, too much oxygen, and birth defects. It was also generally recognized that rats born to mothers with vitamin A deficiency had congenital eye defects. This was extrapolated to humans and for a while vitamin A deficiency was considered to be a cause of RLF. Subsequent studies showed no protective effect by giving vitamin A to newborns. The atomic bomb was also being developed at this time so it was natural to blame radiation and atomic fall-out. Those new-fangled infant formulas were just coming on the market so they too came under suspicion. Many investigators felt that the high amounts of salt in cows milk, as compared to human breast milk, was the culprit. This was never proven. Dr. Terry himself once stated that of "all the probable causes, precocious exposure to light is considered the most tenable." Studies coming out in 1949, 1950 and 1954, disproved Dr. Terry's thesis; infants whose eyes had been shielded from light still developed the disease.

The scramble to find the cause and cure for RLF produced some awkward results. Many learned investigators completely ignored their own findings about RLF and continued their research on light, vitamins and iron believing all the while that oxygen was good and necessary for babies. Nothing could shake that conviction. In their defense, it should be noted that science had already established the cause and effect relationships between oxygen restrictions, brain damage, and death in newborn babies. Almost everyone was hesitant in accepting the notion that there could be such a thing as too much oxygen. As a matter of fact, one of the foremost scientists of that time felt he had conclusive proof that RLF was due to an insufficiency of oxygen and strongly advocated treatment with supplements. Fortunately this advice was offset by other investigators who kept saying that the disease tended to occur more often in babies exposed to high oxygen concentrations and advised caution.

How and why oxygen caused RLF proved to be a hard nut to crack. For one thing, examinations of eyes performed soon after birth usually were normal. Signs of RLF tend to appear a little over a month later. Mild RLF often resolves completely and the eye shows normal development. Only in the more severe cases is there scaring, nearsightedness and retinal detachment with blindness. To complicate matters still further, TOXINS DO NOT BEHAVE IN A REGULAR MANNER. For instance most people taking aspirin do not develop stomach ulcers, but that does not mean that aspirin never causes ulcers. Similarly, oxygen, under special circumstances, behaves like a toxin, therefore its effects can be exasperatingly variable.

Nonetheless, by the middle of this century scientists began closing in on the culprit. Bertha A. Klien, MD, the internationally recognized ophthalmic pathologist who had received her education in Vienna but fled to the United States at the approach of Nazism, began collecting and examining RLF eyes in her Chicago laboratory at the Illinois Eye and Ear Institute. By 1949 she concluded that a lack of oxygen or interruption in its delivery to tissues of the eye set into motion a number of devastating events that led to RLF. In deference to her strict scientific training she hesitated to come right out and state this in a scientific journal because absolute proof was as yet not available.

That same year, however, Drs. Kinsey and Zacharias listed all the possible factors that existed in the Boston Lying-In Hospital, between 1938 to 1947, that could possibly be related to RLF. By checking for differences between the era of no disease and the era of rapidly-rising disease, only increased oxygen therapy stood up against scrutiny. But this did little to clear the air. Other reports showed that there was no correlation between oxygen and RLF. In 1951, for example, studies conducted jointly in two New Orleans medical schools failed to find an increase in cases of RLF despite liberal use of oxygen. And in another study, a year later, oxygen restriction had not led to a decrease in RLF cases.

These conflicting results can be explained by the fact that oxygen, in those days, was delivered in an unregulated manner, mostly by hit and miss procedures. Concentrations of oxygen varied in different settings. Some incubators were air-tight, others were not; oxygen levels fluctuated as incubators were opened for feedings and other procedures. Kate Campbell, MD, in Australia, reported in 1948 and again in 1951 that with the introduction of "a more efficient oxygen cot" the incidence of RLF zoomed upward. She insinuated, rather strongly, that too much oxygen was the probable cause of RLF. Shortly thereafter, in 1952, another Australian physician, H. Ryan, MD, stated that RLF had not occurred in his hospital prior to the introduction of "an efficient oxygen cot."

Dr. Kate Campbell and Mary Crosse, MD are often credited with having established the relationship between oxygen and RLF. Their findings were ultimately proven to be correct but warnings were not heeded at the time when they were made because, as other well-meaning colleagues pointed out, observations were made without appropriate controls, the amount and manner of oxygen usage was not exactly measured, and the study did not have proper statistical design. Basically, the Campbell-Crosse observations did not provide justification for exposing infants to the risk of increased mortality and brain damage that might follow oxygen reduction.

The first controlled clinical study to test the association of RLF with excessive oxygen was begun in January 1951 in the Gallinger Municipal Hospital, Washington, DC by ophthalmologists Patz, Hoeck and DeLaCruz. Infants were divided into high and low oxygen groups and oxygen levels in incubators were measured at eight hour intervals with electronic analyzers. The nursery care of the two groups was otherwise identical. The incidence of RLF in the high oxygen group was about 60% compared to 20% in the low oxygen group. Although the study was supposed to end in June 1955 it was interrupted a month earlier because it was deemed inexcusable to expose infants to toxic doses of oxygen.

Next came a controlled clinical trial conducted under the supervision of a coordinating committee consisting of the leading investigators of the disease. This study, started June 30, 1954, showed, conclusively, that premature infants kept in oxygen-rich environments were prone to develop RLF; but equally important, it indicated that there was no concentration of oxygen in excess of that in air which was NOT associated with increased risks.

Concurrent with clinical studies, top researchers in the United States and Great Britain diligently peered through microscopes and experimented with animals in attempts to unravel the mysteries of RLF. They were seriously hampered in their research because, not having a reliable animal model for the disease, they only had intermittent peeks at what was going on. They were in the same position as a motion picture director trying to put together an understandable movie but only allowed to see one out of every thousand frames of film.

Dr. Isaac Michaelson, the avuncular iconoclast from Israel's Hadassah Medical Center and London University's Institute of Ophthalmology, pushed aside this roadblock by providing scientists with an experimental model, the kitten. He painstakingly traced the development of retinal circulations in the human fetus and in fetal and young kittens and proved, in 1948, that the two were analogous. Dr. Michaelson, incidentally, alluded to oxygen as cause of invidious circulatory changes in the developing eye but made no mention of RLF in his reports.

Dr. Arnall Patz and co-workers at the Wilmer Eye Institute of Johns Hopkins University exposed many species of animals to high oxygen concentrations—opossums with pouch young, rats, mice, kittens and puppies—and reported, in 1953, eye changes that could be interpreted to be equivalents of RLF. But it remained for Professor Norman Ashton at the University of London's Institute of Ophthalmology, and his co-workers, to carry out the ground-breaking work that definitely linked high oxygen exposures to RLF and the same time explained how and why the condition developed. His experiments were so ingenious that they are recorded here in Dr. Ashton's own words, (except where popular language replaces highly technical terms):

We then decided to submit baby animals to high doses of oxygen, but prematurity was the problem (we considered Cesarean section in pregnant apes and marsupials; but kangaroos are not convenient laboratory animals). Now it so happens that the kitten, when born at term, has a retinal blood supply which is still developing and at a stage exactly comparable to the premature baby. A retina from a one-day old kitten (injected with Indian ink, then removed from the eye and laid flat on a slide), shows a growth pattern and stage of development like that of the premature baby.

In our first experiment we placed the mother cat and her litter of kittens in a converted bacteriological incubator. Into this we piped pure oxygen for a test period of three days. The experiments require continuous supervision by day and night. This was the secret of our success—because had we switched off and gone home at night, the retinae would have recovered by the morning. At the end of three days the kittens, and in some cases the mother cat, were anaesthetized, injected with Indian ink and the retinas and other blood vessels, especially the brain, were examined microscopically.

To our great surprise—and even today I recall the moment with considerable excitement—we discovered in this first experiment that the growing retinal vessels of the oxygenated kittens were completely uninjected, only a few branches remaining at the disc (where the optic nerve enters the retina), whereas the retinal vessels of the cat were fully injected as were all other vessels of the eye and throughout the body in all experimental animals.

We had made the fundamental and entirely original discovery that excessive oxygen, if continued and prolonged, selectively destroys growing retinal vessels—and only growing retinal vessels; we called this oxygen blood-vessel-obliteration..... This was the secret of (how and why) RLF developed; for kittens allowed to survive in

Figure 5.1 Prof. Jerold F. Lucey. M.D.

air, being now without a retinal blood supply, developed a wild and abnormal frenzied growth of vessels into the retina and liquid of the inner eye, as seen in retrolental fibroplasia in the premature baby. It is frequently said that this fundamental discovery was made independently in America. This is completely untrue: but it was confirmed a year later in America and in Sweden.

Dr. Ashton's research report paper, which appeared in scientific journals in 1953, essentially proved that high concentrations of oxygen produce severe constriction, then obliteration, of developing blood vessels in immature retinas. A little over a decade later Dr. Ashton repeated his experiments but this time he injected extremely small glass beads into the blood stream of kittens, effectively blocking blood circulation in the eye. He reproduced the exact same RLF seen in human prematures.

The inner layers of the sensory retina of the eye in the developing fetus have no blood vessels until the 4th month of gestation. At that point vessels begin to grow outward from the optic nerve. The nasal portion develops first and is in place by the 8th month of development; the remainder gets its full compliment of vessels a month later, even shortly after birth. This explains why RLF most often affects the outside portion of the retina, the side away from the nose. When the development of retinal blood vessels is interrupted, either by oxygen-obliteration, disease, or any other insult, the body tries to compensate by making a maze of interlaced small arteries, veins, and capillaries. These tend to bleed, paving the way for retinal scars then detachment. Once blindness occurs there is no hope for restoration of sight. This series of events is not absolute. It can stop or stabilize without progression at any point. There may actually be regression of retinal damage in one or both eyes. Nearsightedness and focusing problems stay with those who retain useful vision.

As more and more incontrovertible evidence accumulated about the relationship between RLF and exposures to high oxygen concentrations the New York City Health Department, in April 1954, issued a warning against the routine use of oxygen in nurseries. Shortly thereafter Colorado General Hospital recommended that oxygen concentrations should not exceed 40%. In June 1955 the California State Department of Health recommended that physicians give oxygen only for specific reasons and in concentrations not to exceed 40%. The American Academy of Pediatrics concurred with these recommendations and stipulated that oxygen should be kept below 40%. RLF almost disappeared when concentrations of oxygen in incubators was reduced. But this resulted in a Cadmean victory because other problems soon replaced it. By the 1960's there was an unmistakable rise in deaths and cerebral palsy in premature babies. Articles flooded the scientific literature during the 1960s and 1970s calling attention to the fact that brain damage went up as RLF went down. Neonatologists and pediatricians were presented with a dilemma: withhold supplemental oxygen and risk death or brain damage or give oxygen and risk blindness.

In the 1970's, the incidence of RLF began to rise again. Some investigators attributed this resurgence to the increased use of mechanical ventilators and devices that used high pressures to push oxygen into the lungs of prematures, a common practice in intensive care facilities. Others suggested that exchange transfusions with adult highly oxygenated red blood cells might be the cause. Nonetheless, the impetus for further RLF research lagged until Jerold F. Lucey, MD, (see Figure 5.1.) Professor of Pediatrics at the University of Vermont, in Burlington, shook up the world by stating that there were other players in the RLF game besides oxygen. RLF, he emphasized, was a disease of multifactorial origin and not simply a disease caused by excessive use of oxygen. Singling out oxygen as the sole cause of RLF is an over simplification which has resulted in numerous unjustified malpractice suits", he said. "It is clearly time to stop pretending that we know the cause of all RLF—we do not."

Dr. Lucey was in effect reminding science and society that they were paying a draconian price for progress. Over the previous two decades pediatricians and neonatologists came to rely on new high levels of technology in sophisticated intensive care units as they tried to save prematures, some weighing as little as two pounds. Success depended upon the administration of 'higher than usual' levels of oxygen and this precipitated an avalanche of lawsuits alleging malpractice. "Too much oxygen blinded my baby," became the outcry of disgruntled parents. Attorneys for plaintiffs maintained that physicians should have restricted oxygen from the day they read Dr. Kate Campbell's report in the Medical Journal of Australia, published on July 14, 1951—a report that appeared two years before Dr. Ashton's findings on kittens and three years before results from The Cooperative Study became known. Many babies who had allegedly been injured by oxygen received large awards. But many died or ended up with cerebral palsy when physicians found themselves between the catastrophic end-points of death or brain damage from too little oxygen, and blindness from too much oxygen. Both extremes carried the threat of lawsuits.

Dr. Lucey felt that by focusing attention solely upon oxygen, investigators actually clouded and obscured other factors equally important in producing RLF. To drive home this premise Dr. Lucey retrieved Dr. Patz's 1950 - 1951 study and pointed out that many babies who received high oxygen did not develop RLF whereas some of those in low oxygen did. Even in the subsequent Cooperative Study of 580 low-birthweight babies there were 27 cases of RLF that were not related to high levels of oxygen. In the three decades after the indiscriminate use of oxygen was proscribed and the 40 percent rule went into effect we have witnessed a decline in RLF but never its complete disappearance. But we have also seen a rise in infant death rates and another rise in the incidence of RLF.

Dr. Lucey and co-workers at University of Vermont hooked up the babies in their neonatal nurseries to computer controlled oximeters that measured blood oxygen levels through the skin and continuously monitored and recorded these values. Cases of RLF still appeared. "Even with continual sampling it has not been possible to avoid all RLF", Dr. Lucey said. He felt that the old type of RLF caused by exposure of low-birthweight infants to excessive oxygen is becoming rare today due to the control of oxygen therapy and a new type, more common, occurs in sick, small infants, who formally would not have survived. Studies on full term and preterm infants who developed RLF not related to excessive oxygen revealed abnormalities in the blood supply and circulation to the brain and retina, to lack of oxygen in the uterus during pregnancy, and congenital heart disease. "The retina probably reacts to disturbed oxygen delivery regardless of its cause in a stereotypical way—all of which we call RLF", Lucey stated. In deference to this quixotic behavior the name of the disease was changed to RETINOPATHY OF PREMATURITY, ROP.

Today the incidence of RLF in a neo-natal intensive care unit ranges between 3% to 20%. The rate rises to 32% in babies weighing less than three pounds and shoots up to 75% in even smaller babies. The lower the birthweight the higher the chances of acquiring RLF. If we take the lowest incidence of the disease, 3%, and multiply it by the number of infants weighing less than 3 pounds who ultimately survive, it becomes evident that at least 2,600 to 4,000 babies are still affected by the disease in the United States every year; about one-fourth of them go blind. These figures are virtually the same as those seen at the height of the so-called epidemic of RLF in the 1940s.

Oxygen is not entirely and exclusively at fault, otherwise how can we explain such cases as these: A full term infant developed RLF after less than two hours of oxygen therapy. Another full term baby that died from heart disease was found to have RLF during autopsy even though it never had supplemental oxygen. RLF was reported to occur in only one eye. Some still-borns had RLF. Doctors Aranda and Sweet reported that 20 premature infants never developed RLF although their blood oxygen was extremely high for 10 hours; this was substantiated by constant measurements. And a premature infant that weighed slightly less than three pounds at birth developed RLF after receiving oxygen for 4.25 hours during general anesthesia and surgery for a blocked intestinal tract, whereas its twin sister, who weighed even less at birth, required 79 hours of oxygen in her battle against an immature respiratory system, never developed RLF.

In view of the erratic behavior of ROP there is no universal paradigm on how to treat it. At one time or another water soluble vitamins, iron and cortisone were tried, the latter possibly doing more harm than good. Recently, researchers examined the medical records of 90 infants who weighed less than 2 pounds at birth and found that cortisone treated infants had a propensity to develop severe ROP. Some doctors recommend early surgery, some late surgery, and some cryotherapy or freezing of the retina, a way of creating a ring of scar tissue that slows or stops the growth of abnormal vessels. These all work, to a degree, but since severity varies and there is no way of knowing which eyes are going to regress spontaneously, it often becomes an ethical and moral decision as to when to intervene. Surgery is accompanied by complications. Might not nature do the job in a much better manner if left alone? But waiting too long can be equally dangerous because any treatment, to be effective, must be instituted before retinal detachment and irreparable blindness sets in.

The greatest quandary surrounds the use of Vitamin E as a treatment or preventative of ROP. Remember, earlier in this chapter, the work on Vitamin E in the decade around 1948 by the husband and wife team, ophthalmologists Owens and Owens, was noted. They observed that it was common practice to feed prematures a low fat diet since such babies could not digest fats very well. Water-soluble vitamin supplements were given routinely but fat-soluble vitamin E was not included because of its indigestibility. And to make matters worse, these prematures were usually given supplements of vitamin A and iron, which we have since found out, increases the need for vitamin E. Although initial trials with Vitamin E supplements at the Harriet Lane Home seemed promising other studies failed to confirm these observations and by 1953 vitamin E therapy faded from the scene. It remained an obscure oddity until the early 1970's when Dr. Lois Johnson, Director of Neonatal Research at the Pennsylvania Hospital and professor of Pediatric Research at the University of Pennsylvania in Philadelphia, concerned with the persistence of ROP despite very careful oxygen monitoring, reinvestigated it. In 1974 she reported that giving premature infants Vitamin E shortly after birth reduces the severity of ROP and in some cases prevents it. These studies were repeated and confirmed by microanatomist Frank L. Kretzer and pediatric ophthalmologist Dr. Helen Mintz Hittner at Texas Children's Hospital, Baylor College of Medicine, in Houston, Texas, and also by Neil N. Finer, MD and his group from the Departments of Neonatology, Pediatrics, and Ophthalmology, Royal Alexandra Hospital and University of Alberta Medical School, in Edmonton, Canada. Dr. Hittner's 1981 report stated that "the severity of retrolental fibroplasia was found to be significantly reduced in infants treated with Vitamin E as compared to controls." Dr. Finer's report concluded that "vitamin E therapy significantly improves visual outcome."

Later that year at a conference on ROP held in Washington DC, two reports suggested beneficial effects of vitamin E and one study showed no statistical difference between treated and untreated babies. Nonetheless, physicians held-off using vitamin E. It seems that disturbing side effects often accompanied its administration: calcium was deposited in muscles where vitamin E had been injected; some babies sustained severe damage to their intestines; and some developed blood stream infections. During his momentous 1984 presentation on the multifactorial basis of ROP, Dr. Lucey told his audience that "The risk of adverse effects renders widespread use of vitamin E in low birth weight infants unwise at this time ..... Unfortunately there is no standard dose and no guarantee of safety or effectiveness."

It is estimated that currently 37,000 infants are born every year in the United States weighing less than 3 pounds; 22,000 will survive. Of these 1,500 will have significant visual problems and 500 will be blinded from ROP. The remaining 20,000 infants will not get ROP. The question arises, Is it ethical to give vitamin E to 20,000 infants who actually would not be developing this disease?

Scientists are now struggling with this enigma and answers will not be forthcoming until a paradigm for antioxidants is established—a project that will take at least several decades. Vitamin E is a naturally occurring antioxidant that prevents or minimizes oxygen injury through its known role as a free-radical scavenger. These may sound like strange, new terms, but the phenomenon they describe has been with us a long time. Free radicals are electrically neutral atoms. Free oxygen radicals have a propensity to attack fats and oils and make them rancid. They received little notice outside the food industry until recently when they were identified as important players in cancer, aging, heart, brain, eye, and blood vessel diseases. Now they are the hottest item on the research agenda. Free radicals are found in polluted air, tobacco smoke, radiation, and herbicides. Our bodies also generate free radicals which give rise to super oxygenated toxins that seek out and attack fats throughout the body. When the outer membranes and the nucleus of cells are assaulted by free radicals they lose their ability to transport oxygen, water, and vital chemical elements. This inevitably causes premature aging and cell death. Attacks on chromosomes and alterations to the genetic code putatively cause cancer. At places where blood vessels have been damaged by free radicals, platelets and cholesterol get stuck and begin the processes that culminate in hardening of the arteries, heart attacks and strokes.

The body recognizes and deactivates free radicals by means of enzymes, antioxidants, and white blood cells. The most effective antioxidants are the minerals selenium and manganese, and the vitamins A or Beta carotene, vitamin C, and vitamin E. Beta carotene prevents the formation of a highly active free radical by removing singlet oxygen. Fat soluble vitamin E protects the cell membranes by working in conjunction with antioxidant enzymes and manganese to scavenge free radicals before they attack fat molecules. Selenium is the essential ingredient of an important antioxidant enzyme that converts free radicals back to polyunsaturated fats. The water soluble Vitamin C helps guard the body against reactions within the watery medium of each cell.

The foregoing assaults and counterattacks continually take place throughout the body, but in the eye, light seems to tilt the playing field away from defense and in favor of oxidative damage. The retina, rich in polyunsaturated fatty acids, may be more vulnerable to free-radical damage than tissues with less fat. Antioxidants might provide a shield.

Having said this much about the role of free radicals the question arises: Why not give everyone supplements of vitamins and minerals until we are sure that they are indeed helpful? Unfortunately this idea is not without drawbacks. In a recent study undertaken by the National Cancer Institute in a trial in Finland, Vitamin E, Alpha Tocopherol, caused a non-significant reduction in the incidence of lung cancer and beta-carotene actually caused significant increases. High doses of vitamin A have been linked to birth defects and to liver disease in adults. And, as we have already noted, the side effects in small babies can be disastrous.

As things now stand interventional studies are taking place in order to assess laboratory and epidemiological findings that can be translated into firm recommendations with regard to diet and vitamin supplementation. Sooner or later the U.S. Food and Drug Administration will have to decide whether to allow food and vitamin manufacturers whose products contain antioxidants to claim that they can reduce or prevent heart and blood vessel disease, cancer or eye problems. Perhaps, when all the answers are in we will get a working paradigm for ROP.

PARADIGM POINTERS

The way to establish a paradigm about a disorder is through randomized controlled clinical trials—half of the subjects undergo some sort of experiment while the other half act as controls. Optimally, the investigator's bias never enters the picture, criteria for evaluations are clear-cut and uniformly applied, the data lends itself to statistical analysis, and results are interpreted impartially.

Unfortunately this approach is less than perfect and its weaknesses show up in the studies with a multifactorial disease, like ROP, where many agents are at work, serially or concurrently. To understand this, imagine what would happen if an authority declared that all headaches were caused by drinking alcohol. Alcohol would then be banned, and surely, the number of headaches would go down. Immediately thereafter all the other causes of headaches which went unnoticed, such as poor vision or tight neckties, would become painfully obvious. And as former alcohol drinkers substituted drugs for alcohol new conditions would arise.

Now compare this with ROP. Let us assume that several things in addition to oxygen contribute to ROP, but toxic amounts of oxygen act the fastest. By cutting down exposures to oxygen the incidence of ROP drops, whereupon other factors, presumably immaturity, poor respiration, hemorrhages, even poor kidney function, get a chance to show up—explaining the rise, fall and rise of ROP.

There is another analogy that applies particularly to the investigations on antioxidants and Vitamin E. Suppose a savant declared that 'sleeping under bridges causes pneumonia'. This would indeed be a logical premise because it would be based on examining all the people who slept under all the bridges and finding that they suffered from pneumonia more often than people in nice warm houses. But here again, attributing the malady to bridges rather than to poor diet, depression, improper clothing, not enough rest, and vitamin and mineral deficiencies, could easily lead to the wrong conclusion. In essence, many of the attributes of Vitamin E may be due to variables that are yet to be understood.

As Oscar Wilde said, The truth is never pure and rarely simple. The next chapter also deals with the eye but in quite a different context.

POST SCRIPT

One of the first controlled clinical trials was performed by Dr. James Lind, in 1747. Disturbed by the ravages of scurvy among British sailors who lived in the squalid holds of the queen's ships for months at a time, he descended into H.M.S. Salisbury's dank infirmary, and "took twelve patients in the scurvy, on board the Salisbury at sea. Their cases were as similar as I could have them.... They lay together in one place in the forehold, and had one common diet, water gruel sweetened with sugar in the morning, fresh mutton broth often times for dinner; and for supper, barley and raisins, rice and currants, sago and wine, or the like."

Dr. Lind divided the sailors into six groups, two sailors to a group, and treated them as follows:

Group 1. Received One quart of cider daily

Group 2. Received 25 drops of elixir vitriol three times a day plus a vitriol gargle

Group 3. Received Two spoonfuls of vinegar three times a day with vinegar gargle

Group 4. Received Half a pint of seawater daily

Group 5. Received Two oranges and one lemon daily (treatment lasted only 6 days because of limited supply)

Group 6. Received A combination of garlic, mustard seed, horseradish, balsam of Peru, gum myrrh, and barley water with cream of tartar daily.

"The consequence was that the most sudden and visible good effects were perceived from the use of oranges and lemons; one of those who had taken them, being at the end of 6 days fit for duty.... The other was best recovered of any in his condition."

Dr. Lind's bold randomized controlled clinical trial was purposeful and free of apparent bias in favor of one treatment over the other. But it was prophetic in two ways. First, the treatments tested were not new; the Dutch had found out 150 years earlier that citrus fruits and juices were of benefit to sailors on long voyages, but this knowledge had never been properly tested. Secondly, Dr. Lind's results were not accepted during his lifetime. The British Admiralty delayed 40 years before adding lime juice to seamen's rations, hence the name Limy.

chapter 6
FORESIGHT, NEARSIGHT AND INSIGHT

In the movie SLEEPER, Woody Allen awakens in the 22nd century after two hundred years of frozen sleep and finds that he, as well as everyone else in the world, is wearing heavy horn-rimmed glasses. While this scenario was funny, even believable, when it was filmed, it is entirely possible that a relatively new surgical procedure will virtually eliminate the need for eyeglasses and contact lenses, especially in people with mild, uncomplicated, nearsightedness. That procedure, which resulted from a series of fortunate, or unfortunate, accidents, depending on how you look at them, is called *Refractive Keratectomy*. Keratectomy simply means that the cornea, the transparent tissue just in front of the lens, is incised; and Refractive implies that the focusing of light within the eye is altered. When the operation was first devised it was called *keratotomy* because only punctures or incisions were made in the cornea. When corneal tissue was actually removed, the name was changed to *keratectomy*. Both terms are used interchangeably.

The cornea and the lens bend light rays that strike the front of the eye so that they focus sharply on the retina, the inner side of the back of the eye that interprets or 'sees' images and transmits them to the brain. When an image falls short, either because of too much curvature in the cornea or when the eyeball itself is too long, we end up with a less-than-sharp image or nearsightedness.

Refractive Keratectomy corrects nearsightedness by creating a new shape in the cornea, one that is flatter in the center and steeper around the outside. In practice the eye-surgeon cuts a tiny amount of the superficial layers of the cornea. This allows the biomechanical forces in the corneal tissues to produce a gaping of the incisions and repositioning of the uncut parts, which in effect alters its shape. Wound healing holds this new corneal contour. More about this later.

Myopia means nearsightedness. It comes from the Greek word meaning squint. Apparently the ancient Greeks discovered that nearsighted people could usually see things rather well within arms-length but had to squint to recognize objects farther away.

Nearsightedness is quite common. Full-term babies usually have normal eyes but premature infants tend to be myopic at birth. Nature tends to correct nearsightedness in the newborn, so much so, that it is rarely seen by the sixth month of life. There is a sharp rise in the incidence of myopia during the ages of 5 to 20 years followed by a gradual decline throughout middle and advancing age. The greatest number of cases of nearsightedness are seen during adolescence, reaching approximately 25 percent in that age group. This, of course, explains why so many teenagers wear glasses.

There are marked differences in the prevalence of myopia among racial and ethnic groups. It is most commonly seen in Oriental races, reaching 50 to 70 percent in Chinese and 30 percent in Japanese. Eskimos, native Americans, and blacks usually have very little nearsightedness. It affects twice as many whites as blacks. In Europe the incidence seems to be 10 to 20 percent of the young adult population while world-over the incidence in individuals between 12 to 54 years of age hovers around 25 percent. And things seem to be getting worse, largely because of television, computers, and eye strain. People with advanced educations and high annual incomes are ten times more likely to have myopia than the rest of the population. But the inexorable spread of myopia is not limited to technologically advanced sectors of society. Recent studies show that it is becoming widespread even among Eskimos.

Myopia is among the five leading causes of blindness in the world but not all myopia is due to simple focusing errors of the lens and cornea or elongated eyeballs. There is such a thing as pathologic myopia which is the eighth most frequent cause of severe visual impairment and the seventh most frequent cause of legal blindness in the United States. Retinal detachment and optic nerve disease are often associated with myopia, however, it is not always possible to tell which is cause and which is effect.

Medical science still cannot explain all the underlying causes of both simple and pathologic myopia. Numerous theories have been propounded, the latest and most intriguing, that of doctors David H. Huble and Torsten N. Wisel, who observed, serendipitously, that an eye deprived of vision from birth begins to elongate and becomes myopic. This suggests that proper development of the eye depends upon timely signals from a normally developing nervous system.

Spectacles can be traced back to ancient China. Since they worked so well, little thought ever went into improving vision by any other means, that is, until scientists began looking at vision critically. In the early 17th century Kepler observed that in myopic eyes parallel rays of light were focused in front of, rather than exactly on the retina. By the 18th century anatomists knew full well that myopic eyes were longer than normal and this is why light rays fell short of the retina. They also knew that severe degrees of myopia went hand in hand with excessively elongated eyes.

When the Dutch physician Cornelieus Donders, in Utrecht, 1864, systematically described the optical and clinical aspects of focusing errors in the eye, he unknowingly started the movement to correct eyesight surgically. By showing how parts of the eye could be manipulated to bring images into better focus he virtually invited surgeons to try to correct eye problems by changing the light focusing characteristics of the eye. And in no time at all, spurious eye specialists descended on an unwitting public. Some meddlers removed part of the fluid from the inner eye in hopes of shortening its long axis or changing the refraction of the lens, but sadly, in so doing, they collapsed and ruined many good eyes. Around the turn of this century Dr. J. Ball and Company advertised a device that promised to "restore eyesight and render spectacles useless!" It consisted of a spring loaded mallet suspended over the eye that pounded the cornea flat through closed eye lids. In other travesties, surgeons removed normal lenses as a treatment of nearsightedness. Admittedly, today's ophthalmic surgeons remove lenses but only if they are seriously affected by cataract or disease and can be replaced with an implant.

After a long series of such debacles most ethical physicians concluded that spectacles were the safest and most effective means of managing nearsightedness and all attempts to surgically correct myopia fell by the wayside. However, when the victims of two disparate accidents met—one lost his leg under a trolley car, the other almost lost an eye cut by broken glass—the quest for man-made corrections of nature's mistakes with vision was rekindled, and this time it succeeded.

In 1972, during a fight in a Moscow school-yard, 16-year-old Boris Petrov's eyeglasses were shattered and shards of glass cut the surfaces of his eyes. As the injuries healed, Petrov's pre-existing nearsightedness diminished to the point where he no longer needed glasses. This case naturally came to the attention of the Soviet Union's outstanding ophthalmic surgeon, Svyatoslav Fyodorov, MD. He immediately sensed that here was a way to achieve a permanent cure for myopia, a way to get rid of spectacles. As he put it, "We began to think that if a boy without any ophthalmic knowledge can treat myopia with his fist, may be... we could also treat myopia."

Slava, as he is known to his friends and colleagues, was more than an ordinary doctor; he was an accomplished eye surgeon, researcher, businessman, inventor, politician, and epicure. He was born into a military family in 1927. His first love was flying and his only purpose in life was to become a pilot. He was accepted by the Soviet flying academy at age 16 but this career was interrupted when, while trying to catch a trolley car, he was dragged under its wheels and lost a leg. He then switched to medicine and, since he already harbored a great interest in photography, he gravitated toward ophthalmology, the specialty devoted to eye problems. Slava once observed, quite philosophically, "the loss of my leg may have been one of the best things that ever happened to me."

After receiving his medical degree in Rostov he took a job at a small eye clinic in the town of Cheboksary. Here he happened to read about an American surgeon who had successfully implanted an artificial lens into an eye after cataract surgery and thought about doing the same thing in the Soviet Union. Since there was no way of getting such lenses in Russia, Fyodorov proceeded to manufacture them in his kitchen and then implant them into rabbits. In 1956 he implanted his first intraocular lens in a schoolgirl. Instead of accolades for his triumph he met outcries of protest from the Moscow medical establishment and the municipal officials at Cheboksary. Fyodorov fought back and ultimately persuaded the Deputy Minister of Health in Moscow to support his work. But when the opposition showed no signs of abating, Fyodorov moved to a hospital in remote Archangel on the White Sea. Here he encountered failure after failure as he tried to develop a safe and acceptable intraocular lens. Only later, when he found a skilled craftsman in Leningrad capable of producing technically superior lenses, could he boast about successful operations. In 1967 he moved to Moscow to oversee the construction of a new hospital, Number 81, which later became the Moscow Research Institute of Eye Microsurgery.

Dr. Fyodorov was a well-trained, highly disciplined, medical investigator. Before embarking on his innovative course to surgically correct focusing errors of the eye he scoured the backlog of scientific and medical literature on corneal surgery, then carried out his own agenda of animal experiments. The first surgical operations on the cornea, Dr. Fyodorov found out, were performed by Dr. Lan in the early 1800's. Results were less than satisfactory and Lan's procedures were abandoned. Nothing further was attempted until Dr. T. Sato, in Japan, spurred on by the impending war with the United States, launched a crash program to find a cure for nearsightedness. He had carried out extensive experiments on the corneas of rabbits and had performed many operations on people. Dr. Sato's procedures and results were described in scientific papers which Dr. Fyodorov latched onto with great enthusiasm. Interestingly, in 1960, shortly after Dr. Sato's death, Dr. Fyodorov attended the Japanese Ophthalmological Society Conference in Niigata, Japan, where he learned, first hand, about Dr. Sato's pioneering research and surgical techniques. Upon returning to the Soviet Union, Fyodorov tried Dr. Sato's method of doing keratectomies on four people with myopia, simple nearsightedness due to an excessively curved cornea. He also tried Dr. Sato's recommended surgical procedure for astigmatism, that is where irregularities in the lens of the eye give rise to asymmetrical or distorted images. In both instances Fyodorov found it difficult to place Sato's recommended number of incisions on the cornea, therefore he could not produce the desired refractive changes. He put the procedure aside and went on to more promising pursuits, that is, until the fateful meeting with Boris Petrov.

In 1974, Dr. Fyodorov began performing systematic Refractive Keratectomies in humans. At first, operating freehand, he used a razor blade fragment in a blade holder and checked the depth of his incisions with a depth gauge. Later, as soon as they became available, in 1978, he espoused crystal blades. The following year, 1979, Dr. Fyodorov perfected a special set of marking instruments and a micrometer knife that advanced the blade a fixed amount past a flat foot plate that rested on the surface of the eye to better control the depth of incisions. Fyodorov also introduced circular patterns,

When the American ophthalmologist, Dr. George O. Waring, visited Svyatoslav Fyodorov in the Soviet Union in 1980, he encountered a gregarious and charming colleague who insisted that they gallop on horseback through the birch forests surrounding Slava's country dacha. "There was no way of telling that he had an artificial leg," Dr. Waring said. "You'd swear that we were in pursuit of Yuri Zhivago or being chased by Red Guards." In Moscow, a black limousine with Fyodorov's personal driver was placed at Dr. Waring's disposal. "The driver," Dr. Waring mirthfully recalled, "had refractive keratectomies performed in both eyes—I hoped successfully."

While making plans for the eventuality of a war with Japan, American strategists felt that we had little to fear from an air attack. "The Japanese are so nearsighted they couldn't hit the side of a barn," was the prevailing logic. Although Pearl Harbor quickly changed this stereotypical misconception the basic premise was not altogether wrong; the Japanese people had one of the world's highest rates of myopia.

As the inevitability of war approached, Japan's foremost ophthalmologist, Professor T. Sato, felt obliged to do something about this insidious handicap. As a patriot he felt duty-bound to find a way to eliminate myopia and as an ophthalmologist he chose the surgical approach to change the shape of the cornea of the eye. It was this pioneering work that later gave Dr. Fyodorov such an enormous headstart. Wars seem to escalate research and accelerate discoveries.

Prof. Sato initially unearthed the procedures and findings of Dr. Lan then went on to establish the basic principles of modern keratectomy. He designed an extensive series of experiments on rabbits and then assembled a team of top-notch scientists and eye specialists to carry them out. He also directed the surgical trials on Japanese patients who exhibited excessive nearsightedness, doing many operations himself. He proved that changes in the cornea were related to the depth, number, and length of incisions. In books and scientific papers he published figures and patterns that would surgically reshape the cornea to almost any desired configuration—configurations that would theoretically correct nearsightedness, farsightedness, and astigmatism.

As the consummate scientist, Dr. Sato also explained how and why his techniques worked. When radial incisions, like the spokes of a wheel, are made into the cornea, there is a release of its circular tensions. This creates a gaping of the cornea which results in a relative flattening in the middle and a steepening around the outside edges. The eye becomes more farsighted. When the cornea is incised in a circular manner, like in a bulls-eye, its spoke-like tensions are released, the middle part of the cornea protrudes and becomes steeper. The eye now becomes more myopic which corrects the original farsightedness. Crossed incisions were tried in an attempt to correct astigmatism but these created dangerously large scars that failed to heal therefore Dr. Sato cautioned against them .

To his credit, Dr. Sato demonstrated the potentials of corneal surgery and described the basic patterns that proved most effective. It would remain for Soviet surgeons in the 1970s to modify these principles and introduce variations in surgical technique tailored for the individual eye—innovations that would improve the predictability of outcome.

Now, with hindsight, we can see that Sato and his colleagues made some serious mistakes, the most egregious being the wrong choice of experimental animals. The rabbit, Sato's primary experimental animal, is generally considered a poor model for refractive surgery. Firstly, the flexibility of the rabbit cornea is quite different than that of humans. Secondly, *Bowman's layer*, an important source of nutrients in the human cornea, is absent in the rabbit. Yet, despite this built-in handicap, it is nonetheless remarkable that Sato's team could deduce as much valid information from the rabbit as they did.

Dr. Fyodorov and surgeons of the western world found that superficial, frontal incisions of the cornea gave the best results. This is the procedure in use today. But when Dr. Sato performed his original surgeries on humans, after the rabbit experiments, he concluded that incisions only on the front of the cornea were generally ineffective. This happened because his surgical instruments were not as refined as the ones we have today. The knife used for eye surgery by Dr. Sato and his colleagues, the Okamura trachoma knife, even as modified by Sato, was probably not sharp enough to create consistent incisions in the cornea, especially in soft eyes. Dr. Sato's central clear zone that was designed to keep incisions outside the pupil was also too large. On some patients Dr. Sato made incisions on the back-side of the cornea, the area just in front of the lens, which in effect, damaged the layer of cells that nourishes the cornea. This led to painful, vision-distorting swelling of the cornea. Rabbits did not show such devastating changes because the regenerative capacity of their corneas is far superior to that of humans. Also, the follow-up period was less than one year and corneal swelling appears much later. At the time of these experiments the biology of the back-side of the cornea was unknown. In essence then, Sato's experimental design in humans was flawed. He collected and studied too few cases for accurate statistical analysis and these were observed for too short a time. He fell into the common error of premature enthusiastic promulgation of a surgical technique.

After the end of the Japanese-American conflict, researchers tried to round up and evaluate all of Dr. Sato's former patients. Although the total number of people who underwent eye surgery at Dr. Sato's Juntendo University was unknown, by March 1986, 170 eyes and 103 patients were found for study. They had been operated on between 14 years and 40 years of age. Ninety-nine out of these 103 patients had decreased visual acuity. Almost all of them had swelling of the cornea. One hundred twenty-one or 71 percent of those eyes that could be examined revealed complications of the cornea or lens. Most of the cases had painful, bloodshot eyes with blurred vision. Symptoms were particularly distressful in the morning. The eyes in which incisions had been made in the back part of the cornea for astigmatism turned into disasters. By 1987 it became apparent that two-thirds of the patients operated on between 1945 and 1955 were near-blind as a result of scarred corneas.

The surgery devised by Dr. Sato and his colleagues wilted and died in Japan because of the war and ensuing poor results. It would probably have become an extinct curiosity, very much like dinosaurs and dodo birds, were it not for Boris Petrov's accident and Svyatoslav Fyodorov's intuitive instincts. Initially Fyodorov repeated many experiments on rabbits that had been performed by Sato and proved that by controlling the depth and spacial arrangements of incisions he could get persistent flattening of the cornea. His rabbits showed complete healing of the corneal wounds in three months. Then, with his associate, Valerie Durnev, MD, Fyodorov carried forward the ideas established by Sato. He made superficial corneal incisions that interrupted its circular tension and thus changed its shape and light focusing powers. Together they proved that the outside corneal curvature steepened and the central corneal curvature flattened after surgery. Fyodorov made keratectomy less haphazard by devising a mathematical formula that could be applied to various curvatures of the cornea in different individuals. Surgeons from all parts of the world flocked to the Moscow Institute for Eye Microsurgery to learn, for a fee, how to do keratectomies. Additionally, every such student had to sign a licensing agreement with strict terms and conditions that prohibited any alterations in technique or teaching it to others. The agreement also insisted upon the use of Fyodorov's special instruments. It exacted royalty payments on income from the operation for seven years but did make allowances for free charity cases. Howls of protest against such ethics arose worldwide and acrimony tarnished the merits of the surgery.

Fyodorov, unfazed by this turbulence, continued to push for controllable high quality and high volume eye surgery. This led him to use assembly line techniques for surgery so that the surgeon is only essential for about 10 percent of the operating time. He placed patients on a series of stretchers that moved from one station to another. Personnel at early stations performed preparatory parts of the operation, the trained surgeon did the most critical parts, and personnel at the later stations completed the case and applied the final dressings. His institute has now become a complex comprising eye centers in 12 cities from Khabarovsk in eastern Siberia to Krasnaodar in the Crimea and there was talk of franchising clinics around the world. Patients can even receive eye surgery while aboard special Cruise-hospital-ships that now ply Russian waters. Enterprising travel agents offer week long packages that combine eye surgery at Fyodorov's Institute with sightseeing in Moscow and Leningrad. The Fyodorov Sputnik intraocular lens was exported around the world until newer designs supplanted it. Through licensing agreements with Bausch and Lomb in the United States, Fyodorov has made available absorbable collagen contact lenses as well as sets of surgical instruments.

In addition to his eye work Dr. Fyodorov has persistently campaigned for improvements in medical economics. Taking a dangerous entrepreneurial stance, at a time when Communism still reigned in the Soviet Union, he proposed a system of earning money for his Eye Institute based on the number of patients treated. The hospital would receive the normal allotment of state funds but would keep all savings that arose from decreasing costs and increasing volume. Employees benefited through an incentive fund based on profits, a foretaste of the economic reforms instituted in the 1980's by Mikhail Gorbachev. As Fyodorov's fortunes grew so did his fame. He became a politician and handily won election to the national Presidium. The exigencies of war with Germany, not unlike those of Japan with the United States, prompted Soviet military surgeons to experiment with corneal eye surgery and a lot of valuable data came from operations on Russian soldiers. But it was Fyodorov who compiled all the facts and gave this type of surgery its legitimacy. He also introduced a name-change. Since the incisions in his operation were made in a radial manner, like the spokes on a wheel, the procedure became known as Radial Keratectomy. And, as we shall see, with the advent of the excimer laser, where coherent ultraviolet light is used to sculpt the cornea, the terminology changed once again to Phototherapeutic Keratectomy (PTK) or Photorefractive Keratectomy (PRK). Phototherapeutic keratectomy refers to corneal laser surgery to remove scars or irregularities. Photorefractive keratectomy is similar surgery designed to change the focusing characteristics of the cornea.

When Radial Keratectomy was introduced into the United States in the early 1980's about 24 incisions of various lengths per eye were made by means of razor-blade fragments whose extension was determined by sighting on a linear scale block. Refinements soon followed: diamond-bladed knives with micrometers to set the depth for incisions replaced the razor blade and horizontal cuts were introduced to correct astigmatism.

Acrimony and controversy immediately cast a pall of suspicion around the operation. Invidious essays called attention to the crudeness and unpredictability of the technology and the instability of resulting refractions. Then there was a question of ethics. Who in their right mind would cut a normal eye? With broader trials, reports cited complications such as leaking fluid through the incisions, pain, eye swelling, light sensitivity, fluctuation in visual acuity, over-correction, under-correction, glare, corneal surface problems and induced astigmatism. Rarely, but tragically, corneal infection, cataract formation and visual losses were reported. Although these were exceptions rather than the rule, they nonetheless fomented confusion and distrust.

To stem the tide of criticism, several major eye clinics in the United States were empowered by the National Institutes of Health, NIH, in March 1982, to impartially evaluate such surgery and review the results. The diamond-bladed knife for eye surgery was just coming on the scene, and with improved techniques and in the hands of a select group of surgeons, complications were virtually non-existent. By 1990, according to a report in the Journal of the American Medical Association, radial keratectomy became the most widely applied and most predictable refractive procedure in the United States. The American Academy of Ophthalmology, in an Information Statement dated February 17, 1990, gave tacit approval but qualified its position: "Initial results of PRK are encouraging, but preliminary...long-term results and complications are unknown."

In the meanwhile the excimer laser was being developed. Marguerite B. McDonald, MD, clinical professor of ophthalmology at Tulane University, New Orleans, who pioneered the field recalls, "We ran lots and lots of trials on lots and lots of eyes in monkeys before we were sure of its safety and capabilities. In 1987 we used the excimer laser on a human sighted eye." Excimer lasers proved to be exceptionally accurate with a high degree of safety. They could remove submicron amounts of corneal tissue with minimal adjacent damage, they maintained the structural integrity of the cornea, and there was no tendency for a long-term shift to farsightedness. Accordingly, the Food and Drug Administration, FDA, in March 1995, approved the excimer laser for Phototherapeutic Keratectomy, (PTK) and the management of some other corneal conditions. Thus the operation of Lan, Sato, and Fyodorov entered the realm of modern, high-tech, computer controlled, surgery.

The word LASER is an acronym for Light Amplification by Stimulated Emission of Radiation. Visible light consists of waves that span a fixed range of energies or wavelengths. A laser separates these waves making them monochromatic and coherent, in other words, all the wavelengths are the same size, they move in the same direction, and are in phase or in sync with each other. All lasers have three components: an energy source, usually electricity, an active medium which is a liquid, solid or gas that is made from atoms or molecules which are easily excited by the energy source, and a resonant cavity, where energy packets are collimated or lined up to produce a powerful, bright, focusable beam. The active medium in the excimer laser is a combination of Argon and Fluorine, called a Dimer. Excimer is a term derived from excited + Dimer. It emits coherent ultraviolet light capable of removing extremely small amounts superficial tissues, hence is ideally suited for sculpting applications to the cornea of the eye.

The easiest to understand and most useful action of a laser is its thermal effect or burn. A case in point is the Argon-laser, a thermal injury producer, that seals leaking blood vessels or removes diseased retinal tissue in eyes damaged by diabetes. Another laser-induced effect, called an acoustic transient, induces a pressure wave that disrupts tissue. Additionally, lasers can induce intense electric fields that cause vaporization, optical breakdown and ionization of tissues. In summary then, a laser can burn, cut or obliterate living tissue.

Phototherapeutic Keratectomy and photorefractive keretectomy with the excimer laser is an outpatient procedure. Eye drops are used for anesthesia. The characteristics of the cornea are measured and entered into the laser's computer. The surgeon then directs the laser at the cornea and sculpts it by making micron-sized defects in its surface. In an exact and controlled manner twenty five micromillimeters of tissue are removed with every pulse; an average of 300 to 400 laser pulses, 10 thousandths of a thousandth of a second long, are used; laser time is one to two minutes. The entire procedure takes 35 to 45 minutes. Post-operatively, the patient may be treated with an antibiotic eye ointment until the cornea heals, about three days. Typically, one eye is treated in a session with the second eye treated three to six months later.

The spin-offs and other applications made possible by the pioneering studies of Sato and Fyodorov on keratectomies are exciting and promising. Now, for instance, the excimer laser is being used to treat several eye conditions which heretofore required corneal transplants. This new approach is safer and simpler than corneal transplants. Hospitalization is no longer necessary. There is no prolonged period of vision instability, no rejection of donor cornea, and no need for glasses, contact lenses or second operations to correct improper focusing.

When used for corneal unevenness or scars the excimer laser knocks down very small changes in the cornea, to a level where nature's own healing process can take over and smooth out the corneal surface. A blocking or masking agent may be applied intraoperatively to protect the valleys and selectively expose the peaks to destruction. Myopic patients make excellent candidates for all the foregoing procedures inasmuch as PTK flattens the cornea so that some myopia is corrected when scars or other damage is removed. Physician and patient must be absolutely certain that there is no viral or herpes infections, or any other active disease present in the eye at the time of laser surgery. PTK might stir them up.

The advent of the excimer laser will probably erode the profits of Dr. Fyodorov as well as the manufacturers of lenses and eye-glasses but at the same time it will enrich other sectors of the eye-care industry. Consider: World-wide there are roughly 193 million people who might be eligible for radial keratectomy. Indeed, in the Soviet Union today, assembly-line techniques for performing radial keratectomy in major cities is the rule; a single clinic may perform 30 to 50 operations daily.

Here in the United States roughly 65 million people are myopic. Technical and industry analysts project the PRK market at 3.4 to 12 million eyes, with about 360,000 being treated annually. At $1200 for each procedure, PRK could generate $430 million dollars a year. But it will not all be gravy. At the 1994 annual meeting of the American Academy of Ophthalmology, Jeffrey J. Machat, MD, who has operated the Excimer Laser Surgery Center in Canada sagely observed,

"There will be a lot of players but few winners." For openers, the cost of an excimer laser, about $500,000 is formidable. If leased, instead of being purchased outright, there are rental costs and manufacturer's royalties. Attached to both options is the cost of maintenance, disposables, staff, administration, and optometric co-management fees. Nonetheless there is much posturing at the moment on how to split this pie. Ophthalmologists, who hold medical degrees, view optometrists and entrepreneurial entities, who lack medical training, as intractable rivals.

On the medical side, eye physicians within The American Academy of Ophthalmology have formed a sub-group, Refractive Surgery Interest Group, RSIG, which, according to Marguerite B. McDonald, MD, a founder and editor of its quarterly newsletter, will closely monitor "this fast changing area of ophthalmology.... and keep members informed about all clinical, practice-related and regulatory developments in the area."

Commercial, for-profit, companies are springing up and alliances are being formed in hopes of winning an early edge in the effort to bring PRK to the American public. In England, where government approval of PTK has been in effect for some time, walk-in centers are already operative in London, Birmingham, and Edinburgh. A person walking-in off the street, without any physician referral, becomes a candidate for laser surgery.

At present insurers and managed health care sources regard PRK as a discretionary expense; they will not pay for it. This however, does not seem to faze proponents who know that a prodigal public, inured to sticker-shock, seldom concerns itself with costs when bodily functions stand to be improved without dependence on external prosthetic devices.

As for the future of PRK, the best prognostications come from two directions. Ivan Schwab, MD, Professor of Ophthalmology at the University of California at Davis recently said, "when used as a phototherapeutic tool, let's say for superficial corneal scars, PTK will do much to replace the need for corneal transplants and contact lenses. But controversy will always attend photorefractive uses. Granted, the complications rate is only one or two per cent, but that's a lot when compared to zero, with glasses." Dr. Marguerite B. McDonald feels that "PRK will become like a rite of passage: braces at 10, a car at 16, PRK at 22."

PARADIGM POINTERS

A. It is essential to use the right experimental animal or other module for research. This book cites the happy choice of mice by Dr. Gerhard Domagk with the sulfa drugs, by Howard W. Florey and Ernst B. Chain with penicillin, and by Rita Levi Montalcini and Stanley Cohen in their pursuit of growth factors.

Dr. Sato made the wrong choice. Had he used non-human primates for his experiments, instead of rabbits, the disastrous effects of cutting the posterior layers of the cornea would have shown up and much grief and disability averted. This point is explored further in Chapter 11, CONCEPT OUT OF CHAOS.

B. Experiments must involve a statistically significant number of subjects and observations must be carried out for prolonged periods of time. The history of medicine is replete with popular methods of therapy that were later found to be ineffective, unnecessary or unsafe. Had Dr. Sato kept his patients under observation beyond his one year time-frame he would have noticed that his surgery was fraught with complications and poor results.

This same mistake was repeated with the drug Thalidomide. Short-term tests in Europe, in 1950, suggested that it was a safe non-addicting sleeping pill. But when used over a longer period of time in pregnant women, as a treatment for morning sickness, it was found to be the cause of severe birth defects.

POST SCRIPT

1. The cavalier approaches to surgery on human eyes that took place in Japan and Russia are no longer acceptable. Standards, now in effect world-wide, outlaw such experiments.

    The movement to protect humans involved in drug trials and medical experiments started shortly after World War II when the abhorrent use of humans for experiments by the Nazis came to light. The Nuremberg Code in 1947 and the Helsinki Declaration in 1964, insisted "that all experimentation with human subjects (must) be conducted by scientifically qualified persons yielding fruitful results for the good of society," and all participants must give their voluntary consent.

    In the United States the two main watch-dogs over human experimentation are Institutional Review Boards (IRB) and Human Rights Committees (HRC). Although both are dedicated to protecting the public, the IRB gets its orders from the Food and Drug Administration (FDA) which is charged with supervising investigational studies of medications and biomedical devices. The HRC focuses on the physical, psychological, behavioral and economic needs of people used in experiments, and makes sure that they are respected and treated with dignity and compassion.

    All PROTOCOLS, the detailed plans of proposed experiments, are scrutinized and screened by IRC and HRC who demand satisfactory answers to questions: "What is the rationale behind this particular study? Has it ever been done before? What procedures will be used and does it have to be done this way? How long will it last? What are the expected risks? Will this study produce real, new, usable knowledge? A protocol dealing with the type of surgery tried during the investigational phases of keratectomy would, today, be shot down.

2. Advocates of 'the quick-cure' relentlessly accuse the Food and Drug Administration (FDA) of dragging its heels in approving new drugs and devices. Yet, when we look at Dr. Sato's travesties and all the shams and spurious products that hucksters continue to promote, we must commend and support FDA's restraint and caution.

chapter 7
CAPTAIN OF THE MEN OF DEATH

Pneumonia, the indiscriminate killer of young and old, rich and poor, was aptly called *Captain of the Men Of Death* by Dr. William Osler in his famous 1909 medical textbook. You might think that now, with antibiotics, vaccines, better housing, better nutrition and better medical science, that this designation no longer applies. Statistics, however, prove otherwise. Pneumonia has not disappeared; if anything, it has gained new vigor. This persistence can be explained, to some extent, by the fact that people travel widely and spread their infections quickly over greater distances—the global village effect. In the final analysis, however, it is the vast number of different types of pneumonia bacteria, the ever-shifting characteristics within those types, and the emergence of resistant strains, that make it so difficult to eradicate.

Two men, Louis Pasteur, in France, and Dr. George M. Sternberg, in the United States, almost simultaneously discovered the germ responsible for pneumonia. And in both cases, accidentally. Pasteur, it seems, was chasing down rabies when he came across the pneumococcus. Sternberg was casting a wide net in a study designed to find out which germs lived in the mouth. Both scientists had the presence of mind

153

to record their observations and pave the way for our current knowledge about the germ and the disease it produces.

Pasteur came upon the pneumococcus this way. When, on a midwinter day a five year old boy was bitten by a rabid dog, he was taken to Sainte-Eugenie Hospital to await an inevitable death. Louis Pasteur, one of the few men in the world who knew anything at all about rabies, was summoned to the hospital on December 11, 1880, whereupon he began a meticulous examination of the patient. One of his studies entailed taking blood from the thigh and mucous from the mouth of the child and injecting these samples into rabbits. The rabbits all died. Upon examining the blood and tissues of the rabbits, Pasteur observed dumb-bell shaped microbes and, in terms still used by bacteriologists today, noted that "each one of these small microorganisms is surrounded, as can be detected by proper focusing of the microscope, with a sort of mucous substance areola that really seems to belong to it." Pasteur soon found that these germs were absent in rabies cases and present in normal people, concluding that they were not the cause of rabies. But his description of their shapes and unusual translucent capsules leaves no doubt that they were pneumococci. He had no way of testing these bacteria on humans so he never found out that they caused pneumonia.

At the same time, Dr. George M. Sternberg, Surgeon General of the United States and the father of American bacteriology, injected his own saliva into rabbits. Later, upon examining their blood and lungs he observed the germ we recognize today as the pneumococcus. The development of pneumonia in these rabbits was a fortunate coincidence because Sternberg had to be a healthy carrier of the pneumococcus—a state that is the exception rather than the rule. His descriptions in scientific papers of the appearance and characteristics of the new bacterium are considered classics.

Neither Pasteur nor Sternberg knew that the germ they had discovered was responsible for pneumonia in humans. Five years later, in 1886, Dr. Albert Fraenkel, in Germany, found the Pasteur-Sternberg germ in pneumonia patients and declared that this indeed was the cause of pneumonia. But the German physician Dr. Carl Friedlander took exception to this because he had found a different germ that he believed caused pneumonia. To make matters worse, there was no agreement that the organisms isolated by Pasteur, Sternberg, and Fraenkel were even the same. This confusion can be explained by the fact that the pneumococcus can exist in several forms. Commonly two bacteria are linked together, the diplo form, as used in diplococcus. Occasionally it is seen as long chains, the strepto form, and rarely, it appears as a single cell. Each bacterium is enveloped in a clear, translucent, sticky, capsule, which contains a special polysaccharide, a complex carbohydrate not unlike starch or cellulose, that protects the germ and enables it to invade other cells. Some pneumococci do not have a polysaccharide capsule, but, when grown in the presence of DNA extracted from a species that does have a capsule, they become encapsulated pneumococci of the latter type. Keep these capsules with their special polysaccharides in mind because they are important in the understanding of the treatment of pneumonia, the making of antisera, the induction of immunity, and the final breakthrough with an effective vaccine.

The germ's 'hide and seek' behavior and chameleon form-changes created much confusion among early bacteriologists. Bitter controversy and rancor divided the world's scientists until exact typing methods evolved and proved that Pasteur, Sternberg, and Fraenkel had seen the same organism. It remained for Dr. Albert Weichselbaum to establish the exact roles of both the Friedlander bacillus, now labeled *Klebsiella pneumoniae*, and the Fraenkel pneumococcus: they are different germs, but both have a tendency to settle in the lungs and cause pneumonia.

The pneumococcus tends to grow in chains like pearls on a necklace, but when handled the chains break, not into single pearls but into many two-pearl segments. This is what Weichselbaum saw so he officially dubbed the bacterium diplococcus pneumoniae. The name was adopted by American taxonomists in 1920 and has persisted until a decade ago, when it was changed to streptococcus pneumoniae, which better describes its properties. Back in 1897, Dr. Sternberg felt that there was an incongruity in Weichselbaum's terminology and said as much: "I object to the name diplococcus pneumoniae because this micrococcus in certain culture media forms longer or shorter chains and it is, in fact, a streptococcus." Recent studies have proven that Dr. Sternberg was not only right but that the bacterium bears a genetic relatedness to alpha streptococci. For simplicity's sake , however, the terms pneumococcus and diplococcus will be used throughout this chapter; streptococcus pneumoniae will be used when it is part of a recent written or verbal quote.

The leading causes of death in the 19th and early part of the 20th century were pneumonia and influenza, with pneumonia in the lead most of the time. Epidemics regularly swept through military installations, homes for orphans and the aged, prisons and other closed populations. President Theodore Roosevelt, on a trip to Panama when the Canal was being built, was appalled at the death rate of black workers from pneumonia. He said, "The least satisfactory feature of the entire work to my mind was the .... very large sick rates (from pneumonia) amongst the Negroes, compared with the whites." Housing and hygiene were improved after this rebuke, but these measures probably did not help much because pneumonia, unlike other diseases, is not spread by unsanitary conditions. On the contrary, it is an egalitarian affliction, striking king and commoner alike—claiming such victims as Lord Lister's wife, Agnes, and at age 32, movie star Rudolph Valentino.

Before the discovery of antibiotics there was little physicians could do for the patient with pneumonia: intravenous fluids prevented dehydration; oxygen tents made breathing easier; and antiserums, which often caused allergic reactions, helped stem the infection—but only if they were type-specific for the organism causing the infection. One of the best descriptions of the disease and its treatment, in pre-antibiotic days, comes from Dr. Lewis Thomas' book, THE YOUNGEST SCIENCE:

> The diagnosis was usually the simplest part; the patient complained of the sudden onset of chills and fever, cough, sometimes with blood tinged sputum, and pain in one side of the chest... it was an acute illness lasting ten to fourteen days, with a high fever each day, more chest pain and more cough, perhaps with alarming manifestations of exhaustion and debilitation near the end of this period, and then, suddenly and as triumphantly as the bright sunshine after a thunderstorm, one of the great phenomena of human disease—the crisis. On one day or another, after two weeks of his seeming to come closer and closer to death's door, the patient's temperature would drop precipitously within a few hours from 106 degrees to normal, and at the same time, with a good deal of sweating, the patient would announce that he felt better now and would like something to eat, and the illness would end, like that........The cause was the pneumococcus, a bacterium which was stained dark blue by the gram stain, always as two round, paired cocci. The capsule of this organism contained a polysaccharide, a carbohydrate that endowed it with its invasive properties and protected it against being engulfed and killed off by the host's white blood cells. There were about forty different types of pneumococcus, each with its own special type of capsular polysaccharide.... Around the tenth day (the patient got better upon) the mobilization of an effective defense by the patient's own antibody, chemically designed to fit precisely with the molecular configuration of the polysaccharides of that strain of pneumococcus and no other. Once this happened, and the levels of circulating antibody in the blood were high enough to have combined with all of

the polysaccharides, the pneumococcus was the loser. When it combined with the antibody it was immediately swept up by the leukocytes and killed, and the disease was over. This event was the crisis, the sudden drop of the temperature, the sweats, the return of appetite, the end of the game...
Some types of pneumococcus were more virulent than others, needing quicker treatment. Patients known to be alcoholics were much more vulnerable to... overwhelming infection than normal people. Pregnant women were more susceptible and at greater risk of dying. Old people were the greatest risks of all.
The treatment (included)... the intravenous administration of type-specific antibody directed against the polysaccharide of the particular pneumococcus... it was the intern's first and most urgent task to identify the pneumococcus so that the proper serum could be used... One simply added samples of various antipneumococcal sera to bits of sputum and stained them with methylene blue; if you had the right serum, the capsule around each of the paired organisms would be swollen and dark blue. If you were lucky... and had the right diagnostic serum at hand, you could make the diagnosis within a few minutes... If not lucky, you had to wait... (do) blood culture... or you could inject the sputum sample into a white mouse... One way or another, the intern had to find out the type; there could be no going to bed until that job had been successfully done... it was not enough to know that the organisms were pneumococci; you needed to know next whether it was a type I, or a type III, or a type whatever, or there would be no way at all of treating the disease, nothing at all to do beyond watching the illness run its natural course.

These were the conditions that prevailed from the time pneumonia became a clinical entity until sulfanilamide and penicillin came into common usage, in the 1940s. Obviously, identifying the pneumococcus and establishing its type were the most important elements in the management of pneumonia; they were essential for diagnosis, treatment, and the preparation of antisera. But the means to do this, to identify and type the pneumococcus, were initially non-existent.

Robert Koch laid down the rules, called KOCH'S POSTULATES, for establishing the causal relationships between diseases and their putative causative agents: The germ must be isolated and cultured in pure form; when injected into man or animal it must produce the disease; and the identical germ must be retrieved from the infected person or animal. This was all carried out by Dr. Albert Weichselbaum in the 1890's but as practical knowledge it was of little value to the average physician since the pneumococcus could only be identified by its appearance under the microscope and its characteristics in cultures—long, tedious procedures, subject to error in inexperienced hands. At the turn of this century, in 1900, however, the tools needed to identify the pneumococcus in a precise and definitive manner and then to type it simply, directly, and reliably came to medical science through chance, serendipity, and some keen detective work on the part of bacteriologist Franz Neufeld (1861-1945).

Neufeld showed that the pneumococcus could be identified by the peculiar phenomenon whereby bile from the liver selectively dissolved the organism, only the pneumococcus, no other bacterium. This strange link was discovered in a roundabout way. Neufeld apparently knew that Robert Koch had been able to immunize cattle against Rinderpest, a uniformly fatal disease, by injecting them with bile from an animal that had died of the disease. In Koch's own words, "if one injects subcutaneously a healthy cow with a certain quantity of bile from an animal that died, it sustains a very mild illness and is thereafter immune to this otherwise almost always fatal illness." The bile contained either the virus or toxin responsible for Rinderpest, or a living attenuated modification of the virus, and this produced a mild disease which conferred immunity. Neufeld then began to study the action of bile on other bacteria, particularly pneumococcus, anthrax, cholera, typhus, staphylococci, diphtheria and germs associated with blood poisoning. None of these were affected by bile, except the pneumococcus, which was dissolved by it. In essence then, while Neufeld was looking for something in bile that might possess immunological properties, he discovered a way of identifying and also separating pneumococcus.

Neufeld then demonstrated that bile from rabbits regularly dissolved pneumococci but that heat-killed pneumococci resisted dissolution. Next, by making extracts of bile he proved that bile salts acted like soap and detergents, that is, they altered the surface tensions of fats, oils, and greases, so that water could wash them away. In effect, when bile removed the fat in the germ's protective capsule, the organism dissolved.

In subsequent experiments Neufled encountered complete and dismal failure when he tried to immunize rabbits with the bile from animals that had died from pneumococcal disease; but at the same time he stumbled upon an even greater discovery . In those experiments he injected pneumococci into laboratory animals then tested the animals' serum for antibodies. This was done by mixing, on a microscope slide, serum from each animal with some of the same pneumococci that had been previously injected into them. While peering at this slide under the microscope he noticed that the bacteria had become swollen and tended to clump together. Basically, this provided bacteriologists with an entirely new diagnostic concept: take a bacterium that has a capsule and mix it with serum from a person or animal that is immune to that bacterium and its capsule will swell. The German word for swelling up is *quellung* hence this reaction has become know as the quellung reaction. It is the foundation upon which bacterial typing is predicated. Neufeld described his phenomenon in his handbook of pathogenic microorganisms (1928) but seemed reluctant to recommend it for general use. The quellung reaction remained ignored and neglected until 1932 when it was resurrected in England by Drs. Armstrong,

Logan, and Smeall, who used it to pinpoint pneumococci in the sputum of patients. Albert Sabin, of poliomyelitis fame, was largely responsible for introducing the quellung reaction into the United States in 1933. He describes learning the technique from Kenneth Goodner at the Hospital of the Rockefeller Institute following the latter's visit to Neufeld's laboratories in Germany. The method was rapidly adopted by most laboratories and over the next 15 years was employed almost universally for the identification of pneumococci as a prelude to serum therapy. Animals injected with a certain strain of pneumococci, each with its own specific capsular polysaccharide, produced antibodies only to that strain; 14 different stains produced 14 different types of sera. When unknown pneumococci were mixed with these sera, which, of course, were labeled by type, it was easy to see which serum caused capsular swelling in the germs—its type became evident.

The test most widely used today to type pneumococci depends upon Optochin or Ethyl hydrocupreine, a substance that also came into being as a result of serendipity. We know, from Chapter I, that quinine's ability to reduce fevers and to alleviate malaria was recognized for a long time, so it was only natural for doctors to try it in the treatment of pneumonia, where fevers were excessively high. In 1911 there were reports that it worked but no one could tell whether quinine cured pneumonia or merely abated the fever. Mice infected with pneumococci and then treated with quinine all died. When Optochin, a derivative of quinine, was tried it seemed to have a protective effect, but it too fell by the wayside when germs developed resistance to the agent—probably the first report of bacterial resistance to a chemotherapeutic agent. Because pneumococcal infections were so lethal during the second decade of this century, Optochin was retrieved at the Rockefeller Institute and tried, experimentally, as a possible treatment for pneumococcal pneumonia. It failed to cure and was again consigned to the junk heap. But somewhere along the line someone noticed that the resistance of pneumococci to Optochin had increased 20 fold in patients that had been treated with it, and this prompted others to test it on live germs in cultures in the laboratory. When added to a petri dish in which several kinds of bacteria were being grown Optochin produced clear rings around the pneumococci but did not affect others. By a quirky detour bacteriologists discovered that Optochin dissolves only pneumococci, handing science the Optochin Disc Method which provides a presumptive identification of pneumococcus with 90% to 95% accuracy. Laboratories in this country depend primarily on the Optochin Method because high quality diagnostic sera for typing pneumococci are expensive, time consuming and hard to come by in the United States. Today, Dr. Erna Lund and the Danish State Serum Institute are the world's main resource for specific typing reagents.

Once the pneumococcus and its many types could be exactly identified, it was only a matter of time before vaccines would be made from them. In 1911, the great vaccinator, Sir Almroth Wright, tried to stamp out pneumonia in native South African Gold Miners by vaccinating them with heat-killed pneumococci. His results were ambiguous because he used too few bacteria and too few types to generate immunity. Wright also had an antipathy toward mathematics in science and blocked all attempts to apply statistical methods toward evaluating his results.

At this same time, half way around the world, in the Panama Canal Zone, William Crawford Gorgas, M.D. was also fighting pneumonia in laborers, especially blacks from the West Indies. He had no antibiotics or vaccines to help him, but nonetheless succeeded in reducing the attack and death rates of the disease. Gorgas recognized that pneumonia struck its victims in the early months of their employment and the disease gradually tapered off as recruits built up immunity. He did two things: firstly, he improved sanitation and diets, secondly he stopped the practice of housing all newcomers together in large barracks. He dispersed all new workers into small huts where they were mixed with a cadre of seasoned people who were immune to the pneumonia. This latter measure was a stroke of genius for it immediately stopped the spread of pneumococci among susceptible individuals while simultaneously giving them small exposures and time to build up immunities. Over the next four years, from 1910 to 1914, the annual pneumonia death rate fell from 18.7 per thousand to less than two.

Gorgas became known as the man who made the canal zone free of disease and allowed the Americans to accomplish what the French failed to do, namely, build the Panama Canal. Based on this enviable record, Gorgas was invited to South Africa to help the miners there. On his way to Africa he stopped in England and met with Sir Almroth Wright, where he must have been regaled with the virtues of pneumococcal vaccine. Gorgas stayed in South Africa for three months. He found many similarities to the Canal Zone and accordingly recommended the same measures of isolation, nutrition, and sanitation. But here, the biases and reactionary forces were too formidable so nothing was done. Only much later, when polyvalent polysaccharide vaccines were developed, was the pneumococcal scourge contained.

Pneumococci are normal inhabitants of the upper respiratory tract in five to sixty percent of the population, depending on age and sanitary conditions. They travel from the nose and throat down to the lungs and can involve the covering of the lungs and cause pleurisy, attack the covering of the heart and cause pericarditis, or, as in children, migrate to the ears and set up abscesses behind the eardrum, the cause of middle-of-the-night earaches. In 25 to 30 percent of the cases, where the body fails to localize the germs, they invade the blood stream and infect the coverings of the brain and spinal cord, resulting in meningitis. When they invade the joints they cause arthritis; when they invade the heart they cause endocarditis; and they commonly cause intra-abdominal infections. Mortality is four times higher with blood stream infections as compared with plain lung infections. Pneumococcal disease causes irreversible damage during the first five days of infection irrespective of antimicrobial therapy. Natural recovery from infection is signaled by the crisis.

Pneumonia is as unpredictable and erratic today as it was when first recognized. Patients with terrible looking lungs on x-rays, with lungs full of fluid and very little air space, tend to survive, while those with minimal x-ray changes often die. Dr. Osler was well aware of this capriciousness and back in 1897 speculated that poisons produced by bacteria kill faster than lungs full of fluid. He wrote: "The poisonous factors may develop early and cause from the outset severe cerebral symptoms and they are not necessarily proportionate to the degree of the lung involved.... The gradual failure of strength or more rarely a sudden death, is due...to this common disease, and unhappily against it we have as yet no reliable measures at our disposal."

The toughest job confronting modern day physicians is separating the harmful pneumococci from the casual passengers in a population. This can only be done through elaborate procedures: by recovering germs from the blood, by placing a needle directly into the bronchial tube or into the lung and capturing specific species, by the detection and identification of deadly capsular polysaccharides in the blood, urine or other body fluids, or by demonstrating an immunologic response or antibodies to infection. Presently we know about the existence of 83 pneumococcal capsular polysaccharide types, but 78% of all pneumococcal pneumonias are caused by only 12 of them. The pneumonia vaccine, which will be described shortly, makes use of this information: it is composed of 23 capsular types, which cover at least 80% of pneumonias seen in the United States and Europe.

For a while it appeared that sulfanilamide would halt the pneumonia scourge and later, when penicillin and broad-spectrum antibiotics came on the scene, pneumococcal pneumonia was generally believed to be an extinct volcano. Deaths from pneumonia had dropped 50% in the decade from 1940 to 1950; nonetheless, it still ranked as one of the 10 leading causes of death in the United States. The Centers for Disease Control estimated that between four hundred thousand to five hundred thousand cases of pneumococcal pneumonia occur annually in the United States and the over-all case fatality rate is five to ten percent, even with antibiotics. Pneumococcal meningitis, infection of coverings of brain and spinal cord, often hits young children between one month and four years of age and about 40% of these children die; pneumococcal meningitis is also a major complication of skull fractures.

In 1952 a study of pneumococcal infections in a large New York City Hospital was begun, and everyone naturally assumed that its incidence would be low. Instead researchers found that admissions for the disease occurred with approximately the same frequency as had been observed before the introduction of penicillin and antibiotics. It was also noted that approximately 20% of the patients in medical wards had pneumococcal germs in their blood, similar to the pre-antibiotic days. The fatality rate was 17% to 18% in patients who had pneumococci in their blood as well as in their lungs; the rate rose to 25% in persons 50 years of age or older—despite penicillin, antibiotics and advanced therapeutics. The driving force behind this 1952 investigation of pneumonia was the then 36 year old Robert Austrian, MD, who wanted to know why pneumonia persisted as the only infectious disease among the ten leading causes of death and ranked as the fifth highest cause of death for all diseases, as documented by the US Public Health Service.

Medical peers, quick to criticize the man and his statistics, argued that most of the pneumonia deaths were caused by viruses. But Dr. Robert Austrian (see figure 7.1) had the strength to ignore the conventional wisdom of his peers, whose dogmas conflicted with biological and epidemiological data, and unflinchingly pointed an accusing finger at the ubiquitous and far from extinct pneumococcus. "The preponderance of people who die," Dr. Austrian maintained, "are those who are irreversibly damaged by infection very early in

Figure 7.1 Prof. Robert Austrian, M.D.

the illness even though antibiotic therapy killed the invading organism. They die cured of their infection."

Dr. Austrian's credentials made him ideally suited for the job he tackled. After completing his residency and medical training at Johns Hopkins University he accepted an appointment in 1952 as Associate Professor of Medicine at the State University of New York; later he became chairman of research medicine at the University of Pennsylvania in Philadelphia. His father, Dr. Charles Austrian, had been a superb diagnostician who had spent his entire medical career as student, professor and practitioner at Johns Hopkins University Medical School and, like his son Robert, maintained a major interest in lung diseases.

Besides calling attention to the annoyingly persistent pneumonia problem, Dr. Robert Austrian went on to solve it. He had to start almost at the very beginning by resurrecting the old, once-standard, bacterial typing techniques and classic laboratory bacteriology procedures because they had been abandoned when antibiotics came on the scene. In those days it was easier for doctors to prescribe an antibiotic than to waste time pinning down the exact cause of an infectious disease. After all, if penicillin did not work there were other antibiotics that surely would. But Dr. Austrian knew that antibiotics blunted pneumonia without getting rid of it, so he continued to probe until he was able to produce vaccines made from the capsules of pneumococci. In trials on humans they proved to be more efficient and less injurious than those made by using the whole organism. But then things ground to a halt. The relatively cheap and plentiful supplies of antibiotics put further vaccine-trials on hold. Disheartened but not defeated, Dr. Austrian kept fortifying his vaccine with polysaccharide antigens of the 14 most prevalent types of pneumococci and, by 1967, put pneumococcal vaccines back on the map. Ten years later, after clinical trials in this country and abroad, the FDA approved Dr. Austrian's vaccine, vindicating his vision that immunization was the best way to curb pneumonia.

But we are getting ahead of our story. In order to appreciate the length of time and amount of preliminary investigation it took before Dr. Austrian could even consider making a vaccine, we must go back to the 1920's when Dr. Michael Heidelberger discovered that the capsular substance of the pneumonia bacterium consisted of polysaccharides, which, when injected into animals, gave rise to highly specific antibodies. Capsular substance could immunize. By the mid 1930's scientists learnt how to inject rabbits with these capsular extracts, wait until they made antibodies, then bleed them to make pneumococcal antisera for treatment of the commonest forms of lobar pneumonia. The sera were expensive and difficult to prepare and sometimes caused fatal allergic reactions in patients already debilitated from their infection. But they also produced outright cures in many cases.

The key to making a vaccine from the purified capsular polysaccharides of pneumococci came from Louis Pasteur who speculated that vaccines could be made from components of bacteria or their toxins. He noted, for example, that the spinal cords of rabid animals that had been heated in the process of making his rabies vaccine were noninfectious but could still immunize. On January 29, 1885, he wrote: "I am inclined to believe that the causative agent of rabies may be accompanied by a substance which can impregnate the nervous system and render it unsuitable for the growth of that agent." This momentous premise became the basis for a general theory which led Emile Roux and Alexander Yersin to recover a toxin produced by diphtheria germs that was later used by Emil Von Behring, in 1892, to make diphtheria antitoxin, the first successful treatment for diphtheria.

These discoveries led to the invention of the Schick Test, and the Schick Test led, indirectly, to the making of the vaccine that finally subdued pneumonia.

Although diphtheria antitoxin was life-saving it made recipients terribly sick. When diphtheria epidemics swept through large populations, susceptible children usually died from suffocation and heart disease; children who were immune from previous mild exposures escaped unharmed. Pediatrician Bela Schick set out to find a means of identifying immune individuals so that they might escape needless injections with the dreadful antitoxin. He succeeded: Inject a tiny amount of diphtheria toxin into the skin; vulnerable individuals would show a big red inflammatory reaction at the site of injection but immune subjects, because of neutralizing antibodies, would show no reaction. This became the Schick Test for diphtheria, the gold standard for establishing the presence of immunity to diphtheria. Would a similar test using pneumococci work in proving susceptibility or immunity to pneumonia? To test this theory pneumococci were suspended in a solution and then injected into the skin of patients suffering from pneumococcal pneumonia. The investigators found that the skin reactions were not definitive, but much to their surprise antibodies developed to the several types of polysaccharides from the capsules of the bacteria that had been used in the test material. Dr. Maxwell Finland and his colleagues recognized the potential importance of this discovery and in the early 1930's carried out a series of investigations that proved that the injection of pneumococcal capsular polysaccharides into healthy humans stimulated the production of corresponding antibodies.

This immune response to pneumococcal capsular polysaccharides had been observed and recorded in 1897, in England, by Dr.A.Z.Auld, and lay forgotten and unused for 20 years. He had prepared germ-free filtrates of pneumococci and declared that there was some soluble substance in them that stimulated immunity to the disease. Around 1917, scientists under the leadership of Dr. Oswald T. Avery at the Rockefeller Institute in New York, isolated what they called specific soluble substance or SSS. They proved that it was a component of virulent pneumococci and when it appeared in the blood or urine of pneumonia patients, it almost always signaled death.

Working alongside Dr. Avery was a young biochemist, Michael Heidelberger, who had come to Rockefeller Institute as a fellow in 1912, a year after getting his doctorate in Organic

Chemistry at Columbia University. In 1923 doctors Heidelberger and Avery published their breakthrough report that opened the door for the development of bacterial vaccines, made not from whole organisms, but only from outer capsules of bacteria. They reproduced Dr. Auld's germ-free extracts, and by exemplary biochemical analyses proved that Auld's soluble substance was a specific type of polysaccharide; they also proved that such polysaccharides were contained in the thick capsules or slime layers around the rigid cell walls of pneumococci and similar encapsulated infectious bacteria.

Between 1922 to 1927 Michael Heidelberger continued to study the relationship between chemical structure and immunological specificity, mostly with cellular polysaccharides, and thereby earned the laurel FATHER OF IMMUNOCHEMISTRY. In 1928, leaning heavily on Heidelberger's work, Oscar Schiemann at Robert Koch Institute in Germany, , proved to the world that capsular polysaccharides from pneumococci could stimulate immunity. Later, between 1933 and 1937, Dr. Thomas W. Francis, Jr. and Dr. William. S. Tillett, both at Rockefeller Institute, inoculated over 32,000 youths in the camps of the Civilian Conservation Corp (CCC), during America's great depression, and proved that purified pneumococcal capsular polysaccharides could stimulate the production of specific antibodies in man.

In the early 1940s, as the United States was heading into World War II, Heidelberger together with Colin M. MacLeod, a Captain in the U.S. Army Medical Corps, developed a successful polyvalent pneumonia vaccine, so called because it contained polysaccharides from four strains of pneumococci. When tried on 17,000 adult servicemen in 1944, it proved to be extremely valuable in interrupting epidemics of pneumococcal pneumonia among troops. It prevented many types of pneumonia but failed to protect against two types not included in the vaccine. The war ended, troops were dispersed, and there was no way to assess the long-term effectiveness of the vaccine. Squibb Pharmaceutical Company produced and marketed two polyvalent capsular polysaccharide vaccines for pneumococcal pneumonia from 1945 to 1947, but by then penicillin was working so well that doctors saw no need for immunizations and production was voluntarily discontinued. At this juncture the pneumonia vaccine most certainly would have died an untimely death if not for the vision of Michael Heidelberger and Robert Austrian who put capsular purified polysaccharide vaccines back into modern medicine.

The Albert Lasker Clinical Medicine Research Awards for 1978 honored Michael Heidelberger, Dr. Robert Austrian and Dr. Emil Gotschlich: Heidelberger was recognized "for his elegant studies in immunochemistry which laid the groundwork for the development of capsular purified polysaccharide vaccines for the prevention of pneumonia and meningitis," and, of course, for elucidating the chemistry of pneumococcal polysaccharides. Doctors Austrian and Gotschlich received their award for "perseverance in the development and clear demonstration of the efficacy of a purified vaccine of capsular polysaccharides in the prevention of pneumococcal disease."

As things now stand, penicillin is still effective in most cases of pneumococcal pneumonia but this is bound to change as new resistant species arise. Streptococcus Pneumoniae, type 19A, which emerged in Durban, South Africa, in 1977, now resists penicillin, ampicillin, carbenicillin, methacillin, cloxacillin, erythromycin, clindamycin, gentamicin, streptomycin, fusidic acid, cephalothin, chloramphenicol and tetracycline, the major antibiotics in our armamentarium. Although there is no urgent need to vaccinate everyone at present, it stands to reason that this thinking will change if resistant germs arise faster than we can come up with effective antibiotics.

The vaccine currently approved contains purified capsular material from 23 types of pneumococci, those responsible for at least 80% of all the pneumococcal lung and blood infections seen in this country. It is recommended for the elderly, residents in schools, nursing homes and hospitals, military personnel, and people with poor immunity due to AIDS, sickle cell disease, diabetes, liver and kidney disease and cancer. Children under two years of age respond poorly to the vaccine.

Since many antibiotics cannot get past the blood-brain barrier they are less-than-effective in pneumococcal meningitis. In the event of such an epidemic, vaccination would be the best preventative.

In special populations the vaccine continues to prove its value as an effective control measure. For example, natives in the New Guinea highlands suffer from appalling attacks of pneumonia largely as a result of living in poorly ventilated smoky huts and being exposed to a cold wet climate that damages respiratory systems and induces chronic lung disease. When the vaccine was given to 12,000 adults it reduced the number of deaths from pneumonia and maintained a steady reduction over the three year period of the study. It kept the pneumococci out of the blood stream but had little effect on their invasion of the lower respiratory tract. In South Africa gold miners a single dose of the vaccine reduced the incidence of pneumonia by 80%. They suffer from a particularly high incidence of pneumonia, not because of working in mines but because of **recruit disease** which occurs when young adults from widely dispersed geographic areas congregate in barracks.

Dr. Austrian summed things up perfectly when he said, "Until the nature of the injury produced by pneumococcus is understood and ways devised to repair it, vaccination will remain the only means of protecting those at high risk of a fatal outcome."

PARADIGM POINTERS

A. This chapter has taken us over a wide range of research and to far-flung places on the globe: experiments on Rinderpest in African cattle, sanitation trials in the Panama Canal zone, bacteriological studies by Neufeld in Germany, immunochemistry by Michael Heidelberger at the

Rockefeller Institute in New York, and to hospitals, clinics and field trials where Dr. Austrian's vaccine was perfected. If nothing else, these events prove that changes in paradigms arise less from individual genius than from the collective reasoning power of many brains. This truism was eloquently addressed by Sir Howard Florey, the co-discoverer of penicillin, who, during a speech to the Royal Academy said, "Our increasing knowledge depends on the activity of thousands of our colleagues throughout the world who add small points to what will eventually become a splendid picture, much in the way that Pointillists build up their extremely beautiful canvases."

B. In the Aristotelian world, learned men were taught that a vacuum could not possibly exist; there was no point in looking for it so no one did. Similarly, when errors creep into our paradigms they become stubborn obstacles to progress. Fortunately young undergraduates and medical students who do not have to unlearn presumed gospel are free of orthodox constraints and can see what older, structured, scientists have been taught not to see. Robert Marston in *Medical Science and The Advancement of World Health* put it this way: "Creativity...mysterious as it is...depends ultimately on the dreams of the young... (to) search for new knowledge." Doctors Heidelberger and Austrian, like many of the subjects in this book, were relatively young when they initiated their work.

POST SCRIPT

The Nobel prize has been indicted as an obstruction rather than a stimulus to research. Scientists anxious to move up the academic ladder are inclined to pursue whatever seems hot, whatever might lead to a Nobel prize. This, of course, precludes thinking about the esoteric and the unknown. Hence it is the young, the novice, the dreamers, the altruists, and women who make the most sensational discoveries. Women scientists, whose competitive drives are generally less rigorous, often brave the unbeaten path to produce radically new ideas.

In the same vein, the incessant search for grants plus the Publish or Perish mind-set tend to act as roadblocks to established investigators. Many live by the unstated rules: stick to the safe, the probable, the quick; or conversely, no positive finding, no published paper, no grant, no job.

The traditional avenue of progress by inspired younger scientists may be coming to an end. Rich corporations are luring young brains and talents away from federal laboratories, universities, and academia, and directing them into entrepreneurial, money making pursuits. Today, half of the nation's health research is done in private companies. The National Institutes of Health, the principal supporter of biomedical research, contributed only 32 per cent to the health research effort. The rush to develop practical, marketable, products will adversely change the financial and cultural environment of research and this will dampen, if not curtail, free, inspired, research by young scientists.

PERSONAE

MICHAEL HEIDELBERGER was born in New York City on April 29, 1888. He earned the degrees of Bachelor of Science and Master of Arts in Chemistry at Columbia University. By the age of twenty-two, together with his professor of chemistry at Columbia, he published two scientific papers on quantitative chemical analysis. He earned a Ph.D. in chemistry by 1911, also at Columbia University, then studied abroad. Friends who knew him in Zurich, Switzerland, attest to the fact that he often worked steadily for eighteen hours a day. Upon his return to the United States he could not find a job. The University of Illinois offered him an appointment but he just could not reconcile himself to being buried in the Midwest. While pondering career options he was notified of an opening at Rockefeller Institute provided "that he bring his butt to New York at once." This he did and by 1912 was working alongside his mentor Dr. Avery.

Michael Heidelberger was an accomplished clarinetist and often played with small ensembles. During the summer of 1915, at a camp in Maine, Dr. Heidelberger at the clarinet accompanied a friend at the piano playing Giovanni Pergolesi's composition "Nina." A lovely girl appeared. The pianist stopped playing and said, "Meet Nina." This was Nina Tachau whom the young chemist married.

By 1923 Heidelberger and Oswald T. Avery, at the Rockefeller Institute, proved that capsular polysaccharides could trigger the immune response and their scientific report paved the way for the development of the capsular vaccines. Until the end of his life Heidelberger continued to unravel the chemical structures of immune substances.

Although he ruefully recalled that he came awfully close to discovering the germ killing properties of sulfanilamide, he took great pride in acknowledging that he was awarded the prestigious Albert Lasker Clinical Medicine Research Award not once, but twice. Heidelberger's first award came in 1953 for developing the techniques, indeed, a complete new subscience, in the chemistry of immunology. He was sixty-five years old at that time and maintained a full schedule of teaching and research as Professor of Immunochemistry at Columbia University. He left that position in 1964 to became Adjunct Professor of Pathology and Immunology at New York University Medical School, and at age 90 he was still seen traveling daily to his laboratory at the medical school.

Dr. Michael Heidelberger's son, Dr. Charles Heidelberger, distinguished himself as professor of biochemistry and pathology at the University of Southern California and director of basic research at the Los Angeles County - USA Comprehensive Cancer Center. His grandson Philip Heidelberger received his Ph.D. from Stanford and is engaged in research.

ROBERT AUSTRIAN, MD is the John Herr Musser Professor of Medicine and former chairman of the Department of Research Medicine at the University of Pennsylvania School of Medicine. Besides numerous honors and academic appointments Dr.Austrian has served as a Visiting Scientist in the Department of Microbial Genetics at the Pasteur Institute, Paris, France, and he acted as advisor to the World Health Organization. He published 165 articles in medical journals and served on the editorial boards of many prestigious scientific publications.

Lewis Thomas, MD, in his Forward to Dr. Austrian's book, *Life with the Pneumococcus: Notes from the Bedside, Laboratory and Library*, wrote this:

The major figures in American biomedical research come in several quite different classes. There are those who shift swiftly from problem to problem... And there are those who pick out the one problem that will preoccupy them for an entire career of hard work and then just keep at it, year after year. This may seem the safest way to live a life in science, but it is actually, in real life, the chanciest of all gambles, like putting all your chips on a single number, play after play, until all your money runs out.

Robert Austrian's career has been this last kind. He became fascinated by a single microorganism, streptococcus pneumoniae, long ago, and simply stuck with it. As the years went by, some of his colleagues came to believe that he was simply stuck with it. Finally... as the uncommon reward for steady, meticulous, logical experimentation, he got what he was after: a polyvalent vaccine against pneumococcal infection . (This is) a nice historical record of how science goes when it is going slowly but going well... it is also a lesson in what most savvy investigators take on faith: that if you can learn enough new things about living things at a fundamental level, sooner or later you may have the chance, as Austrian has had, to turn basic science into applied science and, at last, into a useful product.

chapter 8
THE OXFORD INCIDENTS

The final decade of the 19th Century saw the introduction of aspirin, antisepsis, x-rays, radium and the conquest of tetanus and diphtheria. Everyone was excited by such spectacular innovations in science and technology but it was the youth, the would-be scientists of the future, who were most enraptured by these advances. One such inspired dreamer was twenty-three year old Ernest Duchesne who enrolled in France's Army Medical School at Lyon in 1895, the same year that his idol, Louis Pasteur, died. While still a medical student, Duchesne experimented with molds to see if they would kill harmful bacteria. In 1897 he reported that certain penicillium molds could indeed do just that—kill bacteria. He had discovered penicillin.

But forty eight years later, in 1945, the Nobel Prize in medicine was awarded to Alexander Fleming, Howard W. Florey, and Ernst B. Chain for the same discovery. By rights, then, who really deserved the award?

Technically, neither Duchesne nor Fleming, Florey, and Chain were first. We have evidence that the Chinese cured abscesses with moldy bean curd 25 centuries ago and numerous other civilizations had used molds in one form or another for cures, rituals, and rites. In a recorded interview for the Canadian Broadcasting Company Alexander Fleming himself said, "Now, I've sometimes been accused of inventing penicillin. No man invented penicillin. A mold has been making it from time immemorial, only we knew nothing about it. I brought it to the attention of man. I invented the word."

The first person to really bring penicillin "to the attention of man", and to scientifically document its germ killing properties, was the British physicist, John Tyndall, the man who succeeded Michael Faraday as superintendent of England's most prestigious think-tank, the Royal Institution. When, in 1877, Tyndall read about Pasteur's revolutionary theory, how bacteria and molds float randomly in the air, he devised an experiment to prove that there really was such a 'diffusion of atmospheric germs', as he referred to them. Tyndall set up 100 open, broth-containing, test tubes in a square box, 10 tubes across and 10 tubes down, to capture and grow any microscopic bits that might rain down from the air above. By focusing a beam of light through the tubes and observing the light scattering-effects he could literally 'see' his captives. Some of them happened to be Penicillium molds. In his famous essay *Floating Matter in the Air*, he wrote:

> The Penicillium was exquisitely beautiful. In every case where the mould was thick and coherent, the bacteria died
> ...Of two tubes placed beside each other, one will be taken possession of Bacteria, which successfully fight the mould
> ...while another will allow the mould a footing, the apparent destruction of the Bacteria being the consequence.

There it is. Bacteria killed off in test1 tubes by penicillium mold. But, inasmuch as he was primarily interested in the physical properties of light, Tyndall failed to grasp the importance of his finding. His strict objectivity and detachment actually got in the way of perceiving an unexpectedly revealed truth. Yet, for all intents and purposes, almost the same thing happened to Alexander Fleming in 1929. He too came face to face with penicillin but was so intent on trying to solve a problem with his laboratory culture media and ill-tempered bacteria that he neglected to chase down the antibiotic. But Fleming was lucky, lucky enough to share the Nobel Prize for the penicillin discovery and then enjoy all the prestige and glory that went with the award, despite this failure to appreciate penicillin's true potential. But, then again, fortuitous coincidences and lucky accidents have always figured prominently in Fleming's life.

Alexander Flemming (see figure 8.1) was born in Lochfield near Darvel, Scotland, on the 6th of August in 1881, four years after Tyndall had noted the bacteria-killing effect of penicillium, sixteen years before Duchesne was to observe the same thing. Fleming's childhood and young years were quite ordinary; there was nothing to indicate any superior talents except that he was always considered to be a bright student. When he decided to become a surgeon he applied for and received a scholarship from London University. His medical training was completed at St. Mary's Hospital, where he remained and spent his entire professional career. This latter choice was something of an accident inasmuch as Fleming was more interested in the hospital's rifle shooting team, which he joined, than its propensity to turn out surgeons.

Good fortune surfaced long before Fleming earned his professional degree. It undoubtedly started when a small boy named Winston Churchill fell into a swimming pool and would have drowned had he not been rescued by a nearby gardener, Fleming's father. When Lord and Lady Randolph Henry Spencer Churchill asked how they might reward him, the gardener replied that his son, Alexander, wanted to go to college and someday become a doctor. "He might find some assistance useful," the gardener added. The Churchills paid for much of Alexander Fleming's education.

180 THE OXFORD INCIDENTS

Figure 8.1 Sir Alexander Fleming, M.D., (1881–1955), (Portrait by Konrad Hack, Courtesy of Mead Johnson and Company, Evansville, Indiana 47721.)

Several decades later, when pneumonia felled England's Prime Minister, Winston Churchill, the King asked for the best doctor available. Alexander Fleming answered the summons and gave instructions for the use of the new wonder drug, penicillin. "Rarely", said Sir Winston, after his recovery, "has one man owed his life twice to father and son."

Following his graduation from medical school Fleming became a bacteriologist in St. Mary's inoculation department. The buildings were jammed between Paddington Railway Station and the wharves of the Grand Union Canal, one of the seediest sections of London. The inoculation department consisted of four floors facing Praed Street, the main thoroughfare to Paddington Station. Ronald Hare, one of Fleming's fellow scientists described the ramshackle facilities: "Privacy was completely unknown. The department was almost poverty-stricken; we even had to buy our own microscopes. The nearest approach to a refrigerator was a wooden box about the size of a tea chest into which someone inserted a block of ice."

Despite its limitations, St. Mary's had the eminent Sir Almroth Wright at its helm. This attracted the notables of the scientific world who brought with them the latest news and developments within their fields. St. Mary's emerged as a sounding-board for the latest scientific theories and a proving-ground for the newest products from the laboratories. Ehrlich himself came to St. Mary's in 1910 bringing with him samples of 606, Salvarsan, the first successful treatment for syphilis. The arsenic in it made sores on the skin and unless every drop was injected into a vein, it caused tissues to slough. Since Fleming was quite proficient with an injection needle he was handed the job of administering the weekly doses of Salvarsan to syphilitic patients. He thereby became the first British Doctor to use Ehrlich's magic bullet and one of the leading venerealologists in London. It also enabled Fleming to acquire a sizable income.

Wright had absolutely no faith in chemotherapy including Ehrlich's Salvarsan. He contended that the drug merely showed where spirochetes were hiding so that white corpuscles could gobble them up. This disdain for chemotherapy was instilled into Fleming who found it convenient to accept such a pedantic view since it more or less agreed with his own past experiences.

Fleming had been sent to France as a captain with the Royal Army Medical Corps at the outbreak of the first World

War. There he observed first hand that pouring antiseptics into wounds made casualties worse. Although it meant bucking the establishment, he fearlessly expressed his conclusion that antiseptics killed good white blood corpuscles that had massed to devour invading germs, and this left the field open for the rapidly multiplying infectious bacteria. The gory statistics on gas gangrene with limb amputations corroborated Fleming's assertions.

In a later episode, the celebrated political figure, Lord Balfour, developed a sore throat during a visit to America. He gargled with Mercurochrome, a mercury-containing antiseptic, and made such a marvelous recovery that, upon his return home, he initiated a huge and costly research project to prove the virtues of this wonderful drug. Fleming demurred and lent his support to the group of scientists that denounced the scheme. He went further to prove that Mercurochrome was ineffective in the test tube and even harmful in the body.

Fleming came to distrust the use of chemicals within the body. Besides the indoctrination by Wright, Fleming had been educated and trained in an era when the science of immunology was burgeoning. Understandably, he firmly believed that substances in the serum related to immunity, mainly anti-bodies, were solely responsible for fighting off diseases. This concept proved advantageous in dealing with antiseptics and war-wounds, but later, it proved blinding and obscured recognition of penicillin's great potentials.

If we discount the near-drowning episode with Winston Churchill, we can say that Fleming's lucky streak actually started in 1922 when he accidentally discovered the antibiotic substances in secretions, called *lysozymes*. We have his own words as to how this happened: "I had this terrible cold with a drippy, running nose, and some of the mucous must have gotten into one of my Petri dishes. Bacteria which had previously been planted in the dish showed no growth around the mucous."

"Now that is really interesting," Fleming added,' "we must go into it more thoroughly."

After his cold went away Fleming squirted lemon juice into his nose and eyes in order to get enough mucous for his lysozyme studies. Since this was hardly enough material for experimentation he enlisted students and laboratory personnel in a ritual called ordeal by lemon, to donate tears. PUNCH, the English magazine noted for its satire, called it "tear therapy". Fleming coined the term lysozyme to describe this germ killing substance in mucous.

Fleming's reports on lysozymes in the scientific literature attracted wide attention and stirred up laboratories all over the world to look into this strange substance. Indeed, Christian de Duve, Claude and Palade shared the 1974 Nobel Prize in Medicine or Physiology for proving that lysozymes can favorably affect cancer, can retard the narrowing of arteries, and curb infections. Had Fleming stuck with lysozymes he might have earned two Nobel prizes. But Fleming himself soon dropped lysozymes and went on to other things. That was his nature. "He was a high-class theoretician and a damn good technician," his colleague Ronald Hare once said, "but he never could follow through on anything."

To outsiders, Fleming appeared to be the archetypal absent-minded professor. The cigarette always dangling from his lower lip was usually construed to be a forgotten appurtenance although in reality he was a chain-smoker. He was a short man with deep blue eyes and a boxer's flat nose. People who worked alongside Fleming said that he was taciturn, often curt, abrupt, and extremely direct. Tact was never one of his strong points.

He just never felt at ease in conversations, either socially at parties or professionally at work. To strangers he appeared asocial if not antisocial; at gatherings he would seldom exchange banter and almost never spoke to guests. His inability to field even simple questions made him seem arrogant. These traits spilled over into his scientific papers which are totally devoid of anecdotal interest. Social ineptitude apparently did not hamper Fleming's professional attainments for in 1928, the year of his penicillin discovery, he was appointed

Professor of Bacteriology and Deputy Director at St. Mary's Inoculation Department. Sir Almroth Wright, of course, continued his autocratic rule.

Inwardly Fleming was a caring sort, somewhat introspective, and he even displayed a little whimsy from time to time. For example, he grew different colored bacteria in the form of pictures on his culture plates, literally 'painting' scenes. Above all Fleming was scrupulously honest. He never padded his results, reported anything that could not be confirmed by peers, or claimed credit for things done by others. In his speech at the time of the Nobel award Fleming said, "I might have claimed that I had come to the conclusion (about penicillin) as a result of serious study of the literature and deep thought...- That would have been untrue and I preferred to tell the truth that penicillin started as a chance observation."

The exact moment when something clicked in Fleming's brain to make him aware of the fact that he had stumbled upon a new antibiotic—that a contaminant mold was killing the bacteria in one of his culture plates—is not known. It is generally agreed that the penicillin saga began on a Monday in late August or early September, in the closing weeks of the 1928 summer. Merlin Pryce a young bacteriologist studying staphylococci at St. Mary's, had sauntered to Fleming's laboratory to go over some facts and figures for a report in progress. While Fleming busied himself shuffling papers, Pryce picked up one of the many Petri dishes haphazardly strewn around the laboratory.

"Hey, look! There's some mold on this one," Pryce remarked.

"Aye, that is not unusual," Fleming answered. "As soon as you uncover one of these dishes something is bound to fall in. The air is full of mold spores."

Pryce, then noticed that there was a clear ring around a colony of bacteria. "That's interesting," said Pryce "did the mold kill the colonies?"

"Hard to tell. Maybe. I'll look into it —."

In his laboratory notebook Fleming wrote, "I was sufficiently interested to investigate this further. I noticed that for some distance around the spot the microbes were undergoing lysis." By lysis, Fleming inferred that the bacteria had been killed and were disintegrating. Since he himself had no special knowledge of molds Fleming asked a mycologist at St. Mary's, C. J. LaTouche, to identify it for him. LaTouche considered it to be *penicillium rubra* but later, the American mycologist, C. Thom, proved that it was in fact a very similar mold, *penicillium notatum*. Fleming's historic culture plate, incidentally, has been preserved and can be seen in the British Library, still with green mold, yellow colonies of bacteria, and clear areas free of bacteria around the mold.

Fleming then grew his 'wild' penicillium mold in test tubes and glass flasks containing the standard growing medium, broth from meat infusions. A thick dimpled carpet of mold soon covered the surface of the broth, first white, then green, and finally black. Fleming noted that the broth changed too. It became a bright yellow suggesting that the mold secreted something into the broth itself. Then Fleming tested both the mold and the broth to see if they had any germ killing powers. To his young assistant, Stuart Craddock, Fleming yelled, "Stuart, come here, look at this!" The broth proved to be the germ killer.

In a series of experiments Fleming went on to show that broth extracts from his mold cultures prevented the growth of staphylococci, streptococci, gonococci, meningococci, diphtheria bacilli, pneumococci, and a few other bacteria not related to diseases in humans. He also found that his 'mold broth filtrate', as he called it, was totally ineffective against the organisms responsible for typhoid and cholera as well as the inhabitants of the bowel and purulent wounds. The former group of microbes, those susceptible to penicillin, are classified as GRAM POSITIVE; the latter, penicillin-resistant bacteria, are GRAM NEGATIVE. Therefore, on the basis of these observations Fleming concluded that the best use for his 'mold broth filtrate' would be in the bacteriology laboratory for the separation of gram positive and gram negative bacteria.

The terms GRAM POSITIVE and GRAM NEGATIVE originated through serendipity. While working in Dr.Friedlander's laboratory in 1883 (see previous chapter) the Danish scientist Hans Christian Gram had shown that certain chemical and structural differences in the cell walls of bacteria determined why some accepted stains while others remained impervious to them. Dr. Gram's original intention was not to differentiate bacteria but to make them more visible in tissues. Back then it was difficult to see bacteria in tissues because cells, connective tissue components, and bacteria were uniformly stained and looked alike under the microscope. Gram found that by treating tissues specimens with solutions of iodine before staining them with dyes, he could decolorize tissues with alcohol, but bacteria remained distinctly stained. In honor of this accomplishment, stain acceptors are placed into a category known as GRAM POSITIVE and those that fail to take up stain are called GRAM NEGATIVE. This method of distinguishing one group of bacteria from the other has served as a valuable bacteriological tool since the day of its discovery.

When Fleming found that his 'mold broth filtrates' worked predominantly on gram positive bacteria it took little deductive reasoning to conclude that the active principle in the filtrate, whatever it was, somehow affected the outer membrane of bacteria, the cell wall. This was the first clue as to how penicillin worked, something that would prove valuable in years to come.

Craddock and Fleming injected some rabbits and mice with their mold juice and noted that it saved them from infections due to staphylococci.

"We next have to try it on humans," Fleming said.

"Why not on me?" Craddock replied.

"On you?"

"I've cultured some of the pus from my nose. I have a sinus infection due to staph. I've proven it."

"All right we'll wash your sinuses out with mold juice."
On this occasion he cleared up Craddock's sinus infection and later cured an infection in Craddock's eye. On January 9, 1929, in his laboratory note book, Fleming commented that the broth extract of penicillium had worked in a human subject without causing any undesirable side effects. Yet, his report on this phase of research contained only one paragraph to suggest that penicillin might be a disease fighter. To himself, however, he mused, " We must try to purify and concentrate this stuff! How about Ridley?" Frederick Ridley was one of the new young post-graduate students with a smattering of chemistry who had been assigned to work in Fleming's laboratory.

Research requires personnel, supplies, and money. Fleming presented his findings to the Medical Research Club in hope of obtaining these but the scientific community was not impressed. One American University turned down a request for $100 to do penicillin research and then threatened to dismiss a professor who offered to finance the studies with his own money. Undaunted by his colleague's indifference, Fleming continued to uncover more information about his mold and its 'juice'. For one thing, he actually assessed the potency of penicillin. When he diluted one part of his extract with as much as 600 parts of water he noted that it still killed staphylococci. He also proved that his new germ-killing substance was not toxic to live animals.

The identification of penicillin had been entrusted to two young men in Fleming's inoculation department, Stuart Craddock and Frederick Ridley. Neither Craddock nor Ridley had any real experience in chemistry; they were bacteriologists not biochemists. Nonetheless, with elementary knowledge of chemistry, under cramped, uncomfortable, physical conditions, with virtually no supplies or support, they almost isolated penicillin in its pure form. By March 20, 1929, Craddock and Ridley had already concentrated an extract of 'mold juice' that was effective even when diluted 3,000 times. By April 10, 1929, Ridley had produced a solution of penicillin in pure alcohol which was active in a dilution of 1 to 30,000. The two men also showed that penicillin was nether a protein nor an enzyme and its solubility in alcohol suggested that it was a small organic molecule. Some sixteen years later, in 1945, while delivering his paper at the Nobel Prize assemblage, Fleming acknowledged this heroic effort, saying "My colleague, Mr. Ridley, has found that if penicillin is evaporated at a low temperature to a sticky mass the active principle can be completely extracted by absolute alcohol." Many years later, shortly after receiving the Nobel prize, Fleming confided to a Cambridge colleague, "I would have produced penicillin in 1929 if I had the luck to have had a tame refugee chemist at my right hand. (The German refugee chemist Ernst Boris Chain isolated penicillin). I had to stop where I did." It was Sir Almroth Wright's hostility toward chemists, scientific clap-trap he called them, that kept chemistry out of St. Mary's Hospital.

The mistakes made by Fleming with penicillin were the same as those made by Ehrlich and Heidelberger with sulfanilamide. Experiments were performed mostly in glass containers rather than in living subjects and the true potential of penicillin was never appreciated. But in light of the instability of those initial extracts and the minuscule amounts available it is hard to blame Fleming for believing that it would never work in animals; blood and the body heat would decompose it too fast.

Craddock, who worked alongside Fleming, stated in a letter to a friend, "He thought it had possibilities in 1928 or early 1929, but the instability, the cost of production and the very impure product convinced Flem that there was very little future in the stuff." Fleming had also been discouraged by his own unpublished observation that staphylococci became resistant to penicillin very quickly.

Fleming's inability to communicate or convey information to the public carried over to his scientific papers. In his now famous report "ON THE ANTIBACTERIAL ACTION OF CULTURES OF A PENICILLIUM WITH SPECIAL REFER-

ENCE TO THEIR USE IN THE ISOLATION OF B. INFLUENZA" which was received for publication on May 10, 1929 and appeared in the *British Journal of Experimental Pathology*, there is only a short paragraph labeled solubility that makes but one casual reference to the attempts in his laboratory to purify crystalline penicillin. The complete details about the contributions of Craddock and Ridley did not come to light until they were published in 1968. Had Fleming related all the work, all the trials and errors, all the successes and failures, that he, Ridley, and Craddock, had encountered while trying to chemically isolate penicillin, he could have saved future investigators years of wasted effort groping through identical mazes.

Penicillin's antibacterial properties might have been discovered long before 1928 except that in the past, bacteriologists routinely discarded culture plates that had been contaminated with mold. Had Fleming done the same and failed to salvage his unusual plate it is difficult to guess if and when penicillin would have been discovered. Yet, despite his astute or serendipitous decision, there is such an improbable series of fortuitous coincidences and chance events associated with Fleming's work as to seriously question his right to the discovery. For instance, in 1964, Professor Ronald Hare repeated the original experiments described by Fleming in 1928. He inoculated standard culture plates with the Oxford staphylococcus, planted some of the spores of Fleming's penicillium mold on them, and stored them at room temperature. Rarely was he able to duplicate Fleming's results. After much thought and speculation Professor Hare concluded that the room temperature determined the outcome of the experiment. He went on to explain, "Fleming had apparently been trying to grow staphylococci without incubating them. When the penicillium mold landed on his plates, they just happened to have the right species of bacteria on them, one of the few susceptible to penicillin, just at the right stage and state of development."

In a post Nobel prize lecture Dr. Chain added, "Fleming did not discover the growth-inhibiting effect of penicillin on bacteria; what he saw was a bacteriolytic (dissolving) effect of penicillin which it exerts under very special limited circumstances not normally encountered in the bacteriology laboratory and only in a very special bacterial species."

Molds and bacteria grow at different temperatures. Bacteria need warmth and are usually housed in incubators while molds prefer room or cooler environments. It is almost impossible to get the two of them to grow simultaneously yet this is what happened. In looking back at the weather during the late summer of 1928 a logical explanation can be found. London experienced one of its most unusual fluctuations in temperatures: a rare cold spell was followed by a sweltering period of heat and humidity. The cold apparently enhanced the growth of mold and the hot weather stimulated bacterial growth, a one in a million chance event.

Also, the mold spores had to land on the plate at an exquisitely appropriate moment. If the plate had contained an old established growth of bacteria the penicillin would not have worked; the antibiotic destroys only young growing cells. Furthermore, it had to be just the right species of mold. There are over 300 different kinds of penicillium molds but only penicillium notatum possesses extensive antibacterial properties.

Fleming had gone on vacation in the midst of his work with staphylococci and upon his return began cleaning the laboratory. Gathering up outdated and useless culture plates, including those stored on the bench, he dumped them into a sink to be washed. Somehow, the important plate, the one with penicillium on it, stayed on top of the heap, barely the thickness of a sheet of paper above the water. It was then that Merlin Pryce rescued it and Fleming decided to hold it out for further study. The rest is history.

During a BBC interview on television Fleming was asked, "What is this mold like? How do you suppose it got into your culture plates?"

Holding a jar of jam in his hand, Fleming proceeded to unscrew the lid to reveal mold along the edges and replied,

"The mold growing on this jam could very well be the same as mine. And it got to me the same way it got to this jar."

Whether this is what Fleming believed or what he thought his audience would understand is uncertain, but it does not fit the facts. Penicillium is a rare species of mold; it is not ubiquitous. Considering the number of innocuous spores in any sample of air and the mathematics of probability the chances of the right spore hitting the right plate are astronomical. Furthermore, it is very unlikely that the mold spores came from the air outside of the laboratory because Fleming was a short man and could not readily open his laboratory window; besides the window was stuck tight from disuse and the clutter of paraphernalia on the window sill blocked all access to it. In all probability penicillium spores came from the floor below, from laboratories of C. J. LaTouche—and it got there by means of a convoluted trail.

Dr. Stormm Van Leeuwen, MD, a Dutch allergist visiting London in 1928, lectured on his theory that common house molds cause allergic symptoms, especially asthma. In the audience were researchers from St. Mary's Hospital, doctors Ronald Hare and John Freeman. Dr. Freeman was so impressed by the presentation that he insisted upon hiring a mycologist. The successful applicant, C. J. La Touche, went to work and scrupulously collected molds from the homes of St. Marys' patients who suffered from asthma. His laboratory soon housed the largest number and variety of molds in the realm. *Penicillium notatum* probably came to La Touche's laboratory by way of a damp basement in a large old house in Kensington or Belgravia. Again, another fortuitous coincidence.

Despite this successful journey of a particular species of mold to a special bacteriological laboratory and the serendipitous observation that it produced an antibiotic, penicillin itself remained elusive. It was too unstable to collect or work with, there were no funds or support available for further studies, and Fleming had his hands full with routine chores and supervision of post-graduate students. He published his paper which discretely described his preliminary findings then turned his back on penicillin, relegating it to a laboratory curiosity.

Why then, one might rightly ask, did the faculty of Sweden's famed Caroline Institute, the Nobel Prize judges, include Fleming with Florey and Chain when awarding its 1945 Prize in medicine for the discovery of penicillin? Florey and Chain rescued penicillin after Fleming abandoned it; they purified it and generated enough interest in it so that it could be mass-produced. "If it had been left up to Alexander Fleming," many historians argue, "there would be no penicillin today."

Some authorities suggest that the judges felt that Fleming's accomplishments were no less than those of Copernicus who found the universe between house-calls, or those of Pasteur, who laid the foundations for his germ theory while searching for methods to improve French beer. Others maintain that Fleming's boss, Sir Almroth Wright, the awesome god-father of British microbiology, undoubtedly did a bit of 'string-pulling' and he was abetted by Lord Beaverbrook, the rich and powerful statesman, publisher, and patron of St. Mary's hospital. Beaverbrook used his newspapers, the *London Daily Express* and the *London Evening Standard*, to exaggerate Fleming's contributions. After all, romantic myths about penicillin would help boost circulation. The British minister of information also failed to dispense credit equitably; Fleming was lionized while everyone else remained in the shadows.

Publilius Syrus observed, the gods scarcely have power over a lucky man, and this certainly applied to Alexander Fleming. Fame, a Nobel prize, and a knighthood were to be thrust upon him, after Oxford University professors, Howard W. Florey and Ernst B. Chain, who had passing interests in lysozyme, ferreted through the scientific literature for promising biologically active substances and came across Fleming's abbreviated report on penicillin. They recognized its potential and resurrected it for study.

Florey and Chain were not on a random 'fishing expedition'. Both men were acquainted with the scientific literature that detailed experiments that had been performed over the past fifty years on antibiosis—where bacteria made substances that interfered with the growth of others. As early as 1887 C. Garre and another French bacteriologist V. Babes, in separate reports, went so far as to suggest that there might be a way of giving people weak germs to kill off deadly ones. "Someday we should be able to get extracts from weak germs that contain an active principle to fight the more deadly ones," they reasoned. In 1903, A. Lode published pictures of petri dishes that had been contaminated with stray bacteria, with clear areas around colonies of infectious staphylococcal germs, very similar to Fleming's famous culture plate. Florey himself, in a 1930 report with Goldsworthy, called attention to a survey of the scientific literature by French scientists that documented some 200 examples of germs killing off other germs or making products that altered bacterial growth.

At the turn of the century C. R. Charrin and L. Guignard grew bacteria in a broth culture, filtered it, and found it to be lethal to certain bacteria but not to normal animal cells. In 1894, N. Pane, found that he could prevent fatal infections in rabbits that had been infected with anthrax by simultaneously inoculating them with a less virulent bacterium, *pseudomonas pyocyanea*. In Germany, R. Emmerich and O. Low used concoctions of these *pseudomonas pyocyanea*, during the first world war, as an antiseptic and oral spray. Later, in 1924 to 1925, F. Wreed and E. Strack, also German, identified, isolated, and purified the active principle made by *pseudomonas pyocyanea*, then went on to figure out its chemical structure and make it synthetically. It proved to have good antiseptic values but, because it was highly toxic, it fell into disuse. Interestingly, pseudomonas pyocyanea and its extract was one of the three possibilities, besides penicillin, specifically named by Florey and Chain in their application for research funds from the Rockefeller Foundation in 1939.

Even before Oxford targeted penicillin, several other scientists had picked it up for investigation. Professor Harold Raistrick at London School of Hygiene and Tropical Medicine on Gower Street, who was studying the biochemistry of micro-organisms, including a mold known as *penicillium chrysogenum*, happened to read Fleming's scientific report and asked for and received some of Fleming's mold. Raistrick, with two colleagues, P. W. Clutterbuck and Reginald Lovell, successfully grew the mold on synthetic medium as well as on broth. They also confirmed Ridley and Craddock's chemical observations and demonstrated that considerable concentrations of penicillin could be obtained by vacuum distillation. They use ether to extract penicillin both from the residue left after distillation and from weakly acid solutions of penicillin-containing mold cultures. But they abandoned further work when they found that penicillin was very fickle, decomposed easily, and disappeared during processing.

Contemporaneously, an American, R.D.Reid tried to isolate penicillin but also put it aside because he reached the conclusion that penicillin was too unstable to ever be of any possible medical use. Fleming also, in 1934, prevailed upon Louis Holt, a chemist, to try to isolate and purify penicillin. But old nemesis Almroth Wright, who innately feared explosions, insisted that Holt use amyl acetate, a common laboratory solvent which is less flammable and less explosive than ether, as an organic solvent for penicillin. Holt subsequently found that when the filtered mold juice was slightly acid and shaken in amyl acetate the penicillin went into the solvent. He also found that it could be taken out of the amyl acetate with a little water and bicarbonate. This process was quite similar, in principle, to that successfully developed some years later at Oxford.

Pharmaceutical companies, always on the prowl for new, worthwhile drugs and antibiotics also spotted penicillin. As early as October 1936, E. R. Squibb Company began to look at the reports about the initial studies on the penicillium mold, and they were soon joined by Eli Lilly, Lederle, Winthrop, Parke-Davis and Upjohn. Actually, when Florey and Chain started to work with penicillin in 1938, investigators on both sides of the Atlantic were scrutinizing it and there was a large body of literature about many of the mold characteristics.

The German word *Zeitgeist*, literally meaning time ghost, really expresses a time-frame when something is destined to happen. Penicillin's *Zeitgeist* had arrived. Chain himself said, in a lecture delivered in 1971 "I believe that the field of microbial antagonism had become ripe for study when we started our investigations in 1938.... We would have started our research program in these substances even if Fleming's paper had not been published... The development of the antibiotics field might have been delayed by a few years but it would, inevitably, taken place with the same final results which we have now."

Penicillin might have remained an obscure laboratory curiosity if not for the brilliant talents and untiring efforts Howard W. Florey and Ernst B. Chain. Both men were equals in intelligence and training but diametric opposites in mood, temperament and personality. Dr. Florey, the senior of the two, was born, raised, and educated in Australia. He received his medical degree from Adelaide University in 1921, to which he returned in the twilight of his career. Upon receiving a Rhodes scholarship to Oxford University the impoverished student had to work aboard ship for his passage to England.

Initially, Florey was haunted by self-doubts, writing to a friend, "As for research I believe I'm a dud after all—but I'll do a year and see." His drive and enthusiasm soon won the respect of his peers who remember him as a rough, tough, Australian, tense as a coiled spring. By 1935 he earned a full professorship in pathology at Oxford. To outsiders, Florey seemed a forbidding character, but those close to him enjoyed his warm and amusing personality. In essence he was the personification of Sinclair Lewis' *Arrowsmith*: entirely motivated and dedicated to research. And, as with Arrowsmith, science proved a selfish mistress and his marriage fell apart. During the turbulent trials with penicillin Florey drifted close to Margaret Jennings, who, as one friend of his put it, "was not only a competent colleague in research but also a confidante with whom he could rely and unburden himself of problems and anxieties that others did not even realize were troubling him." This started a long and tender personal relationship that culminated in their marriage in 1966, only a few months before Florey died.

Florey maintained a strong bond with his research workers, always observing their individual contributions and showing interest in their subsequent careers. He customarily asked for photographs from departing students which he hung upstairs in his corridor. His favorite dictum was, "If you do the experiment you may not be certain to get an answer; but if you don't do it you can be certain not to get one."

Florey had the good fortune of meeting and making friends with a number of Americans who would later achieve prominence and be in positions to help him bring penicillin into production. At the ancient and beautiful Magdalen College, Florey met the American Rhodes Scholar, John Farquhar Fulton who was then advancing his studies in physiology. Later, while on a Rockefeller traveling fellowship in the United States during 1925 and 1926, Florey established other contacts that would later serve to salvage penicillin, especially after the second World War exhausted British resources.

The other half of this unlikely team was a Jewish refugee from Hitler's Germany, biochemist Ernst Boris Chain. Born June 19, 1906, Chain studied chemistry and physiology at Berlin's Friedrich-Wilhelm University. He escaped to England in 1933 when Nazi persecution escalated but his mother and sister, who elected to stay in Germany rather than emigrate with him, were killed in Therienstadt concentration camp. Chain's pioneering research work on the chemistry of enzymes and snake venom, which was appreciated on an international scale, was furthered under the aegis of Sir Frederick Gowland Hopkins in the school of biochemistry at Cambridge. Coincidentally, Chain also studied Fleming's lysozyme and was responsible for discovering that it is an enzyme with unique antibacterial properties. In 1935, he became University demonstrator and lecturer in chemical pathology at Oxford.

During the early part of their collaboration, before the intense pressures of their penicillin research orchestrated a clash of volatile temperaments, Chain and Florey meshed like the gears in a fine Swiss watch. They could regularly be seen strolling through University parks on their way home, Florey to his house in Parks Road, Chain to his flat in Bardwell Road, talking and gesturing wildly as they exchanged ideas. Through their mutual interests in lysozymes they had developed a common interest in the phenomenon of antibiosis. Florey was qualified to do the biological work and Chain, a crack biochemist, was prepared do the analyses. Chain once recalled, "I told Florey about my findings in the literature of penicillin. He appeared to be familiar with the substance and asked me whether I was aware that in 1933 the late Professor Harold Raistrick and two of his colleagues had worked on it and found it very unstable. I had not heard of this paper but read it immediately after my talk with Florey."

Dr. Chain had a mustache and hair very much like Albert Einstein. He was brilliant but extremely excitable. Those who knew him at Oxford said, "He paced up and down while talking. He would hiss loudly and enthusiastically with every major pronouncement. And at times he could be annoyingly dogmatic."

One day, in the faculty lounge of Oxford University's Sir William Dunn School of Pathology, where a team of eight academic scientists were involved with penicillin, Dr. Norman Heatly cornered Chain and said, "You know, Ernst, this penicillin you are looking for could be almost any color; it could even be white or colorless."

"No!" Chain shouted. "It is yellow!"

"How can you be sure?" Heatly asked.

"But I am telling you," Chain riposted.

Despite his volatile temperament Dr. Chain was an astute scientist and the creator of the dictum: "There is no blueprint for discovery." This philosophy must have been at work as he sorted through reams of scientific publications, first, to learn what was already known about antibiosis; second, to determine which organisms seemed most promising; third, to preclude repeating studies done by others and then claiming credit for originality. Thus, by wandering in a maze of scientific reports Florey and Chain stumbled across penicillin and set out to find it. They initially grew penicillium mold in flasks which yielded one part of penicillin in two million parts of nutrient. But it proved just as elusive for them as it did for Fleming.

"It is so unstable," Chain said, "that you lose it while you are looking at it." Three decades later, in a commemorative lecture, Chain recalled, "that a substance of the degree of instability which penicillin seemed to possess according to the published facts does not hold out much promise of practical application. If my working hypothesis had been correct and penicillin had been a protein, its practical use as a chemotherapeutic agent would have been out of the question because of...allergy and shock."

Nevertheless, the pursuit of penicillin continued. Over the next few years many solvents were tried in an effort to increase yields but nothing seemed to work until Dr. Heatly suggested changing the acidity and alkalinity of the extracting solutions. This did the trick; by 1939 the team was getting enough penicillin to experiment with, enough to think about trials on living things.

Britain was on the verge of war and it was a bad time for research. There were no funds, air raid drills interrupted experiments, key personnel departed to the armed forces, and rationing made it difficult to get even the most ordinary supplies. And when the war did start things got worse. Three days after Britain declared war on Germany, Chain and Florey made application to the Medical Research Council for a grant to work on antibiosis—penicillin was only part of the plan—and it was turned down. Heated words and arguments flowed back and forth and irate tempers would probably have permanently scuttled this line of research if not for a chance event. A researcher who was about to leave Oxford had some unused grant money which he transferred to Chain. Nominally this assured Chain of 300 British Pounds a year for three years with 100 Pounds for expenses; however, when the money was claimed only 25 Pounds, roughly 100 dollars, were immediately available. This pitifully small amount was received on September 8, 1939.

Fortunately, another lucky coincidence rekindled the flame of hope. Florey's old friend in America, Dr. H. M. Miller of the Rockefeller Foundation, happened to be in England at that time and called upon Florey. In the course of their conversation the subject of antibiosis came up and Miller suggested that Florey apply for a Rockefeller grant. Three weeks after Dr. Miller's visit the application for a grant, probably drafted by Chain and signed by Florey, was dispatched to Dr. Warren Weaver in New York. It sought support for two research programs: one for a chemical study of the phenomenon of bacterial antagonism and the other for the study of mucinases, the enzyme that destroys mucin, the slimy component of saliva and other mucousy secretions. Florey asked for $7,700 for salaries and expenses plus $4,600 for equipment. The Rockefeller Foundation made a recurrent grant of $5,000 annually for five years instead of the three that Florey originally requested. The penicillin project had been rescued.

Two years later, in the first scientific report about Oxford's triumphs with penicillin, published in Lancet, August 24, 1940, seven joint authors were listed alphabetically, Chain, Florey, Gardner, Heatly, Jennings, Orr-Ewing, and Sanders. The authors acknowledged the financial support from the American Rockefeller Foundation first and the British Medical Research Council second, in the order of the amount of money each contributed. BMRC felt that it had been slighted and this led to an exchange of very harsh words.

Once the Oxford team knew that funding was available it concentrated its efforts on growing mold and extracting the active principal. By the following year, 1940, enough penicillin had been isolated so that it could be tried on animals. The crucial experiment was carried out on Saturday, May 25, 1940, just as German Armies were sweeping through northern France. Florey injected eight mice with a virulent strain of streptococci; four mice were treated with penicillin and four were used for controls.

"We sat up through the night", Florey recalled, "injecting penicillin every three hours into four of the mice." Heatley added, "I began watching the experiment on a Saturday night. All the mice had been infected. The controls looked quite sick; those treated with penicillin still looked very well. I left the laboratory about 3:45 in the morning, that would be Sunday morning. The controls were sick whereas the treated animals looked very well. I felt that penicillin was really going to work."

Florey was equally elated. "I must confess that it was one of the more exciting moments when we found in the morning that all the untreated mice were dead and all the penicillin-treated ones were alive... looks like a miracle!"

The next logical step was to try penicillin on humans but this presented an insurmountable stumbling block. Since man is 3,000 times larger than a mouse much more penicillin would be needed. This meant turning the university into a penicillin factory, a move that would spell disaster, even dismissal, if it failed. Heedless of the risks the team moved on. The student laboratory at Oxford was converted into a 24 hour a day mold growing factory with the resourceful Norman Heatly dreaming up Rube Goldberg contraptions to isolate and store precious penicillin.

Until now the mold had been grown on the large surface area of shallow vessels. Penicillium had to be kept close to sources of oxygen and would drown or suffocate if trapped underneath its liquid growing media. But this limited the harvest to about a 1/2 inch thick layer on the surface. "We tried tins at first," Florey recounted, "and later we found out that the best container was a hospital bed-pan. Radcliff Infirmary was kind enough to lend us their entire supply, 16 of them. As a matter of fact it was on Christmas day 1940." Penicillium adapted nicely to the bed pans. By January 1941 enough penicillin had been extracted for human trials.

That same month Dr. Charles Fletcher tried to find out if Penicillin was toxic. He administered it to a terminally ill woman and noted that it had no adverse effects. This extract was only 4% pure; it could easily have caused a fatal allergic reaction.

The next step was to try penicillin on a human who was not terminally ill. That opportunity came sooner than expected when the plight of an Oxford Bobby, Albert Alexander, came to light. The 43-year-old local constable, critically ill in the Radcliff Infirmary, was suffering from rampaging abscesses and blood poisoning. Even though he received massive doses of sulfanilamide and blood transfusions he continued to have high temperatures, a swollen head with terrible headaches, a stiff neck, and pus oozing from his eyes. His case was so desperate that it warranted trying an untested new drug.

A salt solution of penicillin was injected into the patient. No one had any ideas about dosage. "They just administered all the penicillin they had," Florey recalled. Within 24 hours there was marked improvement: the temperature was going down, pus was less noticeable, and the eyes were clearing. Every drop of urine was collected and sent back to the chemistry laboratory so that its penicillin content could be extracted and reused. Florey said, "It was like trying to fill a bathtub with the stopper pulled out." By the end of the 5th day the patient's temperature was normal; the headaches, stiff neck, and pus were gone—but so was the penicillin. The infection returned and the patient died. Dr. Fletcher presented this summary, "He was a policeman on his way toward death. I collected his urine to extract penicillin and reused it from day to day. The patient seemed to improve as long as the penicillin supplies lasted. He died shortly after they ran out."

Florey showed a great deal of interest in the critical trial and would incessantly disrupt hospital routines with his inquiries. He also recorded everything about this landmark event in the first published clinical study of penicillin which he called Case One. Chain was much more demonstrative, badgering doctors, nurses, anyone with information, "How's the patient? What's the temperature like? What's he doing?" Hope and exuberance reigned as the patient improved but was replaced by chagrin when the penicillin ran out and the patient died.

In those days, as today, there was considerable 'town-gown' rivalry between university academicians and outside private practitioners. Upon the demise of the Oxford Bobby non-university medical mourners were not long in saying, "That proves it's a lot of rubbish from those fellows across the road." Other waggish critics ridiculed penicillin as "a remarkable substance, grown in bed pans and purified by passage through the Oxford Police Force."

This skepticism was not too far off the mark because The Oxford team had been blessed by an inordinate amount of good fortune. Had the researchers used guinea-pigs instead of mice their original experiment would probably have failed. Guinea-pigs generally succumb to infection before penicillin becomes effective; this is not true of mice. Additionally, Florey and his team had observed that penicillin did not work on all species of bacteria, and in some cases where it was initially effective, organisms soon developed resistance to it. They then looked at products from other molds hoping to duplicate or surpass penicillin. In due course they cultured over 900 molds and fungi and assayed their antibacterial activity. All proved ineffective or toxic; none compared to penicillin. A noted historian later stated, "It seems scarcely conceivable that penicillin, their first choice, should be the only clinically effective antibiotic, or even the best. Yet this improbability was in fact not far from the truth; penicillin has few equals and overall has not been surpassed." Florey admitted, just before his death, "We had a bit of luck with penicillin - a great deal of luck... We happened to hit on an antibiotic that worked in man."

After the untimely death of 'The Oxford Bobby' the penicillin team decided that a second trial should be made on a child because it would require less penicillin than an adult. Dr. Fletcher found a small patient with cavernous sinus thrombosis, an infection of the large veins at the base of the brain, which is usually fatal. After two days of treatment with penicillin the child appeared better; after four days it was sitting up in bed, and after a week the patient was out of bed playing with toys. Further tests were planned when, in one of those unforeseen grand gestures, a small manufacturing company in London's Luftwaffe fire-bombed East End came up with 200 gallons of penicillin filtrate a week, all at its own expense.

Around this time there was talk about securing patents on penicillin but British bureaucrats squelched such thoughts because it was considered unethical to patent discoveries in medicine. It was also argued that research grants would dry up if penicillin was patented and this might interfere with further investigations.

Meanwhile, in the United States, Dr. M.H. Dawson at the Columbia-Presbyterian Medical Center in New York, turned lecture rooms into a penicillin factory and made enough crude penicillin to try it on three patients with bacterial infections on the valves of the heart. Either because of impurities or insufficient amounts of the drug, it failed. In 1940 the American pharmaceutical manufacturer Merck obtained a culture of penicillium notatum from Dr. Fritz Schiff of the Beth Israel Hospital in New York and put several of its top scientists to work with it as a possible antibiotic. By the autumn of 1940 the first batch of nine liters of crude penicillin containing culture medium was available for study. It was now time to start thinking about purification and then mass producing it. This however, was easier said than done—it took three more years before penicillin could be made commercially.

Penicillin had already shown great promise as an antibiotic but the exigencies of war effectively halted all experimentation at Oxford University. The allies had been defeated on the continent, the British evacuated Dunkirk, and the invasion of England appeared imminent. Apprehension proved palpable; everyone feared that the enemy would take over the secret of penicillin and confiscate the information that had been so painstakingly accumulated. If, as Winston Churchill warned, fighting would take to the streets, the valuable, possibly irreplaceable, penicillium cultures might be lost. "Then and there", as Dr. Heatly recalled, "everyone at Oxford smeared the fungus in the linings of our coats," hoping to retrieve it when peace and security prevailed. The invasion did not take place but its threat eliminated all hope of getting back to the research laboratory.

What happened next was pretty much hidden from public view because of wartime secrecy. The details of all the clandestine operations surrounding penicillin have only recently come to light as a result of the removal of the information from its classified status.

When it became apparent that studies on penicillin could not proceed in England, the Oxford team decided to shift everything over to the United States which still enjoyed a peacetime economy plus wealth and huge technological capabilities. Like mendicant friars, Florey and Heatley set off for the United States by way of Portugal, leaving behind a disgruntled Dr. Chain. The two scientists traveled by Pan Am's Transatlantic Clipper and stored their fragile penicillium mold in the plane's refrigerator. When the pair landed in New York, on July 2nd, 1941, the temperature was in the 90's which necessitated a breathtaking daredevil taxi ride from the airport to a midtown hotel and its refrigerator. Again, the mold survived.

Once in New York, Florey wasted little time before beating on doors to drum up interest in penicillin. He tried every place and everyone who might be in a position to make just a few milligrams of the stuff. At that time Florey was not interested in the commercial possibilities of penicillin; he just wanted enough for definitive analysis and testing. But his pleas were met by deaf ears; no one was interested.

Florey called upon his old friend, the former Rhodes Scholar from Oxford, Dr. John Fulton, who was now Sterling Professor of Physiology at Yale University. Fulton got the ball rolling again by introducing Florey to the right people at the U.S. Department of Agriculture, the nation's foremost mold researchers. Percy A. Wells, Administrator of the U.S.Department of Agriculture, telegramed Northern Laboratory Director, Orville E. May, July 9, 1941, five months before Pearl Harbor: "Heatly and Florey of Oxford England here to investigate pilot scale production of bacteriostatic material from Fleming's penicillium in connection with medical defense plans. Can you arrange immediately for shallow pan setup to establish laboratory results... ?"

Florey met with Dr. May, who, when he learnt that the problems were related to the growth of molds, sent Florey to see Dr. Charles Thom, a recognized authority on Penicillia type molds. Dr. Thom then turned Florey and the entire project over to Dr. Robert D. Coghill, fermentation Chief at the regional laboratories of the U.S. Department of Agriculture in Peoria, Illinois. "We felt it was important and went to work on it the next day," Dr. Coghill later recalled.

A July 15, 1941 memo from Dr. Coghill set the stage for penicillin production: "First it would be very desirable to study ...... submerged growth in rotary drums and vats. Second ...... Dr. Moyer should undertake small-scale experiments in a study of the effects of changes in the medium, temperature and other environmental conditions on the production of penicillin ....."

Dr. A.J. Moyer, a staff mycologist, knew from experience that corn-steep liquor was rich in nutrients and excellent for the propagation of molds. He tried it with the mold sample from England and it worked. Penicillium grew and flourished and the yields of penicillin shot upward, as did the joy and expectancy of Florey and Heatley who were now 'regulars' at the Peoria laboratory. But just as things seemed to be going well, an unfortunate, vindictive incident erupted, bearing out the sagacity of Aristotle who observed, "As a rule men do wrong when they have a chance." As soon as it became apparent that corn-steep liquor would yield penicillin in quantities large enough for commercial exploitation, Dr. Moyer published a scientific report taking full credit for the process. This, in effect, gave him all the rights to future patents and royalties. The Americans were indignant; the British felt betrayed.

Some Regional Laboratory staff members suggested that in addition to being blatantly anti-British, Moyer may have had other problems. He was usually withdrawn; unexpectedly he become outspoken and acerbic. When anyone entered his office Moyer would invariably lean forward so that his body hid his notes, no one could ever read them or get an inkling about his plans or projects. After Moyer usurped the credit and rights to penicillin he no longer enjoyed Dr. Coghill's friendship; hostility separated the two men for the rest of their lives.

This episode was but a prelude to a larger proprietary war that was to erupt once penicillin became a commercial success. Remember, penicillin had found its way into American hands as early as 1940 when doctors at Columbia-Presbyterian Medical Center and Beth Israel Hospital, both in New York City, ran some preliminary tests with it. Later, it was American know-how that found the best penicillin producing mold and the methods for growing it in large batches. Roughly three years after Florey brought his sample of mold to the United States the American pharmaceutical firms Merck, Squibb, and Pfizer, produced penicillin by deep culture and patented this process. The British press berated its government for forcing Florey into the arms of United States industry. They also cried 'foul' when it became common knowledge that the United States not only appropriated a British discovery but had the impudence to ask British industry pay royalties on it. Britain contended that it freely handed over all of its information about the production and uses of penicillin to the United States when it entered World War II and was entitled to special consideration. A report on this patent controversy, by John T. Conner, general counsel to the Office of Scientific Research and Development, made at the behest of Vannevar Bush, stated: " The payments by British firms to Americans were not for the penicillin itself but in return for costly engineering, scientific know-how, and other services ..... a normal and acceptable practice which all industries, both British and American, used and was agreed to by the Therapeutic Research Corporation representing British manufacturers. Up to the present time no British firm has paid one shilling in royalty to any American firm solely for licenses."

Florey was never assuaged by this argument. Later in life, after he left England and became head of the medical school in Canberra, Australia, he said, "I have had great extension of my experiences as a result of penicillin. I have seen much of the world and have many friends. There is only one serious regret that I have about the whole affair. That is, that I did not on behalf of my colleagues and the laboratory, patent the process by which penicillin was extracted."

But we are getting ahead of our story. We left Dr. Heatley in Peoria, Illinois, where he diligently pursued ways to nurture his mold and encourage large scale production. Meanwhile Florey went to major U.S. drug companies hoping to get a kilo of penicillin for human trials. "I felt like a carpet bagger," he would often recall. But as before, no one was interested. There could be no profit in a finicky mold that was difficult to grow, that made something that disappeared rapidly, that had no track record for safety or effectiveness. No! It was just too expensive.

When the United States decided to undertake the production of penicillin it spent over $26 million that first year trying to mass-produce it. There was talk of cutting back on this fantasy and it might have happened but for the fact that on December 7, 1941, a Japanese commander radioed his taskforce, "Climb Mount Niitaka", signaling the attack on Pearl Harbor. War with Japan freed penicillin from all constraints. "Damn the expense; get us the medicines we'll need for our casualties!" became the slogan of the day.

The exigencies of war spurred American science into action. Experts and specialists from industry, academia, and government studied the penicillin problem and concluded that a better method of growing the mold had to be developed. This called for submerged production or deep culturing techniques carried out in large commercial vats or steel tanks. The pharmaceutical company, Pfizer, soon perfected a method where the mold and their nutrient liquids were constantly agitated while oxygen was pumped into the tanks. Merk & Company also came upon a similar process independently. The U.S. Government had urged all drug and chemical manufacturers to share information and work together but this never happened; everyone was afraid of anti-trust actions, competition, and the theft of ideas.

Even if everyone had cooperated there still would not have been enough penicillin available for practical purposes. The bottle-neck was the mold itself; it was not a reliable penicillin producer. What was needed was a new and better strain of penicillium. So the cry for help once more went back to the mold experts in Peoria, Illinois.

This plea for a new penicillium landed on the lap of Professor Draper who directed mold research at the Department of Agriculture Laboratories in Peoria. He checked the local soils and thousands of cultures already catalogued by the department but could not find a good penicillin maker. He then put out a call for more and varied sources of mold and within months samples of dirt arrived from the Hump in Burma, South America, Africa, Europe, Canada, and every corner of the United States. Everything was tested in the hope of hitting a penicillin bonanza, and then, like the elderly lady looking for the spectacles on her forehead, the ideal strain was found on a moldy cantaloupe in a supermarket in downtown Peoria.

Professor Draper later told the news media, "We employed a girl to bring in anything moldy she could find. One day she brought in a moldy cantaloupe from a Peoria market and this is the one that produced the most penicillin." The last bottle-neck had been removed. The United States had developed the best mold and the best production methods. By the middle of 1944 there was enough penicillin available to treat all the men wounded in the invasion of Europe on D-Day.

The details about the progress of penicillin from a laboratory curiosity to a new, practical, disease fighter, had been kept pretty much under wraps because of wartime secrecy but two unusual events catapulted the entire drama into 1942 newspaper headlines. In March of that year Dr. Fulton scrounged up some of this new, guarded, and still scarce, penicillin for an ailing friend, Mrs. Ann Miller, wife of Yale University's athletic chairman. She was suffering from scarlet fever and despite having received the new sulfa drugs her temperature hovered around 106 degrees Fahrenheit and she was going 'down hill' rapidly. The penicillin worked and the patient left the hospital alive and well. In another episode, later, on August 5, 1942, Fleming himself telephoned Oxford University to ask if they could spare some penicillin for a friend in St. Mary's Hospital who was dying from streptococcal meningitis, an infection of the coverings of the brain and spinal cord. Florey at once took his entire stock of penicillin to London and instructed Fleming on its use. The patient recovered.

The press and radio, both in Britain and the United States, picked up accounts of penicillin's miraculous triumphs and Fleming and St. Mary's Hospital became, in their view, the greatest benefactors of mankind in centuries. Fleming's name, not unlike the morning sun that blots out the stars, dazzled the world but obliterated countless others whose valiant contributions made penicillin a reality. There was no praise too high or effusive for penicillin; it could cure blood poisoning, bone infections, pneumonia, gonorrhea and syphilis which were the contemporary scourges of mankind. In an interview Fleming went so far as to predict that it would one day be used in tooth-paste and lipsticks.

It is debatable as to who benefited most from penicillin, but there are strong arguments stating it must have been the people who had syphilis. This vicious disease not only destroyed the nervous systems of its victims but, as Henrick Ibsen illustrated in his 1881 play GHOSTS, the disease passes through the placenta and carries its malevolence to innocent offspring. Although drugs containing arsenic and mercury provided some remedy, reactions were unpleasant and end results uncertain. An alternative treatment consisted of inoculating patients with malaria. The spirochete Treponema pallidum is sensitive to heat; very high fevers accompanying malaria suppressed syphilis. Remember, back then we had treatments for malaria but none for advanced syphilis so it made sense to give malaria to patients with syphilis. The gruesome therapy was abandoned in 1942 with the advent of penicillin.

Frank Slaughter, the noted science writer once observed: It is not enough to make a great discovery. The discoverer must teach mankind to apply it to their own betterment, and that task is often more difficult than the search for the truth. This was certainly true of penicillin. Across the world in an Indian prison where Mahatma Gandhi had been incarcerated, his wife Kasturbai, the woman whom he had married as an illiterate 13-year-old child, kept him company. Early in February 1944, Kasturbai began to cough, became feverish, and blood flecked her sputum. She was diagnosed as having severe bronchitis that could lead to pneumonia. The British, sensing that her death might have detrimental repercussions, flew a supply of rare and precious penicillin to the prison infirmary. Before it could be administered to the ailing woman, Gandhi inquired, "What are you planning to give to my wife?

"It is a new drug, an antibiotic, one that kills germs. It is called penicillin. It is given by hypodermic injection."

"I do not allow it!" Gandhi said. "Natural afflictions must be cured by nature. And to puncture the skin with a needle is tantamount to the performance of violence on the human body. An injection is inimical to my belief in non-violence."

Despite around-the-clock nursing by Manu, Gandhi's 19-year-old grand-niece, who had been orphaned as a child and raised by Gandhi and his wife as their own granddaughter, Kasturbai Gandhi died on February 22, 1944, her head resting on her husband's lap. It is more than likely that penicillin could have prevented this tragic end.

Heatly stayed on to work with Merk & Co. while Florey returned to Oxford and then went on to treat casualties in North Africa. In a letter dated November 4, 1942, the same day that British Field Marshall Montgomery broke through the German lines at El Alamein and routed Field Marshall Rommel, Florey wrote, "... we sent 5 grams of stuff out to the Middle East... used, so far as we have been informed, very successfully... As always, the great struggle is to get enough stuff and that, of course, considerably hampers our progress."

During the first world war wounds had to be left open and exposed to the air to prevent deep infection and gangrene. In the second world war, with the advent of penicillin, surgeons could sew up wounds, reduce infections, and speed healing; limbs and lives that were formerly lost were now saved.

By the following May, in 1943, Dr. Coghill reported that 17 companies were working on penicillin. "I do not believe that any one or two companies can possibly make the amount of penicillin that will be necessary," he wrote in one letter, "... every possible step should be taken to encourage anyone who will make a try at it." About two weeks after Japan's surrender in 1945 Dr. Coghill said to an industrial colleague, "... we have discontinued our work on penicillin... the whole problem has lost any semblance of urgency... " On

October 10, he wrote Vannevar Bush, Director of the Office of Scientific Research and Development, "This letter is to inform you that I am resigning my position at this laboratory as of October 31st." Like a meteor, Dr. Coghill blazed through the dark sky of nescience, illuminated it brilliantly with knowledge, then faded into the empty darkness.

Dr. Coghill's acumen and foresight paved the way for the pharmaceutical industry to make penicillin on a grand scale. Vat fermenters, 20 to 30 feet high, with capacities of 1,000 to 12,000 gallons, started to mass-produce it. Subsequently, when the mold was exposed to x-rays and chemicals, mutant strains of penicillium were created and these produced 200 times as much penicillin as original samples. Over the years the price fell from 20 dollars to 60 cents for 100,000 units. Huge fortunes were made by the penicillin manufacturers but very little, if anything, went to the people who did the lion's share of the work. When Dr. Chain was told that he had won the Nobel Prize, he said, "Is it true? Are you sure? After all, no one in our group has ever received a penny out of penicillin."

It was a good thing that penicillin arrived when it did, because, as British journalist and historian David Willson wrote, "Penicillin would almost certainly not be allowed to come onto the market if it were discovered today. Bodies such as the FDA in America or the Medicines Commission in Britain would advise against it due to the frequency of allergic side effects." Furthermore, no one at that time was aware of the tendency for bacteria to become resistant to penicillin. Many germs can produce an enzyme, penicillinase, currently termed beta-lactamase, that can inactivate penicillin. This has given rise to many penicillin resistant species and simultaneously spurred on research for other drugs to meet this challenge.

As noted in the previous chapter everyone believed that penicillin would wipe out pneumonia but this turned out to be wishful thinking because streptococcus pneumoniae, like many other bacteria, soon developed resistance to it. In 1941, ten thousand units of penicillin a day for four days cured most infections but today twenty million or more units are required to do the same job; and even then it is not always successful.

By 1982 fewer than 10% of all clinical staphylococcal infections could be cured with penicillin. This is in contrast with almost 100% penicillin susceptibility in 1952. Through the process of conjugation, where two bacteria become linked together and transfer genetic material between them, staphylococci acquire the genes that make beta-lactamase and incorporate them into their own chromosomes. This changes the biochemical composition of their cell walls radically, so much so that they, in a sense, become new and different species; it also enables them to pass on resistance from one generation the next. Later, when methicillin was developed, it worked for a while, but by 1992 about 15% of all staphylococcus strains in the United States were methicillin resistant. Most of these organisms arose within the confines of hospitals.

The versatility of bacteria is illustrated by this example. In 1941 most streptococcal infections were caused by type A, the one responsible for scarlet fever; it was easily cured with 10,000 units of penicillin a day for four days. Streptococcus A and scarlet fever then disappeared, only to be replaced by streptococcus Type B, which, by 1970, became the most serious life-threatening disease to newborns. Then in the 1980's Streptococcus A returned.

By virtue of the expertise acquired during the development of penicillin, the Oxford group emerged as the world's foremost center for explorations in the realm of antibiosis, and the discovery of a new class of antibiotics, Cephalosporins, must therefore be considered a direct off-shoot of penicillin. This segment of discovery, however, starts not in England but in an ancient city founded by the Carthaginians, Cagliari, on the Italian island of Sardinia.

Cagliari prides itself on its small University that had been in existence since 1626. Giuseppe Brotzu, Professor of

Hygiene, acted as Director of the University during W.W.II, then, with the end of hostilities, he went on to become superintendent of Public Health. One of his duties was to supervise the local sewage disposal works. He noted that raw sewage had some capacity for self purification and decided to look into it to see if there were any micro-organisms responsible for this phenomenon. He ultimately isolated a fungus, which he identified as *Cephalosporium acremonium*, then proved that it produced a product that was antagonistic to many bacteria. A crude broth showed no toxicity and it was effective in treating boils and local infections. Encouraged by this, Brotzu tried to arouse interest in the Italian pharmaceutical industry but this failed. During a chance meeting with Dr. Blyth Brooke, whom he had first befriended when Brooke was British Public Health Officer in Sardinia during the war, Brotzu explained his plight. Brooke wrote to the Medical Research Council in London and they suggested that Brotzu should approach Florey in Oxford. Eventually, in September 1948, a culture of *Cephalosporium acremonium* and Brotzu's scientific report reached Oxford.

The religious world believes that victory, success, and discovery come from divine intervention; the secular world asserts that they are the result of intelligence, dedication, and hard work. But, as we observe the tactics of Dr. Giuseppe Brotzu, we must add one more attribute, a sense of humor. When, during W.W.II, Dr. Brotzu could not find anyone to publish his vague report about an insignificant mold that had unknown and less understood properties, he invented his own medical journal, named *Lavori dell' Istituto d'Igiene di Cheliari*. Its one and only scientific paper, *Ricerche su di un Nuovo Antibotica*, in this uniquely singular issue, was sent to Professor Florey.

Professor Brotzu's information and cultures of *Cephalosporium acremonium* went to Heatly who found that the important principle could be extracted from the culture fluid almost exactly as with penicillin. Then, in 1944, under Florey's instigation, a former Royal Navy Research establishment, where

Florey served as a member of the management committee and Heatly was liaison officer, became the center for research on this new arrival. E. P. Abraham, who worked with Heatly and Florey in the early stages of penicillin, directed the research; he was assisted by Kathleen Crawford and H. S. Burton in collaboration with Brandon Kelly and his colleagues in the antibiotic research station at Clevedon which had excellent fermentation facilities. The group first isolated Cephalosporin P, a form active only against so called gram-positive bacteria; later they discovered Cephalosporin N, active against gram-negative bacteria. These, in turn, led to the discovery of a new kind of penicillin, penicillin N.

In retrospect, it seems that penicillin N was the main antibiotic present in Dr. Brotzu's original culture fluid. The Medical Research Council tried to encourage industrial production of this but it came to nothing because no one could see any future in it. Dr. Abraham and research colleague Newton nevertheless persisted in their investigations of and, in due course, detected traces of yet another penicillin-like antibiotic, Cephalosporin C, in their brew. Florey and Margaret Jennings conducted the initial tests on Cephalosporin and proved that Cephalosporin C was less toxic than Benzoyl penicillin and, more importantly, it protected mice from penicillin resistant staphylococci.

In 1950, still smarting from the usurpation of penicillin by United States interests during W.W.II, the Medical Research Council suggested that Oxford should patent Cephalosporins. But once again an unkind fate mitigated complete victory. Cephalosporin C, produced by fermentation, had a chemical side-chain that interfered with its antibacterial properties. The Oxford team spent years trying to remove this trouble-maker but never quite succeeded. Just like the heavily laden branches of the fruit trees that hovered over starving Tantalus, a breeze, stirred by the Gods, always kept the fruit just out of reach, the side-chains eluded the Oxford chemists. Then researchers at the United States pharmaceutical company, Eli Lilly, beat out the British. They not only found an efficient chemical procedure for changing molecular structure but also prepared several semi-synthetic types of Cephalosporins. By 1978 annual world sales of Cephalosporins exceeded those of semi-synthetic penicillin and amounted to more than 600 million British pounds; royalties to the National Research Council came to over 100 million British pounds.

Although Brotzu took no part in the essential research, the University of Oxford conferred an honorary degree on him in 1971.

Like its predecessor sulfanilamide, penicillin also paved the way for another anti-tuberculosis drug, Streptomycin. Selman Abraham Waksman, Ph.D., a Russian born microbiologist who came to this country in 1910 devoted nearly 37 years of his life to finding a cure for tuberculosis. He would always query anyone who would listen, "Isn't it odd that with all the filth and garbage we toss into the earth we never see germs taking over the planet? There must be something in soil that kills them!" With this in mind he studied soil samples from cemeteries, farms, forests, and for miles around Rutgers University, where he worked, and isolated more than 10,000 cultures of soil organisms. As early as 1923, after observing zones of bacterial growth inhibition around colonies of the mold actinomyces, he confirmed that there were associative and antagonistic phenomena among micro-organisms. In 1940, he and his co-workers found their first germ killer but it was too toxic. In August 1943, Waksman isolated a substance from a strange mold-like organism, *streptomyces griseus*, that he had found in a clod of dirt taken from a chicken's neck and went on to prove that it could destroy the tough-coated tubercle bacillus. This new antibiotic turned out to be Streptomycin for which Waksman received the Nobel Prize in 1952—the same year that Isoniazid, an even better anti-tuberculosis agent, was introduced. It was Waksman who coined the term **antibiotic** to describe those chemicals obtained from microorganisms that killed bacteria. Rene Dubos, Waksman's former student, discovered a soil microbe, *bacillus brevis* from which, in 1939, he extracted the antibiotic Tyrothrycin.

In addition to serving as a springboard for many mold related antibiotics and bioactive substances penicillin also led to the discovery of an entirely new drug, Penicillamine, which possessed no antibiotic properties. Hoping to find new and expanded uses for penicillin, or even to synthesize it, Dr. Chain juggled its molecules and came up with a derivative called Penicillamine. But there it sat. It had no antibiotic value and no one could find any use for it.

But once again serendipity came to the rescue when Dr. John Walshe identified Penicillamine in the urine of a patient with liver disease who happened to be receiving penicillin for another infection. He followed-up on this clue and soon discovered Penicillamine regularly in the urine of patients with Wilson's disease who also had been under treatment with penicillin. When he administered Penicillamine directly to these patients their condition improved. Subsequently, in 1963, Penicillamine was approved by the FDA for Wilson's disease and cystinuria, and it is also used in rheumatoid arthritis.

In Wilson's disease excessive copper in the liver and brain destroy these organs. In cystinuria, a metabolic disorder, the amino acid cystine causes kidney stones. Penicillamine possesses a chelating effect, that is, it binds copper and other metal ions and when the Penicillamine is excreted it carries the unwanted ions out of the body. Similarly, it binds harmful antibodies in rheumatoid arthritis and removes them. In cystinuria it makes cystine soluble, preventing it from forming stones in the kidneys.

The drug that we now rely on to treat gout came to us in a similar manner. Back in 1943, when penicillin was still scarce, Merck pharmaceutical company started up its Renal Program in hopes of finding a compound that would conserve precious penicillin by retarding its elimination through the kidneys. Seven years later, when the drug probenecid was finally found, it was no longer needed; penicillin was in plentiful supply. However, the drug happened to increase the urinary excretion of uric acid, thus became an important breakthrough in gout therapy.

Of all the foregoing discoveries that arose from mold research none has had greater impact on world medicine than that of cyclosporin. By virtue of its ability to suppress our immune responses it interferes with the rejection of foreign tissues and makes organ transplants possible. In 1970, the Sandoz Laboratory in Basle, Switzerland, aware of the handsome profits that came from products derived from molds and fungi, like penicillin and streptomycin, undertook the investigation of two fungi *Trichoderma Polysporum* and *Cylindrocarpon Lucidum*, which were discovered in soil samples from Norway and Wisconsin. In due time they came up with several compounds that seemed to have antibacterial properties but when tested they proved utterly disappointing—a bust. But Dr. Jean-Francois Borel, a stubborn biochemist who had joined Sandoz in 1970 as senior scientist in the immunologic division, kept screening these extracts and between January 1971 and 1972 discovered that some could suppress immunity in cell cultures.

Born in 1933, Borel had originally wanted to become an artist, a painter of grand canvasses, but his practical Swiss parents insisted on a more secure profession. So the young man attended universities in Wisconsin and Texas, obtained his doctorate from the Federal Institute of Technology in Zurich, and gradually became interested in the field of immunology. While experimenting with his fungal extracts Borel was able to show that cyclosporin A had weak antibiotic properties; it also acted as a selective immunosuppressant. But many more years would go by before this finding could benefit patients in need of an organ transplant.

To begin with, there were problems of chemistry, production, and dosage adjustment. Then there was a question of commercial feasibility. After promising beginnings, transplantation of organs other than kidneys had come to a virtual halt. Surgical techniques had reached a remarkable level of sophistication but since we had no effective way of counteracting the immune system, rejection crisis set in sooner or later.

Would it make sense, then, to enter this limited market? That question seemed to receive a qualified "Yes" when David White at Cambridge, England, and his associate, the noted transplant surgeon Sir Roy Calne, obtained impressive results with cyclosporin in animal experiments. Their test proved that it was well tolerated and extremely effective in halting antibody production; most importantly, it interrupted graft versus host disease, the cause of transplant rejections. But now another obstacle loomed. Not a trace of cyclosporin could be found in the blood of the first human subjects to be tested with the drug. Convinced that only the formulation was at fault, Borel concocted a mixture of cyclosporin, alcohol, and detergent, which he swallowed himself. He became slightly tipsy but he also attained satisfactory blood levels.

The first time that the public heard about the biological effects of cyclosporin A was after an oral presentation, in 1976, at the meeting of the Union of Swiss Societies of Experimental Biology. Later, Dr. Borel reported his findings before the British Society of Immunology and in 1978, Dr. Calne undertook the first clinical trials in kidney transplantations.

Other surgeons soon followed the lead and cyclosporin was used, successfully, at first in kidney transplantation, later in more difficult surgery, such as liver, single-lung, heart-lung, pancreas and bone marrow transplants. Used in at least 150,000 patients during the first ten years of therapeutic experience cyclosporin has became the gold standard of immune suppression, with a literature of well over 10,000 papers. As for Dr. Borel, he continues his work in immunology and has even returned to painting. But only as a hobby. Incidentally, had Dr. Borel and his staff used a different strain of cells, cyclosporin A would not have been found.

The widespread interest in soil and mold research started by penicillin has by no means waned. Most recently, in 1987, the mold *Aspergillus terreus* gave us an important cholesterol lowering agent, *Lovastatin*.

On October 25, 1945 the Nobel committee announced that Florey, Chain and Fleming were to share the Prize for physiology or medicine. Nominations had come from many quarters including former Laureates. It has been said that prior to awarding the prize the Nobel committee intended to give the entire award to Fleming. Later it was proposed to give him half and share the remainder equally between Florey and Chain, but in the end, all three shared equally. In later years Florey ruefully said that this was the only real capitol he ever received despite his discoveries and an arduous life in science.

The prize-giving ceremony in Stockholm, presided over by the King of Sweden, usually takes place on December 10th, in the late afternoon, to commemorate the anniversary of Alfred Nobel's death in 1896. The 1945 ceremony, the first after the end of the second World War, was a particularly splendid, glittering occasion. In the customary Nobel Lectures, Fleming gave a straight forward account of his discovery of penicillin and implied that Florey and Chain continued that work. Florey described how research on antibiotics depend upon the development of appropriate techniques and the coordination of research. Chain, who spoke about the chemistry of penicillin, made one very erroneous prediction: "All attempts to improve the very small yield of synthetic penicillin have failed", he said, "and it appears improbable that a synthetic process will be evolved that could compete successfully with cheap biological production of penicillin." Today we have both synthetic and biologically produced penicillin and they are about equal in price.

Howard Florey was knighted in 1944 and continued his research with biodynamic substances until his death in 1968. By contrast, Chain, although he had been awarded the Nobel Prize with Fleming and Florey, received not even a minor civil award. He was elected to Fellowship in the Royal Society in 1949 and left England a year later. These snubs were redressed in 1969 when Chain was awarded a well deserved Knighthood—but only after he had been honored by Universities and learned societies around the world.

Florey and Chain went their separate ways after 1948. Florey returned to his native Australia as head of the medical school in Adelaide. Chain became chairman of the World Health Organization and held numerous scientific and honorary positions in Rome and England. Chain found great happiness during the last 30 years of his life. He married Dr. Anne Beloff, an Oxford biochemist, in 1948, who shared his scientific interests. They had three children, including twins, to whom he was devoted. Before he died, in 1979, at the age 73, Chain left the world this legacy of wisdom: "But do not let us fall victims of the naive illusion that problems like cancer, mental illness, degeneration or old age...can be solved by bulldozer organizational methods, such as were used in the Manhattan Project. In the latter, we had the geniuses whose basic discoveries made its development possible, the Curies, the Rutherfords, the Einsteins, the Niels Bohrs and many others; in the biologic field...these geniuses have not yet appeared...No mass attack will replace them..."

The happiness that came to Fleming with the Nobel Prize was abruptly expunged when his wife, Sareen, died in October 1949. Fleming married Amalia Boureka, in 1953, a Greek bacteriologist who had come to work in his laboratory seven years earlier. Sadly, this marriage, like that of penicillin's co-discoverer Florey, was short lived. Fleming died less than two years later at the age of 73. When Alexander Fleming died in 1955, flags the world-over were lowered to half staff, and the flower vendors of Barcelona emptied their baskets at the tablet commemorating his visit to the city.

Duchesne died from tuberculosis at the age of 38, unrecognized and forgotten.

PARADIGM POINTERS

ZEITGEIST, a German word with no equivalent in the English language, means time-ghost or spirit of time. It compels things to happen, or certain ideas to mature, when they do. Paradigm shifts arise as frequently from ZEITGEIST as from accidental discoveries.

When the time is 'ripe' for a discovery, the breakthrough may come from several sources almost simultaneously. It is generally conceded that if Pierre and Marie Curie had not discovered radium when they did, someone else would have done so shortly thereafter. And if Flory and Chain had not made penicillin into a practical antibiotic when they did, some other biochemist or mycologist would, sooner or later, have duplicated that achievement. And if Watson and Crick had not unraveled the structure of nucleic acid, Linus Pauling would have done so.

Perhaps the best example of ZEITGEIST is the coincidental arrival in England, in 1842, of Alfred Russel Wallace's letter from Australia, outlining the theory of evolution by natural selection. It was identical to Darwin's theory even though Darwin had been secretly working on his over the previous decade.

POST SCRIPT

1. LUCKY VACATIONS seem to appear regularly in the history of medicine. The course of events that unfolded while Fleming was on vacation—the contamination of his culture plate with mold and the destruction of colonies of bacteria—would never have occurred under normal working conditions. The contaminated plates would have been discarded or washed before the slow-growing mold could ever get a foothold.

Similarly, Pasteur was also unexpectedly rewarded after a 'lucky vacation'. In the summer of 1880, Pasteur, and his faithful assistants Roux and Chamberland, exhausted from their work with cholera, decided to close the laboratory and take a week's vacation. Upon their return they took cultures of cholera bacilli which had been made before going away and fed them to chickens. To their surprise the chickens lived. "I suppose," Pasteur remarked "that something must have happened to weaken the germs while we were gone. Let's try new cultures on them."

Cholera germs were grown in broth media and fed to the same chickens. Again nothing happened. When Pasteur was appraised of the event, with a flash of insight derived from genius, he concluded that the old germs were too weak to kill the chickens but they nonetheless provoked an immunity that protected them from new virulent strains. He then inoculated animals and people with bacteria so weakened that they could not cause illness. It worked. It induced immunity and resistance to disease and led to the development of vaccines against anthrax and rabies.

2. SIR ALMROTH WRIGHT who we met in Chapter One, where he tried to dissuade Dr. Domagk from pursuing sulfanilamide, and again in the previous Chapter Seven, where he promoted his dubious pneumonia vaccine, reappears, here, again road-blocking Fleming in his attempts to find out more about the strange penicillium mold. This obstructionism was probably a reflection of his convictions rather than perversity. Wright was a product of his times. He received his training just after bacteria were proven to cause diseases and performed most of his research when bacteria were injected into animals who produced antibodies and vaccines. He was a pretentious, strong-minded, individual, described by contemporaries as looking like Tenniel's illustration

of the Lion in *Alice Through the Looking Glass.*

Wright's rise to the peerage came not as a result of contributions to science but from expediency. Wright had developed a typhoid vaccine which the prime minister, Lord Haldane, wanted to use on the Army during World War I. Many scientists opposed this scheme because they felt that the vaccine was useless. Lord Haldane surmised, correctly, that by raising Wright's stature, opponents would be intimidated and quieted.

This is the letter that advanced his plan:

Dear Wright. We must have your Typhoid prophylactic for the Army but I have failed to convince the head man in the Army Medical Service of this. I have therefore got to build you up as a Public Great Man: the first thing to do is to make you a knight. You wont like it but it has to be. Haldane.

The plan worked. All the soldiers in the British army received Wright's Typhoid vaccine and the great vaccinator became Sir Almroth.

chapter 9

ORPHANS AND THEIR RELATIVES

Previous chapters dealt with discoveries related to bacteria. We now come to discoveries about viruses where accident and happenstance have played even greater roles. This is not surprising when you realize that viruses are so infinitesimally small that a hundred million polio viruses, for example, could fit on the period at the end of this sentence. They cannot be seen with conventional microscopes, they cannot be grown in the laboratory by customary means, their components tend to vary, their effects and behavior change with different hosts, and they can even infect bacteria.

Elie Metchnikoff, on his way home from Sweden where he claimed his 1908 Nobel Prize for unraveling the way white blood cells scavenge germs, stopped off at Count Leo Tolstoy's estate, Yasnaya Polyana, near Moscow. Tolstoy, who had his own ideas about the world and society, doubted the direct benefit to man of most scientific theories and said as much to his distinguished visitor. Metchnikoff replied, "Theory is much closer to life than is immediately apparent and many benefits to man are traceable only to complete abstract observations. Microbes were discovered long before one suspected that they had a role in human disease and only later has it become clear that the knowledge of microbes gives one the possibility to fight against human suffering."

In a similar vein, viruses were discovered long before their roles in human suffering were understood, long before we found out that they not only cause infectious diseases, but also trigger the events that lead to cancer, immune deficiencies, and weird, slowly evolving degenerations in the brain. And, as Metchnikoff so sagely observed, when abstract observations on viruses were reduced to factual knowledge, the damage they caused was considerably curtailed.

Early work with viruses led to a situation not unlike that of a Lost and Found Department in a large city. We had lots of items on hand but until they were claimed there was no way of knowing to whom they belonged.

Viruses in search of a disease were originally called *Orphan Viruses*. One such orphan cropped up around 1948 while scientists were groping their way among bowel-movement specimens looking for the causes of poliomyelitis. The virus was labeled *coxsackievirus* after Coxsackie County where the stool specimens originated. When it became apparent that this type of virus was enteric, that is, it entered through the gastrointestinal tract, and it was capable of producing pathological or harmful changes in cells, two identifyers, enteric and cytopathogenic were added to its name. This gave us the terminology we use today, ECHO, which stands for *Enteric Cytopathogenic Human Orphan virus*. Further experimentation showed that there were two strains of Coxsackievirus, group A and group B. Several years later, after many antibodies to viral diseases had been collected, virologists were able to prove that Coxsackie A virus caused meningitis and *Herpangina*, a disease characterized by swollen gums and sores inside the mouth; group B caused inflammation of the heart, meningitis, pleurisy, and an illness where severe joint and muscle pains were the most prominent symptoms.

Another orphan, *cytomegalovirus*, CMV, showed up much earlier, in the 1930's, but it did not find a niche until some twenty years later. Unlike most disease-producers, it was found everywhere, therefore virtually ignored or accepted as a normal fellow traveler, like bacteria in the mouth. By 1950 we surmised that almost everybody who reached middle age or beyond had had contact with CMV. We also believed that CMV infections were invariably fatal to newborn infants. In 1955 scientists were forced to take a second look at this unpredictable adversary when the first non-fatal case in a newborn was brought to light. Since then we have learnt that CMV strikes about one percent of the general newborn population. About 15% to 50% of infected babies sustain complete hearing loss and 25% have serious eye problems. There are 5 or 6 strains of CMV and each strain presents a different picture.

How we captured and identified this quixotic virus, plus some valuable philosophic thought about discoveries in medicine, was presented in a lecture delivered July 18, 1975, by Russell J. Blattner, MD, the eminent virologist and Chairman of the Department of Pediatrics of Baylor Medical College in Houston Texas. Recalling his early days just after graduation from medical school, he related how, "in 1934 Dr. Margaret G. Smith who is in pathology at Washington University (St. Louis), assigned to those of us who were doing the autopsies, the task of collecting salivary glands. This was a rather difficult thing because we were always worried about cutting through the cheek. However, as a result of the study of these specimens, she was able to demonstrate that between 16% to 20% of all young children autopsied, whatever the cause of death, showed the presence in their salivary glands of typical cytomegalovirus inclusions.... We wondered what possible value it could have. But it did mean something, because as the years passed it became apparent that the cytomegaloviruses are species specific and that there are many different strains.... Dr. Smith succeeded in growing the cytomegalovirus in specific tissue culture, a technique which she had learned from Dr. Enders and his coworkers in Boston.

All this was very exciting. Actually it laid the foundation for a much broader understanding of viral disease than we ever had before because it reemphasized the great importance of the host, the age of the host and the host response... In the baby or perinatal period, death or profound injury can result. In older persons we see completely inapparent infection.... This has given us an understanding of what happens in German Measles. If the Rubella (German measles) virus is introduced into the host early during pregnancy... it causes the rubella syndrome (deafness, heart, eye, and neurological deformities). As often happens in the history of medicine and science, a rather meager uninspiring beginning with seemingly unimportant observations has led to a very important concept of disease."

Besides German Measles and CMV, many other viruses act differently in different age groups. For example, in the United States, where hepatitis B is most often acquired in adulthood, the disease is acute and over-and-done-with in a short time. In Asia and Africa, where the virus is usually picked up at birth, it causes lingering, persistent problems or latent infections that can erupt sporadically.

In 1975 another enigmatic orphan came to light. While screening blood for possible hepatitis contamination, a group of virologists in England found some odd-looking viral particles that did not resemble the hepatitis agents they were looking for. These were collected and when tested proved to be a single stranded DNA virus that required rapidly dividing cells for replication. Since they did not fit the characteristics of any known disease causing viruses they were labeled human parvovirus B-19 and put aside as a strange virus without a claim to any disease. Ten years later M.J. Anderson proved that this so-called new virus was actually a very old virus that was responsible for a very old disease, *Erythema Infectiosum*, familiarly known to most parents as *Slapped Cheek Disease*, because children with the infection usually seem to have a light case of measles and develop a rosy-red rash on the cheeks. The disease was first described at the end of the 19th century and recognized in 1926 as a clinically separate entity.

There still are myriad orphan viruses in the world, viruses that have been catalogued but not linked to specific conditions. Viruses are more plentiful than ever suspected, possibly making them the most numerous things on earth. They have been discovered in vertebrate animals, plants, insects, and even in bacteria.

Many of the essential characteristics of viruses were first found by studying *bacteriophages,* viruses that infect bacteria. Around the turn of the century, just a few years after the world learnt about the existence of viruses, Felix H. d'Herelle of the Pasteur Institute in Paris, quite by accident, stumbled upon these sub-microscopic bacteriophages. While studying, of all things, a diarrheal disease fatal to locusts, he isolated the causative microbe and traveled from Argentina to North Africa to spread the infection in front of advancing columns of insects. While growing the germs on culture plates he was perplexed by clear spots in his bacterial lawns but not until he was recalled to France to investigate an epidemic of dysentery did he gain a unique insight into the meaning of these clear areas.

One night, during the 1900's, d'Herelle mixed filtrates from the feces of dysentery patients with broth cultures. The next morning, d'Herelle wrote, "On opening the incubator doors I experienced one of those rare moments of intense emotion which reward the researcher for all his pains; at first glance, I saw the broth culture, which the night before had been very turbid, was perfectly clear. All the bacteria had vanished. They had dissolved away like sugar in water. In a flash I understood what caused my clear spots—a virus that was parasitic to bacteria. If this is true of the germs in the intestines of the sick men, as in my test tube, the dysentery bacilli will have dissolved away under the action of their parasite."

Subsequently investigators discovered viruses that attack Anthrax, Diphtheria, Cholera, Bubonic Plague, and Gonorrhea. Phages, the abbreviated term for bacteriophages, were just beginning to gain a foothold as a form of antibacterial treatment in humans when sulfanilamide and penicillin came on the market; phages just dropped out of sight.

Unlike other microorganisms, viruses are essentially obligate intracellular parasites, that is, they can only thrive within other living cells. They cannot be propagated in the absence of living host cells. Viruses do not reproduce like other lifeforms. When entering a host cell they borrow genetic material from their hosts and make copies of themselves, hence we use the term *replication* to describe viral reproduction.

Since both bacteria and viruses are common disease producers , there is a tendency to consider them as close relatives. Actually they are as different as kittens and tigers, although both are cats. Perhaps the greatest difference lies in the equivocal nature of viruses. Are they alive or dead? Are they inert chemicals, crystals, or do they possess the attributes of living things such as the ability to grow and reproduce? Shortly after the discovery of viruses Friedrich Loeffler (Fig. 9.1) and Paul Frosch produced evidence indicating that viruses were living things yet in 1935 Wendell M. Stanley crystallized them, just like common table salt. Only recently, with the advent of electron and field-force microscopes and super-high-speed centrifuges that cause rapid sedimentation of virions, segments of viruses, have we begun to find out what constitutes a virus and how they operate. But in order to better understand these complexities we ought to first look at conditions surrounding the discovery of viruses and some of the seminal work that gave us our basic understanding of their structure and behavior.

Between 1892 and 1898, two men, working independently and unaware of each others experiments, yet using the same basic tools, discovered the particles that we now call viruses. Dmitry Ivanovsky, a Russian botanist, and Martinus Willem Beijerinck, professor of bacteriology at the Polytechnic in Delft, the town where Anton Van Leeuwenhoek had lived and invented the microscope, undertook the study of an infection of tobacco plants called *Mosaic Disease*. Both men

Figure 9.1 Friedrich Löffler. (From the archives of the Deutsches museum, München.)

squeezed the juice out of mottled, infected, tobacco leaves and looked for germs under the microscope. None were found. Attempts to grow them in the laboratory also failed. The leaf juice was then filtered through unglazed porcelain: When the sap from diseased plants was passed through porcelain filters and then injected into healthy plants it produced Mosaic Disease—suggesting that a filterable infectious agent was present. It was Chamberland, who worked with Pasteur, who discovered that the pores in unglazed porcelain were so small that bacteria could not get through; liquids that did pass were free of bacteria.

Beijerinck went further than Ivanovsky. He realized that a poison in these filtrates could have produced Mosaic Disease so, he reasoned, "if it is just a poison it will kill one plant but not others. All I have to do is take the juice from leaves that have been treated with the filtrate and pass it on to other plants and see what happens." To Beijerinck's amazement leaves of healthy plants treated with such filtrates came down with Mosaic Disease. It was not a poison. It had to be a germ of some sort, a germ that could not be seen or even grown on culture medium in the laboratory but could pass through pores small enough to block bacteria. "There must be a disease without a microbe," Beijerinck concluded, and in 1898 he published his idea of a living fluid . This was later termed filterable virus and subsequently shortened to virus.

Simultaneously, in Germany, the bacteriologist team of Friedrich Loeffler and Paul Frosch discovered that by taking fluid from cattle with Foot-and-Mouth Disease, then passing it through the pores of a porcelain filter, they could induce the disease in animal after animal, proving that the invisible infectious agent was able to reproduce itself within the infected animals. Three years later, in 1901, two Americans, Walter Reed and James Ferrell, showed that Yellow Fever behaved the same way.

Although the world of science would not accept Beijerinck's theory, no one was able to prove that it was wrong. The only thing known for certain was that the great Louis Pasteur could not find the bacterium that caused rabies and went so far as to imply that the disease was caused by a virus. He stipulated that viruses were so small that they could pass through the finest porcelain filters, and that there was no microscope powerful enough to make them visible. In truth, however, a virus had been seen as early as 1887, in Scotland, by John Brown Bust, who had stained some Cowpox liquid and noted strange little particles in it. These were Cowpox virus particles which, being unusually large, could be seen with the ordinary microscope.

In 1908, Karl Landsteiner, the discoverer of human blood groups, injected bacteria-free samples of spinal tissue from human poliomyelitis patients into monkeys and thereby proved that the disease was caused by a filterable virus. Simon Flexner and P. A. Louis at Rockefeller Institute confirmed the viral theory of poliomyelitis by passing polio of human origin from monkey to monkey. But for all intents and purposes, viruses remained unseen until the development of the electron microscope. In the decade after 1932 scientists began to identify viruses by characteristic appearance: mumps virus resembled a bowl full of spaghetti whereas the polio virus was absolutely symmetrical.

One of the most intriguing breakthroughs in the annals of virology took place around 1910 when a perceptive farmer picked up his newspaper and read about a young scientist in New York City, Dr. Francis Rous, who was studying cancer in animals at the famous Rockefeller Institute for Medical Research. The farmer's Plymouth Rock hens were developing tumors and dying. No one from the local farm bureau could offer any help, so he took one of his infected hens to the Rockefeller Institute in New York City and placed it and his problem on Dr. Rous' lap. Rous was fascinated by the story as well as the appearance of these cancers because, in the farmer's words, "it seems like there was an infection going on, almost like chicken cholera." Rous dropped everything and went to work on this new challenge. The first thing he did was to take a small piece of tumor from an infected hen and implant it into another hen of the same strain. Lo and behold! similar cancers developed. Then, in a series of classic experiments, Rous was able to show that the cancer initiating substance was smaller than a cell and harbored many of the characteristics of Beijerinck's *living fluid*. This led him to duplicate Beijerinck's methods: he ground up the cancers and passed the mishmash through filters with pores so small that they would filter out all possible bacteria and cells. The filtered material, when injected into hens, produced cancers. Rous concluded that the cancer causing substance was smaller than a cell, smaller than a bacterium, and it was transferable, possibly a virus. He was awarded the Nobel Prize in Medicine in 1966, four years before he died, 56 years after he made his breakthrough discovery. The reason for the delay was that many researchers unfamiliar with the special techniques needed for viral studies were unable to duplicate Rouse's work. As time went on, however, a new breed of virologist with viral expertise replaced the general bacteriologist, and in their hands Rous's infectious agent, called RSV, Rous's Sarcoma Virus, successfully induced rat and mouse tumors.

The virus that Dr. Rous found did not actually cause cancer; it carried the cancer-activating signal or *oncogene* from one hen to the next. Rous' greatest contribution to science was his explanation of this phenomenon: an oncogene bearing virus was simply carrying a cancer promoting gene donated by a previously infected cell, and infection with this virus could induce cancer. This was later substantiated when Hepatitis B virus was found to initiate cancer in the liver. Similarly, the virus that causes infectious mononucleosis can cause cancer in the throat and cancer in B-lymphocytes. The 1975 Nobel Prize winners, Howard Temin at the University of Wisconsin in Madison and David Baltimore at Massachusetts Institute of Technology, working independently, proved that the rous sarcoma virus, found in chickens, and rauscher leukemia virus, found in mice, are viruses that steal cancer causing genetic material from one cell and initiate cancers when they infect other cells. The viruses responsible for leukemia in mice, chickens, and cats, and the foamy viruses in monkeys and cattle behave the same way.

We will return to virus' ability to cause cancer but in order to understand the mechanisms involved we have to learn more about viruses themselves. This is best done by following the trail of the pioneers.

Back in 1932 Wendell Meredith Stanley and his wife got tired of the hubbub in New York City and decided to live in Princeton, New Jersey. Stanley was working as a chemist at the Rockefeller Institute, and, since the Institute had laboratories in Princeton as well as New York, he asked for a transfer. "If you go to Princeton", he was told, "you will have to work on viruses." Rather than stay in New York, Stanley accepted the challenge, and, as a chemist, attacked the problem of viruses chemically.

He started with Tobacco Mosaic, the same substance that Beijerinck had worked with over 30 years before. His plan was to take infected plants and chemically remove all plant material, which, in effect, would leave only the virus, something he could then analyze. In order to get enough tobacco plants he had to grow them himself, unwittingly becoming a farmer. Stanley grew tobacco plants, infected them with mosaic virus, then harvested and froze them. As the plants froze the liquid in them expanded, their cells burst open, and as luck would have it, Stanley obtained virus laden plant-liquids free from solid material. Additionally, he thawed and mashed up plants, chemically removed all plant material and then tested the residue for viral activity. Slowly, painfully, and systematically plant extracts became clearer and clearer until Stanley had a pure solution of virus in water. One day while examining the purified solution he observed beautiful crystals, like frost on the window pane. When tested, the crystals produced disease; the water from which the crystals were extracted did no harm. In June 1935 he published his startling results: A VIRUS CAN BE CRYSTALLIZED.

Ostensibly, this contradicted the findings of Friedrich Loeffler and Paul Frosch who, in 1898, had shown that the virus of hoof-and-mouth disease was a living thing that could reproduce itself. Now it appeared that viruses belonged somewhere between life and non-life. They could appear as ordinary crystals, like salt, for instance, but they could still grow and spread inside of a plant and retain living and reduplicating properties.

The answer to this enigma lies in the structure of viruses, namely, as blocks of genetic material from living cells. Actually, they are incomplete life-forms that depend on their hosts to provide many of the fundamental life processes that are absent in the virus itself. The free virus particle, or virion, consists of genetic material called ribonucleic acid, RNA, or deoxyribonucleic acid, DNA, but never both, surrounded by a protective coat, or capsid. Some viruses have proteins within the capsid while others contain lipid (fat), and carbohydrate. Some RNA viruses, known as retroviruses, produce an enzyme that can synthesize DNA. The virion is an inert macromolecule that lacks the components necessary for independent life and reproduction. But by using genetic material from host cells it produces new virions, which, upon release from cells, initiate new cycles of infection.

H. Fraenkel-Conrat, working in Stanley's laboratory in Berkeley, California in 1956, demonstrated that it is the nucleic acid in the virus that makes it dangerous. After a virus enters a cell its nucleic acid takes control of that cell; the virus' nucleic acid now directs the host cell's metabolism and functions, thereafter, the host cell starts turning itself into more virus particles. Ultimately the cell swarms with virus particles made out of the cell itself. Some viruses are released by killing the cell while others, the producers of silent infections, may replicate within a cell and leave it without causing any injury.

Contemporaneously with Stanley, two obscure, young scientists at Rockefeller Institute, Albert Sabin, who would later give us the vaccine that conquered polio, and his co-worker Peter Olitsky, also experimented with the polio virus, but unlike Stanley, who specialized in chemistry, these two men were primarily interested in the virus' growing characteristics and how and why it caused paralysis. They achieved a modicum of fame in 1935, the same year that Stanley crystallized the tobacco mosaic virus. While trying to grow the polio virus in various parts of human embryos they found that it would only thrive in the brains and spinal cords. But to make absolutely sure that this was the case, they performed extensive experiments in an attempt to grow the virus in other media and other tissues. When all of these trials failed; the intrepid scientists boldly declared that the *Polio virus will only grow in nerve tissues.* This was an unfortunate episode and the source for later embarrassment. The scientists had chosen a strain of polio for their tests that had been passed through the nervous tissues of laboratory animals so many times that it had forgotten how to live in any other medium.

Some twelve years after this event, virologist John Paul visited the summer home of John Enders in Waterford, Connecticut. In the course of a leisurely afternoon Enders casually remarked that he and his colleague had succeeded in growing the polio virus in human tissues other than those of nervous origin. "For the moment I was stupidly unaware of the implications that this finding held," John Paul later recalled. It not only shot down the Sabin-Olitsky pronouncement, but it paved the way for a successful polio vaccine.

Prior to 1935, when Stanley first crystallized a virus, scientists could only describe the effects of viruses in cells and in whole animals. There was no way of 'seeing' them. Stanley enabled people to see crystallized viruses, but actual visualization of the living, active, virus had to wait until the next decade when the electron microscope became available. And except for Sabin's growth of viruses in nerve tissues, there was no way to grow viruses in the laboratory. A step closer to physically identifying viruses was made eight years earlier, in 1927, when two research scientists, Ernest W. Goodpasture and Charles Eugene Woodruff met at Vanderbilt University School of Medicine in Nashville, Tennessee, and found that they had a common interest in viruses. Woodruff showed

Goodpasture a young chick that had a lump on its head, "the first of its kind", he said. "I did it by injecting the chick with fowl-pox virus." Woodruff was essentially saying that he could cause tumors in chicks by injecting them with a virus. Woodruff already knew that chick cells harboring fowl-pox virus were easy to spot under the microscope. The virus itself could not be seen, but the infected cells built a container around the virus, called an inclusion body, and these could be seen with the ordinary microscope. Woodruff collected these infected cells in a micro-pipette and carefully transferred them to the skin on the head of a chick. "You've just seen the results" Woodruff said to Goodpasture. "That proves that the inclusion body contains the virus."

Woodruff and Goodpasture now set their sights on growing fowl-pox virus in living cells. They knew that Ross Granville Harrison, between 1907 and 1910, working in the Department of Anatomy at Johns Hopkins, had successfully cultivated animal tissues outside the body, and that the legendary Alexis Carrel at the Rockefeller Institute had kept pieces of chicken heart and kidney alive for years in artificial surroundings and came close to growing viruses on the same basis. Goodpasture adopted Carrel's technology and soon succeeded in growing chick kidney tissue in his laboratory.

When Woodruff's wife, Alice, joined the department, Goodpasture handed her the chore of trying to grow fowl-pox virus in tissue cultures made from the cells of the chick kidney. But the virus just refused to grow here. "Well, let's try the chick embryo itself," Goodpasture suggested, "it's made up of living cells, there are no bacteria inside of it, and the egg shell protects it." Whether Goodpasture was aware that this had been tried before is not known, but Emile Roux, who was Pasteur's associate for twenty years, had grown a virus in a chick embryo as early as 1910 but was never able to isolate the virus from the cells. Alice Woodruff, using fertilized chicken eggs, held them up to a strong light and looked for the shadow of the developing embryo. Then, with a tiny blade, she cut a flap in the shell, injected fowl-pox virus into the embryo, sealed the window in the shell with a thin piece of glass held in place with Vaseline, and replaced the egg into an incubator. But this did not work because, almost without exception, bacteria gained entrance through the window and killed the embryo. Everyone put their heads together and agreed that bacteria were somehow being transferred to the embryo together with the virus. Henceforth all fowl-pox samples would be tested for absence of bacteria before transfers were made. Fowl-pox specimens were cultured in broth; if they proved sterile, that is, no bacteria grew in the broth, they were used to inoculate chick embryos. Again the embryos died; death resulted from bacterial infection. Next Woodruff tried, with infinite care, to aseptically transfer pure inclusion bodies from infected cells to embryos but met with no more success than previous attempts. Every time the little group thought about abandoning the experiments someone would say, "let's give it just one more try. If we can only manage to get the virus to grow in one embryo we can use that as a pure source for other experiments. We won't have to go back to inclusion bodies and infected animal cells."

As luck would have it, Alice Woodruff discovered an inoculated egg in which the embryo was thriving; it also had a slightly swollen claw where the virus had been placed. This was the beginning. When the swollen tissue was transferred to another embryo it produced more swollen tissues. Soon the team had a number of embryos that contained pure virus. Under the microscope the swollen tissues showed inclusion bodies. To make sure that this was the same virus they started out with, they took some of the swollen tissue from the embryo and injected it into a grown chick; it developed typical fowl-pox. Now scientists had a way of growing viruses as well as detecting their presence and their effects.

Immediately thereafter, in 1937, Max Theiler of Rockefeller Institute, grew yellow-fever virus in a fertile egg and passed it through hundreds of mice and chicks until he had a mutant that could be used in a vaccine. Theiler's approach to attenuation or weakening was the cornerstone of the research that led to viral vaccines.

At this point in time viruses could be grown in living animals, fertile eggs, and chick embryos, but they still could not be propagated in flasks or test tubes in ordinary laboratories. It remained for Dr. John Franklin Enders, together with his bright, eager, assistants, Thomas Huckle Weller and Frederick Chapman Robbins, to hand scientists the breakthroughs that not only made it easy to grow viruses but also opened up the door to making viral vaccines.

The fact that this trio succeeded at all must be attributed to a series of fortunate meetings, accidents, and a willingness to fish in troubled waters. When Dr. Enders accepted the 1954 Nobel Prize in medicine, he cited Claude Bernard on the merits of unpremeditated science: "Physiologists should not be afraid to act somewhat at random, so as to try—permit me the common expression— fishing in troubled waters."

The three men who came together in a laboratory in Boston's Back Bay grew up in remarkably different worlds. The senior member, John Franklin Enders, was the only one of the trio who had not inherited a bent but instead inherited a fortune. Enders' grandfather was president of the AETNA Life Insurance Company and his father had been president and board chairman of the Hartford National Bank. Together they left an estate of 19 million dollars. When asked what he thought of the money as an inheritance, Enders answered succinctly, "I think it is a good thing." Enders attended aristocratic New England prep schools and cultivated a general interest in the arts and a passion for rowing. In 1915, as World War I loomed on the horizon, Enders joined the Naval Reserve, piloted aircraft, and taught flying. By 1920 Enders had his Bachelor of Arts degree from Yale and, while his brother went into banking, John set himself up in a real estate office. He was not a good salesman; the business failed. John Enders thought he might become an English teacher and enrolled in Harvard's School of Arts and Sciences where he spent four years immersed in

English, Celtic and Teutonic literature. He took a Masters Degree and wrote a doctoral thesis on the use of gender in the grammar of ancient languages.

Amidst disconsolate wanderings in dead languages, Enders met an Australian medical student, Hugh Ward, at a rooming house in Brookline, Massachusetts, who took Enders to his bacteriology laboratory to meet professor Hans Zinsser. The dramatically handsome forty year old Zinsser was not only an engaging teacher and famous author but also a charming raconteur who would discourse authoritatively about music, philosophy, literature, and the future of medicine. John Enders was instantly and permanently enchanted by Zinsser. He dropped literature, entered a program leading to a doctorate in bacteriology and immunology, and began working under Zinsser; this allowed him to join that enchanted circle around Hans Zinsser that functioned like a Socratic symposium. By 1930 Enders had become an expert in serious allergic reactions and was appointed instructor at Harvard Medical School. He married Sarah Frances Bennett and had arrived at a pleasant plateau in his life, professionally and domestically.

By age 40 Enders had found the virus responsible for distemper in cats and later, with William Hammon, developed a vaccine for it. Zinsser was always interested in Typhus and Enders thought of impressing the master by finding a way to artificially grow the responsible *rickettsia*. He made a good start and came awfully close but was beaten out by Haerald Rea Cox, who grew the organism on chick embryonic tissue.

At the time Hans Zinsser died in 1940, Enders had become interested in the nature of viruses and began looking into ways of growing them in his laboratory. When one of his students, Thomas Weller, indicated that he too had an interest in tissue culture work, Enders took him on as an assistant.

Weller's father was chief of the department of Pathology at the University of Michigan and his grandfather was also a doctor, so young Weller gravitated toward medicine naturally. After receiving bachelors and masters degrees at Michigan in 1937, he went off to Harvard Medical School, where he concentrated on parasitology. In the summer between the second and third years at Harvard he received a Rockefeller Foundation Fellowship that took him to Florida, where he performed extensive public health work. When Weller returned to Harvard he was convinced that his future lay in the field of viruses.

Weller's roommate was Frederick Chapman Robbins, a year younger than Weller, who, like Enders in his youth, did not quite know what he wanted to do with his life. He was certain that he would not go into botany, where he would have to compete with his illustrious father, William Jacob Robbins, professor of Botany at the University of Missouri and talent-scout for Rockefeller Foundation Fellowships. He tried engineering then abandoned it and transferred to Harvard Medical School where he fell under the spell of Enders' and Weller's enthusiasm for virology. After getting his M.D. he decided to bypass the competition for internships and instead took a job in the pathology and microbiology laboratories of Boston Children's Hospital; this was available to students right out of school.

Weller, Robbins and Enders might have come together at this time, but war broke out and the three men were driven in different directions. Weller was sent to the U.S.Army's Antilles Department Laboratory that specialized in parasitology, biology and virology. Here he tried to grow the viral pneumonia virus in a mongoose. "The mongoose is not friendly and when you come to get him, he comes more than half way to get you," Weller used to tell friends. Fred Robbins was indoctrinated at Walter Reed Hospital and then assigned to teach medical technicians at a school in El Paso, Texas. By 1943 he began to feel unused and abused, so in response to his complaint he was sent to head up a viral and rickettsial laboratory in bombed-out, disease ridden, starving, Naples. Like Weller, Robbins was also attracted by the large number of cases of what was believed to be viral pneumonia. Through keen investigations he proved that this outbreak was due to Q Fever (so named in honor of Queensland or as an abbreviation for questionable.) It had swept through the German army as well but they called it *Balkan Grippe*.

Professor Hans Zinsser, the man who brought Enders, Robbins, and Weller together, was a renowned historian as well as one of the foremost bacteriologist of his time. In his book, RATS, LICE AND HISTORY, Zinsser shows off his vast medical and scientific acumen by demonstrating how epidemics, rather than military strategy, defeated armies, even causing the fall of the Roman Empire. He wrote: And typhus, with its brothers and sisters—plague, cholera, typhoid, dysentery— has decided more campaigns than Caesar, Hannibal, Napoleon, and all the inspector generals of history.

The truth behind this premise surfaced once again during World War II, when diseases often incapacitated vital segments of most armies. This was particularly troublesome for the Germans in Italy. Laboring under the illusion that they were dealing with some sort of a viral condition, the Germans treated *Balkan Grippe* like a common cold or pneumonia, which, of course, proved futile. But Robbins and coworkers in the field found out that rickettsia were behind the pneumonias in Italy and advantageously instituted specific treatments. Dr. Robert Austrian, who gave us the vaccine against pneumonia (Chapter 7, CAPTAIN OF THE MEN OF DEATH), served in the China-Burma-India theater as part of a team that investigated typhus fever, a rickettsial-type disease not too distant from that studied by Robbins. As a result of superior medical knowledge supplied by men like Robbins and Austrian, plus the backup of sanitary engineers who cleaned out animal and insect disease transmitters, the Americans lost fewer men to illness, giving them the deciding edge in battle. As Zinsser said, "Epidemics get the blame for defeat, generals the credit for victory."

Meanwhile, back in Boston, Enders now in his 40's and too old to fight, looked for ways to contribute to the war effort. His friend, Frederick Russell, a professor of preventive medicine, suggested that mumps might be a good place to start; it had been a serious medical hazard in World War I and might crop up again. Enders plunged into new studies of mumps virus, which, as early as 1934, he had already isolated in embryonic chicks by the Goodpasture technique. But the virus would not grow in non-living media in flasks; he had to rely on the mainstay of viral laboratories, the monkey. Enders then used the monkey salivary glands to produce a vaccine that proved successful, first among monkeys and later among humans. Enders was appointed to the status of civilian consultant on epidemic diseases by The Secretary of War.

The death of his wife Sarah temporarily slowed down the energetic and inquisitive Enders, but as soon as the war ended he headed up the brand-new Infectious Disease Research Laboratory at Boston Children's Hospital. Now he would not have teaching duties and could pursue the culture of viruses full time. The laboratory was set up in an old three story brick building in Boston's Back Bay just about the time when two army medical corp majors, Thomas Weller and Frederick Robbins, were returning home from the war.

Weller, in quick order, married Bostonian Kathleen Fahey, finished a year of Pediatric training at Children's Hospital, then joined Enders in setting up his new laboratory. Robbins would have liked to join the group, but there was no money available for his salary. The National Foundation for Infantile Paralysis was then funding some programs for viral research, so Robbins applied for and received a fellowship on the stipulation that he work only one year with Enders; after that he was committed to go to Australia for a two year tour of duty with Macfarlane Burnet. Included in this team were two beguiling laboratory assistants, Carolyn Keen and Alice Northrup. Within six months after setting-up the Back Bay laboratory, Alice Northrup married Frederick Robbins and three years later Carolyn Keen married John Enders.

Enders gently nudged Weller in the direction of mumps research and by March 1948 Weller succeeded in growing the mumps virus in a soup of chick embryo brain and ox blood.

He then decided to try to grow the chickenpox virus on human tissues. He set up twelve flasks containing human embryonic muscle and skin tissues, poured specimens taken from a chickenpox patient into four of them and put another four aside as controls. Somehow polio virus got to the remaining unused flasks.

One story has it that Weller, upon hitting a snag in his attempt to grow the chickenpox virus in an artificial medium, complained to his mentor, "I'm still trying to grow these damned cells. I've made up a whole series of flasks with samples of skin and muscles cells from newborns that aught to be tender enough for any virus. I have more tissue culture material than I need for my studies; would you care for a couple of these spare flasks to see if you can grow anything on them?"

Another version maintains that Frederick Robbins came to Enders at this same time complaining that he too was running into a blank wall in his experiments with the mumps virus. "Fred," Enders advised, "sometimes we push too hard; nature gets upset and pushes back twice as hard. Occasionally it is better to leave a project and then come back to it. How would you like to work on polio for a while?" Then, as he peered over his half-frame glasses, Enders added, "I've got some polio virus in the deep freeze. Let's give it a try."

Records show that the team added polio virus from Enders' freezer to a culture of cells derived from human embryonic skin and muscles. This was the Lansing strain which came from an 18 year old boy who had died from polio in Lansing Michigan. The virus thrived in the flasks. This triumph led to the long sought-after method of growing viruses artificially in the laboratory; it also contradicted Albert Sabin and Peter Olitskey who had stated unequivocally, a decade earlier, that polio viruses would grow only in nerve type tissues.

Dr. Enders knew that Dr. Robbins had been trying to grow the mumps virus in a culture of cells from the mouse intestine and surmised that here might be a way of enlarging the sphere for polio propagation. Enders approached Robbins and stated his case: "In the early 1930's it was discovered that large amounts of polio virus could be found in the stools of individuals with non-paralytic polio," Enders began. "The polio virus seems to live very well in the human intestine; maybe it actually grows inside intestinal cells. Why don't you take some polio virus and see where it hangs out and what it likes to eat?"

"But you are talking about human intestinal cells," Robbins protested. "I've been working with mouse intestinal cells. Do you think the virus is smart enough to know the difference?"

"We won't know until we try, will we?" Enders replied. And as an after thought he added, "since you already have the mouse intestine cells ready to go this might be a good place to start."

In short order both Weller and Robbins had the polio virus growing in their respective tissue cultures: Weller in human skin and muscles cell cultures and Robbins in intestinal mouse cell cultures. When the viruses were tested for their polio effect, Robbins' mouse intestine culture showed no effect but Weller's human tissue cultures showed sustained growth with no loss of viral potency. Although these advances were heartening, Enders, familiar with the many false leads of the past decade, suggested further experimentation before going public with a scientific report.

Weller and Robbins joined forces to fully explore the means of growing polio viruses in tissue cultures. These contained penicillin and streptomycin as well as both animal and human cells. The virus grew exceptionally well in human intestinal cells, thereby confirming Enders' suspicion that people naturally harbored polio in their intestinal tract. After establishing growing procedures the team then diluted their cultures, prior to testing them for viral activity, and, to their surprise, the virus remained active with every dilution; even when diluted a million, million, million times. Enders,

Weller, and Robbins published their findings in SCIENCE January 28, 1949 and were awarded the Nobel Prize in medicine five years later. It was only a matter of time before other scientists would grow the virus, which of course opened the door to the development of polio vaccines.

At Noble prize ceremonies Enders said, "In 1948 we had no immediate intention of carrying out experiments with poliomyelitis viruses...but knowing that the virus grew in the intestinal tract the decision was taken to try a mixture of human embryonic skin and muscle tissue in suspended cell cultures in the hope that the virus of chickenpox would grow in this. These cultures were close on hand in a storage cabinet where the Lansing strain of poliomyelitis virus was kept. Thereupon it suddenly occurred to us that everything had been prepared almost without conscious effort on our part for a new attempt to cultivate the agent in extraneural tissue."

That declaration in Stockholm in summary stated that the inspiration was in the air, the situation presented itself, the deed was done and never mind who did it. As it turned out the chickenpox virus did not grow in the flask , but the polio virus did. This resulted in two benefits; no longer were monkeys needed and, secondly, a large source of polio virus was now available for study and for vaccine preparation.

The team went back to work and cultivated the Brunhilde strain of poliomyelitis by similar techniques with equally good results. In the years that followed Enders' team isolated and grew at least 13 different strains of polio virus in tissues of human origin.

According to the terms of his National Foundation Fellowship, Robbins had been scheduled to remain with Enders for only one year but the thought of losing him in 1951 was repugnant to Enders, so he urged Robbins to resign his fellowship and Enders agreed to pay him out of his own pocket $7,000 a year; this was $1,000 more than Enders was earning. Weller went on to isolate the chickenpox virus, the cytomegalovirus, and the German Measles virus, the last one coming from Weller's 10 year old son.

Despite these mighty triumphs in the laboratory, poliomyelitis itself continued to run rampant, striking 57,000 people every year and killing 3,300 of them. Sulfa drugs, penicillin, and other antibiotics proved useless. The only hope for many victims whose chests had been paralyzed by poliomyelitis was the Drinker respirator, the *iron lung*, which also came into being partly through design and partly through happenstance. In 1926 Phillip A. Drinker was appointed to a commission at the Rockefeller Institute to develop improved methods of resuscitation. This study was sponsored by the Consolidated Gas and Electric Companies of New York, who were hard pressed by fatalities resulting from gas poisoning and electric shock. Phillip Drinker's brother, Cecil, and a young physiologist, Louis A. Shaw, were studying respirations in cats around this time, so Phillip got his start from them. The device used in the cat experiments was very similar to the one that finally emerged as the iron lung. It was essentially a big box in which the cat's body fitted but the head protruded to the outside. A rubber collar at the neck sealed off the box. Drinker, together with his coworker and Charles F. McKhann, discovered that the changes in volume inside the box exactly matched changes in the animal's chest as it breathed. As the cat inhaled the volume in the box decreased, and vice versa. They then devised a chamber where the opposite conditions prevailed: when the chamber was evacuated the patient's chest expanded and inhalation followed; when pressure was increased in the chamber the chest was forced to contract and the patient exhaled. Polio was almost uniformly fatal when respiratory paralysis set in, so Drinker and Shaw rushed to build an adult-sized respirator in the shortest possible time. The tank was built by a local tin smith; a pair of vacuum cleaner blowers created the pressures, and the necessary valving was salvaged from the laboratory's workshop. The subject was placed into the tank on a garage mechanic's creeper, the rubber collar was slipped over the subject's head using a soaped sheetmetal cone to protect the face, then the endplate that sealed the tank was clamped shut. The entire device was built in a matter of weeks for about $500.

The first three test subjects were Drinker, Louis Freni, who was a laboratory worker, and Shaw. In 1928 the stage was set for a clinical trial when an eight year old girl with poliomyelitis entered a Boston hospital and began to go down hill rapidly. Her breathing was becoming weaker and weaker and death seemed imminent. The machine was brought into her room so she could see it and become accustomed to its noise and, as she slipped into coma from lack of oxygen, she was placed into the machine. Within a minute or two she regained consciousness and a little later asked for ice-cream. News of the success of the *mechanical respirator* spread like wildfire throughout the world. It was not unusual for doctors from overseas to call Drinker at all hours of the day and night and ask for instructions on its operation. At the height of the epidemic in 1931 the demand for respirators exceeded the supply, so a crash program to build them went into effect. By 1950 Massachusetts General Hospital had so many polio patients using ironlungs that an entire floor of one building had to be gutted to accommodate them.

Incidentally, this work on respirators for poliomyelitis laid the foundations for the design and perfection of the apparatus that later provided ventilation and super-oxygenation for premature newborns. This is described in Chapter 5, TOO MUCH OF A GOOD THING.

Clearly there was an urgent need for a polio vaccine. The big break in this direction came in 1952, when several investigators at Johns Hopkins University found that there were three different types of polio virus: Brunhilde type 1, Lansing type 2, and Leon type 3. A committee representing the National Foundation For Infantile Paralysis was sent there to check out these findings. They confirmed the results of the Johns Hopkins group and, in effect, notified researchers that if a vaccine was to be successful it had to contain the three different types of polio virus; immunity to a single strain did not furnish protection. One of the committee members, a slightly built man with a prominent nose and balding on the top of his head that gave the impression of an unusually high forehead, was Jonas E. Salk, a prominent virologist from the University of Pittsburgh. He immediately put this information to use and went on to win the race for a vaccine against polio.

Without the discoveries at Johns Hopkins University and those of Dr. Enders and his coworkers, Dr. Salk might not have had any better luck than his predecessors. In 1935, for instance, Dr. Maurice Brodie at New York University School of Medicine, employing almost the identical methods that Pasteur used with rabies, injected polio virus into the spinal cords of monkeys, ground up the cords and treated them with formalin to kill or weaken the viruses, then filtered the mixture. He inoculated the filtrate into experimental animals only to find that the vaccine did not work; worse still, it provoked serious, often fatal, reactions to proteins present in nerve tissues. Dr. John Kolmer, a bacteriologist at Temple University in Philadelphia, the man who had treated President Coolidge's dying son, also made a vaccine. He claimed that it was perfect. When tested it produced no immunity and in the process killed six children and paralyzed three others.

Armed with this data picked up during his visit to Johns Hopkins University Dr. Salk returned to his laboratory at the University of Pittsburgh, took over several floors of the research building and put 50 people to work growing all three types of the polio viruses in tissue cultures made with monkey kidney cells. From experience Salk knew that formaldehyde inactivated or killed the virus but it did not do it all at once. Each day only part of the virus that was in a batch at any one time was killed: on the first day it would kill perhaps 90% of the viruses, the next day 90% of the survivors, and the next day 90% of the remainder. "But," declared Salk, "I don't dare have even one living particle in my vaccine." After subjecting his polio viruses to repeated doses of formaldehyde the vaccine was tested on monkeys. Later, when there was evidence that the monkeys were responding to the vaccine by making antibodies, they were challenged with live polio viruses. The monkeys stayed healthy, they were immune to poliomyelitis.

Being the consummate scientist, Salk was not satisfied with this triumph, so he proceeded to check the levels of antibodies in the blood of monkeys that had received the vaccine. Levels were high, proving that his vaccine was a powerful inducer of immunity. Dr. Salk now tried his vaccine on children who had had polio and therefore could not get it again. If the vaccine was really good it would nevertheless elevate their polio antibody levels. During the spring of 1952 Salk administered his vaccine to a group of crippled children in Lettsdale, Pennsylvania, and by October found that they all showed significant elevations in antibodies against polio. The vaccine was a success but "dare we try it on people who have not had polio?"

On a fine day in May 1953, 38 year old Dr.Salk drove his wife, Donna, and their three sons from their home to his laboratory where all five members of the Salk family received polio vaccine. There were no harmful effects; it could justifiably be tried on others. The first field trial was on several hundred children in the suburbs of Pittsburgh. Nurses first took blood samples from each child, which would show the basic level of polio antibodies, then Salk injected his vaccine, and after several weeks the children were again tested to see if they made polio antibodies. They did and Salk was sure of the efficacy and safety of his vaccine.

The National Foundation for Infantile Paralysis made preparations for mass field trials. Results would be judged by an independent committee headed by Thomas Francis an esteemed biologist and, coincidentally, one of Salk's former teachers. Most of the virus for these studies was grown in the Connaught Laboratories of the University of Toronto, Canada, and each type was grown separately. After four days of growth the viral fluid was chilled, poured into 2 1/2 gallon bottles and shipped to the United States. Here formaldehyde was added to kill the virus, then the three types were mixed together and tested on laboratory animals. Once safety and efficacy were assured the vaccine was shipped out for mass field trials; these began on April 26, 1954. It took almost exactly a year to sort out all the facts and figures assembled by the Committee, but on April 12, 1955, at 10:20 AM, ten years to the day after the death of Franklin Delano Roosevelt, Dr. Francis mounted the rostrum at the Rickham auditorium, University of Michigan, faced the television and newsreel cameras and announced, "The vaccine works." Salk, commented, "Dr. Enders pitched a very long forward pass, and I happened to be in the right spot to receive it." The Detroit FREE PRESS wrote, "The prayers and hopes of millions... were answered." A passer-by wrote THANK YOU DR. SALK on an office window at the University of Pennsylvania.

It seemed like the Salk vaccine was an unqualified success but on April 24, 1955, the nation's euphoria was interrupted when first cases of paralytic polio due to the vaccine surfaced. This was ultimately traced to the dangerous Mahoney strain of polio virus. The problem was corrected and thereafter the vaccine and its discoverer enjoyed world acclaim.

A chance meeting between Salk and Basil O'Connor aboard the ocean-liner Queen Mary, in mid-Atlantic, in September in 1951, played an important role in the development of the polio vaccine. O'Connor had been the fund raiser for the Warm Springs Hotel, Georgia, where President Franklin Roosevelt enjoyed the buoyant mineral waters for the treatment of his polio, but now was the guiding light of the National Foundation for Infantile Paralysis. Both men were returning from the second International Poliomyelitis Congress in Copenhagen, Denmark. During their conversations O'Connor was overwhelmed by Dr. Salk's extensive background in virology and his particular erudition with regard to polio. What impressed him most, however, was that Dr. Salk was going to use only dead viruses and there would be no chance of his vaccine ever causing the disease. Accordingly, O'Connor threw his weight behind his new acquaintance and saw to it that Dr. Salk and the University of

Pittsburgh were liberally funded. As the initial reports of early successes reached O'Connor he passed them on to syndicated newspaper gossip columnists Earl Wilson and Walter Winchell; the publicity would enrich his National Foundation for Infantile Paralysis.

It is interesting to speculate whether the polio vaccines would have been developed as rapidly as they did if Franklin Delano Roosevelt had not contracted poliomyelitis. Consider this: W. W. Keen, M.D., the famous surgeon who successfully removed a cancer from the jaw of President Grover Cleveland, had become the Roosevelt physician. When 39 year old Franklin became ill on August 11, 1921, at the family retreat on Campobello Island, off the coast of Maine,. Dr. Keen missed the diagnosis; he called it flu and did nothing. Many believe that Roosevelt might not have become a cripple if his true condition had been recognized and treated immediately. At the time he was stricken, FDR was one of the most popular and glamorous figures in the Democratic Party. Had he remained healthy he might well have been the presidential nominee in 1924 and would have been almost certainly defeated in that year of Republican dominance. Had he been renominated in 1928 he would have been trounced as Republican prosperity enthralled the country. Roosevelt's admirers and some of his critics believe that his special talents were in part a result of his affliction. Eleanor Roosevelt stated, "He learned infinite patience and never-ending persistence" as a result of his disability. Lawrence Durrell, wrote in SICILIAN CAROUSEL: Unexpected and fateful is the trajectory which life traces out for our individual destinies to follow. How well the destiny of Franklin Delano Roosevelt exemplified this.

Dr. Salk's success did little to ease the on-going feud with Dr. Albert Sabin. As Salk later said, "It had all the characteristics of a holy war." Dr. Salk, as early as 1949, applied his experience with influenza vaccine, which was a killed virus vaccine, into making his polio vaccine. He was well aware of Dr. Albert B. Sabin's position, that a weakened but living virus would make a better vaccine, but Salk reasoned that a killed virus vaccine would confer immunity without possibly infecting patients, and thus without repeating the debacles of the past. Goaded by Salk's triumph, Sabin pushed harder and within a few years, 1956 to 1958, developed his trivalent oral polio vaccine (TOPV). He succeeded only after he abandoned his 1935 belief that the polio virus would only grow in brain, spinal cord, and nerve tissues. Had he not done this he probably would have met the same humiliation that Dr. Maurice Brodie encountered earlier.

Albert Bruce Sabin was born in Bialystok, Poland, and came to the United States when his family emigrated in 1921. He received his M.D. degree from New York University in 1931, the same medical school which Jonas Salk later attended. He then worked at the Rockefeller Institute and finally joined the staff of the Children's Hospital Research Foundation and the College of Medicine of the University of Cincinnati where, after several false starts, he perfected his oral vaccine. In contrast to the killed virus developed by Dr. Salk, Sabin's vaccine contained all three types of live but weakened polio viruses. Sabin's vaccine could be given orally, Salk's had to be injected. Dr. Sabin administered the first experimental doses to himself and his family, which he ethically felt was his only choice. In later life Sabin's daughters could not forgive their father for the ordeal of being tested with his new, untried, vaccine. By now the United States was committed to the Salk Vaccine, so Sabin had to go to the Soviet Union to try his out. Later, from 1958 to 1960, it was tested in Mexico, Czechoslovakia, and Singapore. When licensed in the United States, in 1963, it replaced the Salk injectable vaccine as the standard immunizing agent.

Recently, CDC, Centers for Disease Control and Prevention, shocked us out of our complacency by reporting that several people contracted paralytic polio from the very vaccine that was supposed to prevent it. While the Sabin vaccine is entirely safe in 99.9996% of cases, that is, it produces immunity without causing the disease, that fact offers little comfort to the 8 to 10 people who acquire vaccine-associated paralytic polio every year in the United States. Not only has this re-awakened the forty-year-old scientific argument between two giants of preventive medicine, Dr. Albert Sabin and Dr. Jonas Salk, but it is prompting CDC and the American Academy of Pediatrics to re-examine their recommendations with regard to polio vaccine.

For 30 years doctors have had the choice of two polio vaccines. Salk's inactivated vaccine, made from killed viruses, cannot cause poliomyelitis, but, although immunized individuals build up good immunity against the disease the polio virus itself can still thrive in their intestinal tracts—a likely way of passing on 'wild type' virus infections to unvaccinated populations. Sabin's oral attenuated, live vaccine provokes a more powerful immune response and also prevents live viruses from growing anywhere in the body; but, under unusual circumstances, may cause poliomyelitis. It was the Sabin vaccine that actually eliminated polio epidemics in America. Today, of the 20 million doses of polio vaccine on the market less than 500,000 are prepared from killed, Salk-type viruses. The vulnerability of a non-immunized population was demonstrated in 1979 when 16 cases of paralytic poliomyelitis occurred among Amish people in the United States and Canada who, on religious grounds, rejected vaccination.

Simply put, the question facing health authorities today is, should we switch to Salk vaccine to make certain that no one ever gets vaccine-acquired poliomyelitis?

Those who favor this idea can show that there have been no home-grown cases of naturally occurring polio in the United States since 1979. The disease has been eliminated in the Western Hemisphere since 1991. Additionally, the World Health Organization's immunization efforts have also decreased the risk of importing the disease into the United States. Since the chances of a major outbreak are remote, the United States could now switch to killed virus vaccine.

The NO sayers point out that one in five American children is still unvaccinated for polio; these children, or anyone with a weakened immune system, could contract the disease brought in by newcomers or silent carriers. In India, for instance, where polio is still smoldering and vaccination rates are low, the Salk vaccine would not protect the population because a large dangerous reservoir of virus persists in people's intestines. Besides, the Sabin vaccine comes in a sweet-tasting liquid that is usually administered on a sugar cube placed directly into an infant's mouth. The Salk vaccine can only be injected and requires at least 3 trips to the doctor. Parents and youngsters will not welcome yet another shot to the already long list of immunizations. Also the injection costs about $5 a dose compared to $2.27 for the Sabin type of vaccine.

After weighing the risks and benefits of the two vaccines, a panel of experts with CDC have indicated that the best course would be to inoculate infants twice with the killed virus vaccine at 2 to 4 months of age and then twice more with the live virus vaccine before the age of six years. With this arrangement infants whose immune systems are as yet not strong avoid exposures to the more dangerous live preparation. Older children, who have tougher constitutions, will get the benefits of full protection. Experts expect that the change could prevent most vaccine-associated cases of polio.

In 1988, the killed-type vaccine was improved by making it from virus grown in human instead of monkey cell lines in tissue culture. This will replace the Jonas Salk vaccine, which was licensed in 1955. Its main use will be limited to certain adults at risk to exposure to wild polio viruses and children and adults with suppressed immunities.

The final chapter on polio vaccines has by no means been written. Dr. Sabin told a scientific congress not long ago, "The scientific challenge is to look at the genetic elements of neurovirulence and see whether these can be deleted without interfering with the ability of the virus to replicate,"—implying that there may still be a way of manipulating the molecules of viruses so that they would become more stable, provide improved protection, yet not cause disease. Genetic engineers can now delete, substitute, and recombine intranuclear molecules; changing just one or two of the 7,500 sequences in DNA or RNA might transform a virulent virus into a harmless agent that could still immunize humans.

At present there are four kinds of vaccines. The first is made from living organisms that have been weakened or rendered harmless, like the Vaccinia-Small-Pox Virus, which, as the United Nations World Health Organization proclaimed in 1979, completely eradicated the disease. The second group of vaccines are made from killed organisms, like the Salk Polio vaccine. The third kind are made from weakened or altered bacterial toxins, such as Tetanus or Diphtheria toxoids. The last and newest category of vaccines called, *Subunit Vaccines* are made by using immunogenic pieces of a bacterium, virus or toxin.

Viral antibodies are extremely specific, so much so that an antibody against polio, for example, would not react against smallpox. By 1968 we knew how to distinguish one virus species from another by testing for antibodies to proteins in viral capsules and this started a cascade of vaccines that depend on using pieces of viruses to stimulate antibody responses and immunity in people. Jonas Salk used this response against outer capsule protein of polio virus as the basis of his vaccine. The vaccine against Hepatitis B, for instance, uses a piece of the Hepatitis B virus's outer protein envelope that stimulates antibodies against the entire Hepatitis B virus.

The Hepatitis B vaccine might never have been developed if not for an interesting bit of serendipity on the part of Dr. Baruch Blumberg. In 1964 while studying Australian aborigines, who had never come in contact with hepatitis, or, for that matter, any of the diseases of civilization, Dr. Blumberg unexpectedly discovered a new and unusual antigenic substance in their blood that reacted with serum from people who had hepatitis. It should be remembered that an antigen is an organic molecule that excites an immune response; and an antigen is also attacked by its specific, antagonistic, antibody. Dr. Blumberg was already aware of the fact that patients with acute and chronic hepatitis made antibodies to their disease in response to antigens liberated by the hepatitis viruses. When serum from these hepatitis patients was mixed with his newly discovered Australian antigen, the antigen was attacked by the patients' antibodies. This reaction was easy to recognize and led to a test that enables doctors to test for hepatitis. Furthermore, by using the antigen to stimulate the production of immune substances in people it became the basis for the development of the hepatitis vaccine. Dr. Blumberg was awarded the 1976 Nobel Prize in Medicine for this work.

Antibacterials came to medicine with the sulfa drug in the late 1930's, but the first antiviral compounds were not developed until the 1960's. One of the difficulties has been that viruses require the active participation of host cells to thrive; any agent that affects viruses is also likely to injure the person harboring the virus. Nonetheless there have been a few successes: AZT, Acyclovir, 3TC, Interferon, Amantadine and Vidarabine have effectively interrupted nucleic-acid replication or disrupted the protective coating of viruses, which makes them more vulnerable to attack by the body's defenses. Vidarabine, incidentally, had been isolated in 1957 from the Caribbean sponge and later from soil microorganisms found in Naples, Italy, but in 1964, French investigators Privat deGarilhe and J. DeRudder, while testing the compound for anticancer activity, serendipitously found that it was active against herpes simplex and chickenpox viruses. Two new drugs that prevent viruses from replicating themselves, Indinavir and Ritonavir, are on the verge of gaining FDA approval.

Finding antiviral drugs is but half the problem; the other half is keeping up with viral changes that make them resistant to drugs as fast as we develop them. Amantadine proved effective for Asian Flu in 1966, but was not effective later, when the virus went through genetic changes and emerged as Hong Kong Flu. That such changes have been going on for a very long time is demonstrated by this notation made by

Thucydides twenty-four centuries ago, when an influenza epidemic raced across Athens and the Mediterranean Basin. "All speculation as to the origin and its causes, if causes can be found adequate to produce so great a disturbance, I leave to other writers,...I shall simply set down its nature, and explain the symptoms by which perhaps it may be recognized by the student, if it should ever break out again." After three years of annihilation, 430-427 BC, the disease vanished as quickly as it had appeared.

The influenza virus is unique among infectious agents of man. By genetically modifying its proteins, a single virus can mutate and create a new strain which possesses characteristics unrelated to those in its previous form. In recent years we have been confronted by influenza strains with names like Hong Kong, Turkey, Wisconsin, Swine, Tahoe, Taiwan, and Singapore, each with its own peculiar method of eluding our vaccines.

In 1910 Dr. Francis Rous at Rockefeller Institute handed us the initial clues about the relationship between cancer and viruses. But many other real and putative causes of cancer, such as toxic chemicals, tobacco smoke, contaminants in food and water, air pollution, pesticides, auto emissions, industrial wastes, and radioactive materials, kept cropping up and attracting scientist away from viruses. However, a strange event in Africa, in 1960, brought attention back to the virus as a cause of cancer. Denis Burkitt, a British physician, discovered that a certain tumor of the lymph glands appeared with an unusually high frequency in Uganda. It seemed to arise in clusters, commonly affecting whole families or entire villages. In his honor the tumor was labeled *Burkitt's Lymphoma*. Subsequently, British researchers, Michael Epstein and Y. M. Barr, discovered a type of herpes virus in cells from Burkitt's Lymphoma patients; it was dubbed Epstein-Barr virus or EBV. When, in 1982, it was proven that EBV caused *Burkitt's Lymphoma*, scientists were forced to accept the fact that viruses could initiate changes in cells that resulted in cancer.

How viruses actually carry out their nefarious deeds was unraveled by Howard Temin and David Baltimore. Working independently, they discovered *reverse transcriptase* an enzyme that regulates basic molecular functions in cells. They then showed how viruses enter cells and use the enzyme to make copies of genes and then insert such segments into another hosts' cells. Essentially, by inserting part of their genetic material into cells viruses take command of the cell. When viruses turn on growth signals for that cell, it causes cancer. Howard Temin, David Baltimore, and Renato Dulbecco shared the 1975 Nobel Prize for medicine for this breakthrough in molecular biology. In 1979 Dr. Robert Gallo at the U.S. National Cancer Institute, almost simultaneously with researchers at Tokyo Cancer Institute and Kyoto University, confirmed that viruses cause cancer in human beings by swapping genetic materials.

Later, between 1976 to 1983, Michael Bishop and Harold Varmus at the University of California in San Francisco, demonstrated that a cancer causing gene found in the rous sarcoma, a form of cancer, was almost an exact copy of a gene normally present in chickens. The normal gene was called C-SRC whereas the viral *oncogene* was V-SRC. C-SRC has since been found in the genetic components of other birds, animals, insects and humans, proving still further how vulnerable we all are to an exposure to oncogenes, genes that can induce cancers. Oncogenes seem to have a limited function in early fetal development, where they accelerate the growth of cells that transform an egg into a baby, but they then retire into the background when they are no longer needed. Apparently some viruses carry genes that mimic oncogenes and when these are deposited within a host's cells they trigger cancers. Cancer and the contagious process sometimes overlap.

A tug of war is always going on between viruses and people. In order to do their damage viruses must pass through skin, intestinal lining, mucous membranes, coverings of the brain and spinal cord, and evade a number of other defenses—like antibodies and white blood cells. To help them, viruses have a variety of proteins in their outer coating which enable them to lock onto cells of their host and prepare an avenue for invasion. Some viruses, like influenza, disguise these proteins so that the human immune system cannot detect them. Others, like the Dengue-2 virus, evades our antibody-immune system by taking control of killer cells sent to destroy them, in effect, making these cells carriers of the virus. This scenario is like that portrayed in the motion picture INVASION OF THE BODY SNATCHERS, where space aliens take over human bodies and there is no way of telling the human-humans apart from the alien-humans. Fortunately, humans and higher animals can and do make antibodies that can recognize and neutralize most foreign viral proteins. When people are initially infected by a virus they develop antibodies which not only get rid of the virus but protect them against further attacks. Children as a rule get light cases of measles, mumps and chickenpox then remain immune for the rest of their lives. And when these children become parents they seldom get the disease from their children.

Over the first three quarters of this century, science discovered the nature of viruses, figured out, more or less, how they cause diseases and cancer, and even tamed a few of them. It was, however, in the field of slow-growing viruses that scientists made their most dramatic advances. Try to imagine how difficult it would be to find the cause of a disease that produced symptoms, one, five, or ten years, after infection. Yet this is the exact situation that doctors faced when they came up against Kuru. Through the fortunate confluence of an inquisitive pathologist, a perceptive veterinarian, two anthropologists, and an intrepid pediatrician, this Gordian knot was untied.

In 1953 an Australian patrol officer in New Guinea, J. R. McArthur, reported that a mysterious disease was killing the primitive Fore people in the wild, largely unexplored, Eastern Highlands. The natives called it Kuru, from their word meaning "trembling from fear and cold." Two years later, in 1955, the Australian government posted Dr. Vin Zigas, a young Lithuanian physician, to its Kainantu hospital, with instructions to look into this phenomenon. He found that the disease often started innocently with prolonged headache and minor loss of coordination; it progressed rapidly so that in less than three months some natives developed tremors or shaked uncontrollably while others could hardly walk because their limbs splayed sideways or gave-out without warning. By the fourth month adults leaned on heavy poles for support or lay on the ground, unable to walk, talk or eat, and died of starvation if pneumonia did not kill them first. Almost invariably every patient died within a year.

Dr. Zigas took a short vacation then returned to the Fore people bringing with him supplies needed to study the disease in depth. He collected samples of blood, sputum, and brain, then sent them to the medical research institute in Melbourne Australia. He also enclosed a long formal report suggesting that Kuru was "probably a form of encephalitis," infection and swelling of the brain. When standard examinations and analyses of the specimens failed to reveal any cause of disease, Sir Macfarlane Burnet, director of the Institute, called for the best expert in this particular type of outbreak, American pediatrician, D. Carleton Gajdusek.

Gajdusek received his M.D. from Harvard Medical School then specialized in pediatrics and neurology. The biophysical research he performed at Boston Children's Hospital provided many answers to problems related to war-time survival. As student, mentor, or coworker he met the most advanced scientists of the time: James Gamble, Michael Heidelberger, Linus Pauling, Max Delbruck and John Enders. In 1952, the young virologist was drafted by the U.S. Army and served in Korea, studying the viral epidemic of hemorrhagic fever. Upon his discharge from the army he joined Dr. Joe Smadel, head of the department of virus and rickettsial diseases at Walter Reed Army Medical Center in Washington D.C. At age 33, he answered Dr. Burnet's summons, departed for New Guinea, and soon was sending specimens to the United States for study.

Dr. Igor Klatzo, a neuropathologist at National Institute of Health, was assigned to do the pathological examination of the brains forwarded to NIH by Dr. Gajdusek. Initial examinations showed nothing. But Dr. Klatzo, who never took NO for an answer, peered at his brain specimens over and over again, first with conventional high-powered microscopes then with a polarized-light microscope and felt, rather than saw, an unusual pattern of star-shaped brain cells, astrocytes. Following up on this hunch he came to appreciate that here was "a strong and profuse proliferation of astrocytic brain cells", but it still did not match anything that he had ever seen before. Kuru was something new. The picture seen under the microscope brought back to Dr. Klatzo's mind something that he believed he may have seen, read, or heard about, during his student days in Germany, so he undertook a comprehensive search of all the neurological degenerative diseases in the German medical literature. He found what he was looking for: *Creutzfeldt-Jakob disease* (CJD), a rare disease with no known cause, characterized by brain changes similar to Kuru. It was a brilliant association but offered no clues as to cause or cure.

As luck would have it, a missing piece of the puzzle was about to be fitted into place. In the spring of 1959, Wellcome Institute, a London medical foundation dedicated to the history of medicine, mounted an exhibition on Kuru. It displayed photographs of the forbidding terrain of the New Guinea Eastern Highlands, disfigured and horribly affected patients seen by Dr. Gajdusek, and slides and specimens prepared by Dr. Klatzo. A young American veterinarian, William Hadlow, dropped in at the Burroughs-Wellcome Museum to see the exhibit. He had left his post at National Institute of Health's Rocky Mountain Laboratory in Montana, and was now working with the British Ministry of Agriculture at Compton, in Berkshire, as a pathologist studying Scrapie, an infectious disease of sheep. He had been invited to England because Scrapie was currently causing great financial losses among sheep ranchers.

Bill Hadlow often explained the vagaries of Scrapie this way: the disease has been known for centuries and as early as 1775 English farmers described how their animals were tortured by itching before dying; the sheep scraped their skins constantly, hence the name. It is an unusual disease. It could take up to four years to incubate, that is, four years might elapse between the time an animal picked up the infection and the time it manifested signs of the disease. Although infected, animals appeared to be healthy. The disease always affected the central nervous system. Cell-free filtrates from infected sheep killed healthy sheep; therefore, a virus must be the cause.

Similarities between Kuru and Scrapie popped into Hadley's mind. He submitted a letter to the British Medical Journal, Lancet, making the bold suggestion that these two diseases might be caused by the same type of organism. Gajdusek was astounded by this observation, which not only gave him a valuable lead but also explained why he and the team at National Institutes of Health, NIH, failed to find their quarry—they did not keep looking long enough. The viruses behind Scrapie and Kuru were slow-growing viruses that took many years to betray their presence.

Gajdusek now embarked on an experimental program designed for slow growing viruses and enlisted the help of Dr. Joe Smadel, his mentor, and Dr. Clarence Joseph Gibbs, Jr., another virologist at NIH. The team ground up the brains of Kuru victims and injected various doses into laboratory animals, primarily chimpanzees at the Patuxent Wildlife Center, 25 miles from Bathesda, in the Maryland countryside. It was an expensive program because it meant keeping a colony of chimpanzees and other primates around for years and years after they had been injected with human brain material.

All the chimpanzees that had been inoculated with ground up brains died two to five years later. Examination of the chimpanzee brains showed changes similar to human

Kuru; pathological changes were identical to those seen in CJD. As luck would have it, a patient with CJD had just died and human brain tissue was available for study. When investigators at NIH injected it into animals it produced the same neurodegenerative pathological changes as seen in CJD, but only after an incubation period of a year or longer. This was the first time that two degenerative diseases of the human central nervous system were proven to be caused by the same filterable virus. Dr. Gajdusek received the Nobel Prize in 1976 for this work.

The final piece to the puzzle was supplied by Dr. Robert Glasse and his wife Shirley Lindenbaum, two social anthropologists, who had gone to New Guinea to study the stone-age Fore people. They lived with them in a one room hut with a single latrine, gathered and ate their food, participated in rites and rituals, and through long conversations and interviews gained their respect and confidence. The first thing they learnt was that Kuru was of recent origin; it could not have been genetically transmitted. They next discovered that women and children were mostly affected by Kuru.

What were the women and young children doing that the men and people in other tribes were not doing? The answer was cannibalism. In the Fore villages the men and young boys did the hunting and they kept all the meat for themselves. The women lived separately. Although they raised pigs they had to surrender them to the males; women and children subsisted on vegetables, frogs, insects, and leftover scraps. In order to compensate for their lack of proteins they resorted to cannibalism. Upon coming across a dead body, older women scraped off some of the flesh and removed Kuru tainted brain from the skull, which they steamed, then ate. Toddlers acquired the virus by eating infected meat or brain, or through open sores or scratches on their skin.

In the 1950's cannibalism was outlawed. Native constables, missionaries, government agents, and doctors went into the Eastern Highlands to discourage the practice. As cannibalism died down the incidence of Kuru began to fall and has been declining ever since.

But that is not the end of this tale because just as this book was going to press newspapers around the world played up the British government's concern over a situation not far removed from that in New Guinea. Health Secretary Stephen Dorrell stunned the House of Commons by stating that there was a connection between MAD COW DISEASE, a form of brain disease found in Britain's cattle, and Creutzfeldt-Jacob Disease. The incidence of the disease, although rare, has been rising steadily in England and poses a threat to the $7.5 billion beef industry.

In the 1970's Dr. Stanley B. Prusiner at the University of California at San Francisco became intrigued by a woman patient in her mid-fifties who succumbed to Creutzfeldt-Jakob Disease. He gave up his thought of becoming a practicing neurologist and went into biochemical research, hoping to find the cause of this strange affliction. Dr. Prusiner initially discovered that extracts from the brains of sheep that had died of Scrapie, even after being heated, irradiated, and treated with chemicals, enough to destroy bacteria and viruses, remained virulent and killed other animals. Prusiner labeled such disease causing agents *Prions*, a name derived from "protein-aceous infectious particle". They are smaller than viruses, lack all genetic materials, and cannot be seen with the electron microscope. Prions appear to be identical to proteins normally present in the body but may differ in how chains of molecules fold up or merely by the types of molecules stuck onto them. Theory has it that there are hereditary and infectious prions, the latter arise as a result of mutations by a change in structure of the protein. Prusiner thinks the small difference between the abnormal prion protein and the normal protein somehow confers infectivity on the prion. The relationship of slow-growing viruses and prions is still being debated by scientists.

In addition to the slow growing viruses we also have virus infections in a persistent state, that is, the virus is continually detectable. Here, neither the virus nor the immune response brings about cell destruction. Measles and HIV viruses invade the cells of our immune system, become part of them, and can be detected for inordinate periods of time after initial infection or recovery.

PARADIGM POINTERS

The theories that go into a paradigm can be compared to stepping stones over a river. Take the viruses for example. They were discovered just a few years before 1900, crystallized in 1935, grown in tissue cultures in the mid-1940s and in the 1970s and 80s we figured out how viruses replicate and cause disease and cancer. Each one of these advances acted like stepping stones that brought us closer to the ultimate understanding of viruses. The 21st century will probably bring more stepping stones that will bring us closer to finding a vaccine or treatment for AIDS and other viral onslaughts.

PERSONAE

JONAS SALK, MD

Jonas Edward Salk was born in Brooklyn, New York, in 1914. He and his two younger brothers were raised there and his father worked in that city's Garment District. He has written, "my earliest recollection of a choice of a career was to be a lawyer. I wanted to go to Washington and be able to pass laws that would correct the injustices I felt existed. In college I became interested in science and medicine and in the possibilities of preventing the injustices of disease." He attended medical school with the idea of concentrating on research, so shortly after receiving an M.D. degree from New York University in 1939, he plunged into studies on influenza viruses. Throughout the 1940's Salk worked on an anti-influenza vaccine, starting out as assistant professor of epidemiology at the University of Michigan then continuing on when he became head of the virus research laboratory at the University of Pittsburgh in 1947. He subsequently became research professor of bacteriology from 1949 to 1954, professor of preventive medicine and chairman of the department from 1954 to 1956, and professor of experimental medicine from 1957 to 1963.

After Salk's poliomyelitis vaccine proved successful, he received the Congressional Gold Medal, the Presidential medal of Freedom, the Jawaharlai Nehru award for International Understanding, the Albert Lasker Award, and the rank of Officer of the French Legion of Honor. World polls ranked him with Gandhi and Churchill as a hero of modern history. In 1963 Dr. Salk established and directed the Salk Institute for Biological Sciences at LaJolla, California. It was designed by the noted architect Louis Kahn, but Salk contributed greatly. The institute provided a haven for outstanding scientists, artists, and philosophers, such as Jacques Monod, Leo Szilard, Warren Weaver, and a number of Nobel laureates. In 1970 Salk married Francoise Gilot, the one time companion of Pablo Picasso, and the mother of two of Picasso's children.

When Dr. Jonas Salk died, on June 23, 1995, at the age of 80, President Clinton said, "The victory of this medical pioneer over a dreaded disease continues to touch many—from the students who study his work to the countless individuals whose lives have been saved by his efforts."

Nobel prize winner, Renato Dulbecco, when he wrote Dr. Salk's obituary, stated: Salk was a very kind person of profound convictions who would not be easily led by other opinions, although he listened to them. He persisted in doing what he thought was right, even in the face of strong opposition.

For his work on polio vaccine Salk received every major recognition available in the world from the public and governments. But he received no recognition from the scientific world—he was not awarded the Nobel Prize nor did he become a member of the U.S. National Academy of Sciences.

The reason is that he did not make any innovative scientific discovery. The fact that a fundamental advance in human health could not be recognized as a scientific contribution raises the question of the role of science in our society."— Dulbecco

ALBERT SABIN, M.D.
Albert Sabin, M.D. died March 3, 1993 at the age of 86. He had received more than 40 honorary degrees and virtually every honor in the U.S. and abroad for his work. Most of his career was spent as Professor of Research Pediatrics at the University of Cincinnati College of Medicine. After 1961 he concentrated on studying the role of viruses in cancer. He remained scientifically productive through his 70s as consultant to the World Health Organization and Fogarty International Center. Dr. Sabin had competitive, outspoken, and dogmatic traits that sometimes led to conflicts.

His obituary in a scientific journal said this: "At a time in which greed seems to be at an all time high and commitment to public health at an all time low, Dr. Sabin's career serves as a beacon for scientists."

Dr. Sabin himself once said, "A scientist who is also a human being cannot rest while knowledge which might be used to reduce suffering rests on the shelf." Yet, Chapter 2, A LESSON FROM THE WIZARD OF OZ, shows that knowledge which might be used to reduce suffering, nonetheless, still rests on the shelf.

chapter 10

JANUS AND THE CANCER CONNECTION

The ten million cells constantly dividing in our bodies usually do so at the right time and at the right place. Occasionally, when things like radiation, viruses, or toxic chemicals disrupt normal genes, the substances that control cell growth become disorientated and cells, like buckets of water in the SORCERER'S APPRENTICE, keep multiplying interminably. Cancer cells have a great advantage over normal cells in the battle for survival. Whereas the latter grow old and die-off regularly, cancer cells persist until they kill the body that made them. Cancer, then, is the general name for a great variety of complex growths that differ in size, shape, origin, structure, and rates of spread. Each follows its own path of destruction and each responds to its own type of treatment.

In much the same way as the two profiles of the Roman God JANUS face in opposite directions, chemicals, physical agents, and radiation act in diametrically opposite ways. They *cause* as well as *cure* cancers.

This perverse faculty was first observed with radiation and later with chemicals. The ability of physical elements like ultraviolet and sunlight to cause cancer was brought to our attention in 1905 when the French physician, Dubreuilh, reported that grape harvesters in the vineyards of Bordeaux often developed skin cancers on the backs of their necks. Other forms of radiation did the same thing. In 1899, four years after x-rays came into common use, a technician who checked newly manufactured x-ray tubes by fluoroscoping his own hand, developed fatal skin cancer. This warning was ignored and most of the first generation of people working with x-rays also died from cancer. Later, in the 1920's, cancer claimed many of the workers who painted watch dials with radium. Yet, it was the very same radium and x-rays that gave us the first weapons against cancer that did not depend on dire removal by surgery or caustics.

Subjecting cancer to radiation and chemicals is a relatively new treatment for a very old condition. The Ancients knew cancer well. It is mentioned in accounts of the Egyptian Pharaohs and in the literature of India, Persia, Greece and Rome. Leonidas of Alexandria in 180 AD extensively dissected out breast cancer, cutting through healthy tissue with knife and cautery, almost duplicating the surgery of today. The first cancer hospital, in Rheims, France, was built in accordance with the 1740 bequest of Jean Godinot, canon of the Rheims cathedral, but the townspeople feared that the disease was contagious so the hospital was moved to the outskirts and renamed St. Louis Hospital. It housed cancer patients until 1846 by which time the fear of contagion had subsided.

Historians, right up to the beginning of this century, observed that cancer mortality was lowest wherever there were shortages of doctors. Blood-letting, purges, herbs and nostrums, often hastened the demise of cancer victims while radical surgery in the absence of sterility, antisepsis, blood transfusions, and anesthesia, dimmed all prospects of survival. Yet, that is all our forbearers had.

During those heady days in the closing decade of the 19th century, when newspapers trumpeted the discovery of new germs, new vaccines, and new breakthroughs in physics and chemistry, everyone expected that the cure for cancer would soon follow. One such hopeful was Germany's Kaiser Wilhelm II who was very worried about his family's predisposition to cancer. Upon learning that Paul Ehrlich had been awarded the 1908 Nobel Prize in medicine the Kaiser invited the famous scientist to a special audience.

"And now, Herr Ehrlich, when will you give us a cure for cancer?".

"It's not that easy, your Highness," Paul Ehrlich replied. He then proceeded to explain, in one of his typically detailed discourses, how much more difficult it was to fire 'magic bullets' at cancer cells than at bacteria. Unprepared for this disappointment, the Kaiser turned away from Ehrlich and said, "Well, if you can't do it, it won't be done." The meeting was peremptorily terminated.

Actually, Ehrlich believed that magic bullets in the form of chemical compounds, antibiotics, or immune antibodies could be devised to selectively eradicate cancer cells. Initially he tried chemical compounds that had been created to kill the spirochete of syphilis but soon found that normal as well as tumor cells were killed. He then proposed a mechanistic approach, whereby antibodies would deliver some lethal substance to the site of a tumor, by-passing normal parts of the body. Unfortunately Ehrlich died before he could work out the practical applications of this premise and the idea died with him. A half century later, on the other side of the world, Japan's Dr. Hamao Umezawa picked up the ball dropped by Ehrlich and carried it successfully over the goal line. This feat is described later on.

Largely through serendipity and chance, radiation came on the scene at the beginning of the twentieth century and the age of chemotherapy appeared a half century later. Radium appeared in 1898 and x-rays, almost simultaneously, in 1895.

While physicist Wilhelm Conrad Roentgen experimented with high voltages in a Crookes tube—electrodes sealed in a vacuum tube, invented by England's William Crookes, now called a cathode ray tube—his laboratory assistant noticed that a fluorescent screen in a distant part of the laboratory was glowing and brought this to Roentgen's attention. Roentgen realized that a new type of ray had been discovered: it traveled in a straight line, it penetrated opaque substances like the black cardboard that had been in place over the Crookes tube, and it was not deflected by a magnetic field. Roentgen presented his discovery to the world on Friday, November 8, 1895. Not knowing the true nature of it, he called it X, hence x-rays.

Several lucky coincidences played a role in this discovery. Had metal or lead been placed over the Crookes tube instead of cardboard the rays would have been blocked and unnoticed. Had there been any covering over the fluorescent screen its glow would not have been seen. And if the Crookes tube had been aimed in a slightly different direction the chances of the rays eliciting that glow would have been next to zero.

There were hundreds of cathode ray tubes in American and European laboratories at the time that Roentgen made his discovery and many experimentalists must have, for some time, been producing these rays without knowing it. Roentgen was also lucky not to have been maimed, killed, or afflicted with cancer as a result of his frequent exposure to radiation.

Nowhere is the Janus connection more startling than in the ability of chemicals to cause and cure cancer. The first link between chemicals and cancer was found by Japan's Katsusaburo Yamagiwa, who, in 1916, produced cancer in rabbits by painting their skins with coal-tar products. Since then we have recognized thousands of compounds that harbor the potential to produce cancer and the list keeps growing. More than 50,000 man-made chemicals are currently in commercial and industrial use and several thousand are put on the market every year. Not all are cancer producers but determining which are and which are not, amounts to a Sisyphusian endeavor. To identify every environmental carcinogen and eliminate each one would require, at least, a redesign of our industrial sector, at most, a redesign of the way we live and dispose of our wastes. Yet chemical compounds, even those most toxic and noxious to humans and the environment, can be trained to do our bidding. This curiosity was brought to our attention fifty years ago by Dr. Cornelius Packard Rhoads.

On December 3, 1943, at 7:30 p.m., on a cold moonless night during one of the bleakest periods of World War II for Americans, the air-raid warning system around Italy's seaport city of Bari inexplicably failed to work during an attack by enemy bombers. The planes had targeted ships in the harbor that were delivering cargoes of fuel and ammunition to the Allies. One of the ships, the Liberty, was blown out of the sea. Besides being loaded down with high explosives it also contained 100 tons of mustard gas in warheads of airplane bombs. The reason behind this awesome cargo was that American intelligence had found out that both the Germans and the Japanese were prepared to use poison gas on a large scale. The mustard gas bombs that had been deployed to Bari would serve as a counter-measure for any eventuality.

Although the poison gas alarm at Bari was belatedly sounded and gas masks were issued, there were many casualties from mustard gas poisoning. Dr. Rhoads, who was stationed at Bari at the time of this disaster and had to care for casualties, was struck by the overwhelming destruction of blood cells in his patients. At first they showed a slight rise in the numbers of white blood cells but this was followed, the next day, by a critical drop in one type of cell, lymphocytes. In succeeding days the rest of the white blood cells disappeared. Immature blood cells often appeared in the circulation, suggesting that the body was trying to replace the losses. In mild cases the blood counts returned to normal seven to twenty-one days later. Serious cases died or were salvaged by heroic blood transfusions.

Dr. Rhoads noted that infections were rare and there was no breakdown of other tissues, regardless of the degree of poisoning. With stunning insight he deduced that this terrible chemical compound might prove useful in controlling cancers of the blood system, notably leukemia, where excessive white blood cells caused death and destruction. After investigating the properties of the nitrogen mustards and the feasibility of using them to fight cancer in humans he introduced them into medical practice—ushering in modern cancer chemotherapy.

Actually a large body of information about mustard gas and compounds related to them, called nitrogen-mustards, was already on hand but no one appreciated the connection to cancer. From experiences during World War I it was apparent that mustard gas possessed great military potential and between the two world wars several studies were undertaken to assess its offensive value. Independent researchers and universities received grants to find out exactly how the mustard compounds worked. By the time Dr. Rhoads came on the scene scientists had figured out how the mustard chemicals changed in the body and what effects the poison as a whole, as well as its individual components, had on enzymes and on various organs. One of the earliest findings was that nitrogen mustards latched on to proteins wherever the two came together. A practical off-shoot of this finding was advice to infantrymen: In an emergency prick the skin and apply a drop of blood to the eye, as a protective film. Since blood has protein in it, it will bind with the mustard gas. This was never tried because gas masks usually got in the way.

Another weird finding during those early investigations into nitrogen mustards was that they also interact with certain plant disease viruses, notably the virus of Tobacco mosaic. This effect is still under investigation.

World War II is past history, the Bari incident has long since been forgotten, and chemical warfare, except for sporadic resurgences, has been replaced with atomic weapons. But phoenix-like, those early seminal studies on mustard gas gave us a better understanding of chemotherapy and paved the way for its rapid introduction into cancer therapy. William Shakespeare had it right when he wrote: There is some soul of goodness in things evil, Would men observingly distill it out.

As soon as it became apparent that chemicals could fight cancer, Leon Ores Jacobson, M.D., a quiet, retiring physician in Chicago, Illinois, carried Dr. Rhoads' concepts to new and greater heights. In March 1943 he used nitrogen mustard on a man with lymphatic leukemia who had not responded to other treatments. No adverse effects were observed. Thereafter he and his research team at the University of Chicago were among the first in the world to use nitrogen mustard compounds in patients with other types of cancer. In 1945 Jacobson and co-workers reported that a patient with Hodgkins Disease enjoyed a lasting remission after being treated with nitrogen mustards —demonstrating for the first time that widespread cancer could be cured with chemicals. This research remained top secret during World War II and was only released in 1946. By then Dr. Jacobson had treated 59 patients with cancer and two-thirds of his Hodgkins Disease patients appeared free of their cancers.

Dr. Jacobson also proved that bone marrow cells could be successfully transplanted. The first attempt to use bone marrow as a therapy was reported in France, in 1890, by D'Arsonval and Brown-Sequard, who tried to cure patients with leukemia and blood deficiencies by giving them oral doses of bone marrow. Needless to say it failed because the vital elements were destroyed by digestion. Between 1930 and 1950 there were many reports of bone marrow transfusions but none survived the test of time. In 1948 Jacobson demonstrated that mice could be protected from lethal doses of whole-body radiation by shielding the spleen where many essential blood cells are stored. Then, in the late 1950s, Dr. George Methe renewed interest in the procedure when he gave bone marrow transplantations to survivors of an atomic accident. Doctors tried to give bone marrow cells intravenously but these were rejected or destroyed by the body's immune system. Major advances in histocompatibility testing and knowledge of the HLA system paved the way to successful bone marrow transplantation in the late 1960s. Here, too, serendipity played an important role, as explained in Chapter 13: MIGHTY LIKE A ROSE.

The domino effect, where tipping over just the first domino in a long row makes them all fall in a rippling fashion, is eloquently epitomized in the pharmaco-chemical discoveries that followed in the wake of Rhoads' and Jacobson's breakthroughs. The contributions of Gertrude B. Elion, George Hitchings, and Howard J. Schaeffer, exemplify this perfectly.

Gertrude B. Elion, the daughter of a dentist, received her degree in biochemistry from Hunter College in New York City but, since serious research was not for women in those days, she had to accept the job of testing pickles and berries for the A&P chain of food stores. Her chance came in 1944 when the men went off to war and there was an opening at the Burroughs Wellcome pharmaceutical company. She was assigned to the laboratory of George Hitchings, the man with whom she was to share the Nobel Prize 44 years later.

At a time when the role of DNA as the carrier of genetic information was barely understood, Hitchings and Elion began to investigate compounds that could block key enzymes needed by cells for the manufacture of DNA. Using the common bacterium found in milk and milk products, *Lactobacillus casei*, as a test organism, Elion uncovered thioguanine and 6-mercaptopurine (6-MP), two anticancer drugs useful in treating leukemia. While trying to modify 6-MP Elion developed azathioprine, commercially labeled Imuran, which was unimpressive against leukemia but proved highly effective in suppressing the immune response. It made kidney transplantation a practical reality in the early 60's and is now used also to treat autoimmune disorders, such as severe rheumatoid arthritis. Further attempts to potentiate 6-MP led to allopurinol, trade named Zyloprim, a compound that became the most effective drug in treating gout.

Gertrude Elion, never one to rest on her laurels, went on to investigate the effects of chemical compounds on viruses. As far back as 1948 it became known that chemicals could affect viruses but the exact mechanisms were still fuzzy. With a new colleague, Dr. Howard J. Schaeffer, who had come to Burroughs Wellcome from the University of Buffalo, many compounds that theoretically harbored antiviral properties were invented and tested. Quite unexpectedly, one of these, acyclovir, proved to be 100 times more effective than its parent in destroying the virus responsible for herpes infections. It was approved by the U.S. Food and Drug Administration in 1981 under the name Zovirax.

George Hitchings, born into a shipbuilding family in Washington State, received his Ph.D. in biochemistry from Harvard in 1933. When he joined Burroughs Wellcome in 1942, as the sole member of the biochemistry department, it was unlikely that any academic institution would have been willing to support his fanciful ideas in biochemistry. Guided by the principle "that every cell type must have a characteristic biochemical pattern, and therefore be susceptible to attack at some point that is critical for its survival and replication," he and Elion explored ways in which drugs that interfere with the metabolism of cells can be made to fight disease. An early application of this strategy led to the development of trimethoprim a compound that greatly enhances the antibacterial activity of the sulfa drug sulfamethoxazole; the combination is found in Septra, a mainstay in urinary tract infections. Another important drug, the antimalarial compound Daraprim, pyrimethamine, was discovered in a similar manner.

"I was 63 years old," Hitchings used to say, "before anyone praised me." But praise, when it did come, was doubly gratifying. In 1988 the Nobel Prize in medicine was awarded to Hitchings and his long-time associate, Gertrude Elion.

The golden age of the pharmaceutical and chemical industries unfolded in the two decades immediately after World War II. In this brief span of time more drugs for the containment of cancer came to light than ever before. Some came, as we have just seen, from modifications of the nitrogen mustards. Some came, as we shall see later in this chapter, from antibiotics.

But the most unique agent, cisplatin, came from a chance observation in an obscure biophysics laboratory in Michigan State University.

In early 1968, biophysicists Barnett Rosenberg and Loretta Van Camp designed a series of experiments to find out which frequencies of electricity destroyed cells most effectively. They chose the bacterium *Escherichia coli* as the 'experimental animal'. It is ubiquitous and is normally found in the human bowel, but it is particularly suited for experiments because it grows readily in the laboratory, it reproduces at an astronomically fast rate, and volumes have been written about its chromosome structure and genetic chemistry. The purpose behind these studies was to explore the feasibility of using certain frequencies to sterilize medical supplies and perhaps as a means of preserving food.

When small electric charges from platinum electrodes were applied to E. Coli, strange, albeit unexpected things occurred. Instead of exploding, shriveling up, or otherwise dying, the bacteria grew to enormous sizes. But they did not reproduce rapidly as might be anticipated with such phenomenal growth. The organisms lost their ability to divide; they were dead-ended giants. Naturally, the first inclination was to attribute this bizarre effect to the electric current but many more experiments, often stymied by false leads and excursions up blind alleys, proved that something on the platinum electrodes was behind this weird phenomenon. "We'd used platinum for electrodes," Dr. Rosenberg recalled, "because it's a good conductor of electricity and it's relatively inert. But our experiments gave us a clue that the effect had something to do with platinum itself rather than with the electric current."

At this juncture Rosenberg and Van Camp could have continued their prosaic line of research but they were astute enough to recognize that their findings had potentials for cancer therapy. They reasoned: cancer is the uncontrolled duplication of cells; this material stops cells from dividing and reproducing; therefore it might interrupt the multiplication of cancer cells. Accordingly, Rosenberg and Van Camp sent their observations and data to the National Cancer Institute. There, many platinum compounds were tested but cisplatin, where two atoms of chlorine and two molecules of ammonia are attached to a core atom of platinum, proved best in stopping test bacteria and cells from reproducing. "It's a chemical compound known since 1845 though its biological activity had never been discovered," Dr. Rosenberg said. "We still aren't sure how it works; there is general agreement that, in some way, it attacks the (genetic) DNA chain in a cancer cell." When tried clinically at the National Cancer Institute and subsequently at the prestigious Institute of Cancer Research in London it demonstrated effectiveness in cancers of the head, neck, thyroid, reproductive, and urinary systems. It also had antibacterial and immunosuppressive properties. It was approved by the U.S. Food and Drug Administration in December 1978 and sanctioned in Great Britain the following year.

Although cisplatin is not a perfect drug—nausea, vomiting and kidney impairment usually accompany therapy—it nonetheless can add months of life to a dying person. And one of the first cancer victims to acknowledge this was former United States Senator from Minnesota, Hubert H. Humphrey, best remembered as the 38th Vice-president of the United States under Lyndon B. Johnson and as the democrat who was defeated by Richard M. Nixon in the presidential election of 1968.

Young Hubert started out as a pharmacist in his father's drugstore in Dolan, South Dakota, but changed careers when the 'great depression' forced the store to close. He received a degree in political science from the University of Minnesota in 1939 and a year later earned his masters degree in political science from Louisiana State University at Baton Rouge. His political career began in 1945 as mayor of Minneapolis, after having been defeated for the post in 1943. In 1948, he was elected Senator from Minnesota and attained prominence as a champion of civil rights, Medicaid, and arms control.

On October 7, 1976, the senator underwent surgery at New York's Sloan Kettering Memorial Institute in an attempt to remove advanced, malignant, cancer of the bladder. Two days later, in an episode still hotly debated—was it good journalism or the height of perfidy—he and his family read newspaper accounts of his condition together with predictions that he would be dead in a few short months. Although this was consistent with statistics of the time, journalists, however, did not take into account cisplatin, a new drug, still in its trial stages, that would extend Senator Humphrey's lease on life.

Dr. Lewis Thomas who was a staff physician at Sloan Kettering Institute at the time Senator Humphrey was undergoing treatment, gives us this vignette in his book THE YOUNGEST SCIENCE:

> A few years ago Hubert Humphrey was a patient, in for treatment of the recurrent bladder cancer which eventually killed him, fully aware of the gravity of his situation, worried and somber on the evening of his admission, quiet and thoughtful, alone in his room at the time of my visit. We talked for a while; he was well-informed about his plight, knew that his chances for survival were slim, almost nil... Over the next few days he transformed himself, I think quite deliberately, into the ebullient, enthusiastic, endlessly talkative Humphrey—not so much for his own sake as for what he saw around him. There were about forty patients on his floor, all with cancers of one type or another ..... some at the end of the line, beginning to die.
> Humphrey took on the whole floor as his new duty. Between his own trips to X ray or various other diagnostic units, he made ward rounds. He walked the wards in his bathrobe and slippers, stopping at every bedside for brief but exhilarating conversations, then ending up in the nurses' station, bringing all the nurses and interns to their feet smiling..... One evening I saw him taking Ger-

ald Ford along, introducing him delightedly as a brand-new friend for each of the patients. Together, Humphrey in his bathrobe and Ford in a dark-blue suit, nodding and smiling together, having a good time, Ford leaning down to be close to a sick patient's faint voice, they were the best of professionals, very high class.

In his article YOU CAN'T QUIT that appeared in the August 1977 Readers Digest, Senator Humphrey, then 66 years old, wrote, "Deep down, I believe in miracles." This was an obvious reference to his year long chemotherapy treatments but easily could have been applied to the near-miracle that produced cisplatin. Senator Hubert H. Humphrey died in Waverly, Minnesota, on Jan 13, 1978.

Platinum has had a long and close association with medicine. Because of its inertness, it was, and still is, customarily used for electrical leads in cardiac pacemakers and in dentistry. An interesting testimony to this inertness took shape on a busy London street in September 1978, when a Bulgarian defector, Georgi Markov, was shot with a small platinum-iridium pellet impregnated with poison. Mr. Markov was hospitalized but died four days later simply because doctors could not detect any evidences of serious injury. Had this missile been made of ordinary metal it would have produced inflammation and doctors might have discovered it. Instead, the inert pellet never betrayed its presence and the patient died. It was found at autopsy.

The prescient feeling that platinum possessed magical if not miracle powers dates back to medieval alchemists who felt that it could transmute lead into gold. The metal was known to the ancient Egyptians who used it in decorative inlays but it was the Spaniards who really brought it to our attention in the 16th century shortly after they conquered South America. While placer mining in the rivers of the Choco region of Columbia, settlers found and cursed grains of platina, little silver, that accumulated in their gold pans.

Painstakingly, one by one, they threw them back into the river to ripen, believing that it was gold not buried long enough to turn yellow. Two hundred years later when sizable amounts of platinum did get to Europe, the technology of the times could not melt it so it was used primarily to counterfeit gold bars and coins. Only in the 19th century did scientists finally figure out that platina was not one metal but six separate elements, palladium, rhodium, iridium, osmium, ruthenium, and platinum, itself. This finding, plus the ability to melt the metal, now opened up new and daring uses for it, foremost of which was the making of fuel cells. These were first invented in 1842 by William Grove, a British electrochemist, then reinvented as new technology for satellites and space craft. Fuel cells depend on platinum to catalyze hydrogen and oxygen to combine to form water, a process that gives up electrons, a prerequisite to generating an electric current.

Almost contemporaneously with the cisplatin discovery scientists stumbled upon another anti-cancer drug in the forests along the Pacific Coast. As part of a large scale plant screening program, sponsored in 1961 by the National Cancer Institute, the pacific yew tree, *taxus brevifolia*, was harvested and its bark, leaves, and pulp were tested for biological activity. Some of the crude extracts contained a substance that changed the way cells behaved but it took ten years to isolate the active agent, TAXOL, and then an additional twenty years to prove that it really possessed anticancer properties. In September of 1991 the National Cancer Institute approved its use in patients with advanced ovarian cancer who failed to respond to standard therapies. A particularly effective form of therapy is one day of taxol followed by one day of cisplatin. The best results with chemotherapy have been obtained with combinations of chemical agents, colloquially called chemical cocktails.

The yew tree has a long and intriguing history. It is found in almost all English cemeteries, Shakespeare referred to it frequently, and sorcerers derived a poison from it. It takes about six trees, each at least 100 years old, to get enough taxol to treat one cancer patient. But there are not that many trees left. Many were lost throughout the United States during periods of indiscriminate logging operations. Research is now being directed toward moving taxol from the forest to the laboratory, that is, make it from basic chemicals. But Nature did not create taxol in *taxus brevifolia* just to fight human cancers. It presumably has myriad functions in a tree and is therefore tangled up in other complex substances. The trick is to find that special effective subunit of taxol, identify its chemical structure, then reproduce it synthetically. Until this happens we must try stop-gap measures like trying to extract taxol from portions of the bark or the flat needles, parts that can be removed without killing the tree.

Taxol is only one of at least ten anticancer drugs derived from natural products. The National Cancer Institute continues to search for new natural products from unexplored marine and tropical sources and has stated that "The diverse and complex structures and mechanisms of natural products in pharmacology are likely to continue to transcend the imagination of human chemists."

Dr. Umezawa, almost single handedly, added antibiotics to medicine's armamentarium in the fight against cancer. This monumental advance, coming from Japan, represents a natural progression of the great biomedical and science heritages of that country. Kitasato was first to cultivate the tetanus bacillus in 1889, he discovered the plague bacillus in 1894, and with Von Behring, defined the role of antitoxin in immunity. Shiga discovered the bacillus of dysentery in 1898 and worked with Ehrlich on syphilis and in chemotherapy. And as noted earlier, Katsusaburo Yamagiwa, in 1916, proved that chemicals could cause cancer.

As soon as he had received his M.D. degree in 1937, Dr. Umezawa went right into research. "Research is my habit", he used to say. "It was the easiest thing for me to continue after medical school." In 1943 he was awarded his Doctorate of Medical Science on the basis of his studies on the effects of chemicals on immunity and viral infections. It was at this point that one of those innocuous, simple, incidents, that springboard great men to pursue great ideas, took place. A scientific report by Rene Dubos at the Rockefeller Institute describing the isolation of the antibacterial *tyrothrycin* from a soil bacillus captivated young Umezawa's imagination. Because of the war, communications with Dubos were out of the question. Instead, Umezawa began looking around to see what organisms might be present in local soils and soon discovered many molds belonging to the *actinomycetes* family from which antibacterial agents could be extracted. Research stopped when he was drafted into the Japanese army at the outbreak of hostilities with the United States. With the advent of peace he was asked by the army medical school, in 1948, to develop methods for the large scale production of penicillin—word had filtered back from Germany that the United States was on the threshold of such an accomplishment. This, plus his earlier work with *actinomycetes* laid the foundations for his subsequent discoveries of the antibiotic kanamycin and the anticancer drug bleomycin.

Immediately after World War II Japan experienced a sharp rise in tuberculosis. Streptomycin was already available but resistant strains were emerging and this prompted the search for additional antibiotics. Umezawa discovered kanamycin in 1956 which proved effective in tuberculosis and was also active against a wide spectrum of other bacilli. In an interview Umezawa said, "Luck played its role in the enormous success of kanamycin." He was referring to the fact that in 1957 staphylococci were becoming resistant to all the mainline antibiotics. "Kanamycin was effective against these and other resistant organisms and so became a life saving drug," Dr. Umezawa laconically added.

In 1962 Dr. Umezawa, now Professor Hamao Umezawa, helped found the prestigious Institute of Microbial Chemistry Research in Tokyo. Grants from the government initiated the project, major financing from the Japan Professional Bicycle

Race Association kept it solvent and well fed, but in the final analysis, it was the royalties from kanamycin that assured its success. Professor Umezawa became its first and foremost Director. Amidst all the confusion connected with setting up and operating this extensive research facility, the government called upon Umezawa to look into the terrible rice plant scourge that had been jeopardizing Japan's basic staple. The disease, caused by the fungus *Piricularia oryzae*, had been kept under control with mercury compounds but this led to widespread poisoning of the waters and marine life surrounding Japan. People eating contaminated fish and sea food developed Minamata Disease, severe deterioration of the nervous system due to mercury poisoning—so named from the disastrous dumping of industrial mercury wastes into Minamata Bay. Professor Umezawa and his research team found an effective deterrent in an antibiotic called kasugamycin, derived from *streptomyces kasugaensis*, not unlike streptomycin that had been discovered earlier by the American, Selman Waksman. A peculiar phenomenon surfaced that gave the researchers many a sleepless night: the antibiotic worked on rice plants in the field but it would not work in the test tube. Only later did the scientists find that the antibiotic only works in an acid environment and works best in the exact acidity of rice plant leaf juice.

Once kanamycin was produced commercially, Professor Umezawa assumed that there would be no further reason to continue research on it, but when, in 1965, strains of bacteria resistant to kanamycin began to appear, Umezawa went back to probing the cause of such resistance. At first he worked under the presumption that weak bacteria were being killed off and stronger ones, with better genes, were taking their places; as they reproduced they would give rise to resistant strains. This proved to be but part of the answer. With great insight Dr. Umezawa was able to show that the chemical structure of kanamycin was being attacked by bacteria in a site-specific manner; certain bacteria could liberate enzymes that broke up the kanamycin molecule and made it ineffective. Together with his brother, Sumio Umezawa, a new derivative of kanamycin, called DKB, which resisted attacks by enzymes was synthesized but it never reached commercial production because penicillin, streptomycin, chloramphenicol, aureomycin, terramycin and second generation antibiotics came on the market faster and cheaper.

While all these undertakings and exploits were under way Dr. Umezawa was quietly and persistently chipping away at one of his earliest concerns, cancer. Back in 1951 he began looking into the possibilities of using antibiotics to treat cancer —they stopped germs from growing, they might do the same for cancer cells. After screening countless filtrates of many different microorganisms and testing them on experimental animal tumors Dr. Umezawa observed that in some cases the tumors stopped growing. In 1953 he published one of the world's first papers announcing that an antitumor antibiotic could be found if deliberately searched for. It took another 15 years before this promise was fulfilled.

"It all started when a man appeared with an intestinal tumor" Professor Umezawa later recalled. "He agreed to an injection of an extract containing actinomycin, still of undetermined structure, and the tumor eventually disappeared. I thought the antibiotic must have been partly responsible." Concurrently, in Germany, Dr. C. H. Hackman also observed that actinomycin was effective against some animal cancers.

Antitumor antibiotics inhibit the synthesis of substances necessary for the life of a cell; they also alter the cell membrane, making the cell more vulnerable to attack. Umezawa was looking for an antibiotic that would accumulate in tumor cells in higher concentrations than in normal cells or an antibiotic that was particularly more damaging to tumor cells than to normal cells. By 1956 Dr. Umezawa and his research team isolated phleomycin, an antibiotic that effectively slowed down cancer. Further tests showed that it damaged kidneys so it was dropped. But the search went on and by 1962 bleomycin, a mixture of ten active components, was isolated and found to be active against certain tumors. Not only was it more chemically stable than phleomycin but it did not suppress immunity nor damage bone marrow and kidneys. Bleomycin is still a major anticancer fighter. It does not work in every cancer but is very active in select cases.

Now, with better biological, biochemical, and immunological technologies on hand the main thrust in cancer research is to perfect Paul Ehrlich's magic bullets for cancer. The most promising approaches lean heavily on immunotherapy, tumor specific antibodies, liposomes (tiny packets of drugs enveloped by fatty material), and viruses to deliver chemotherapeutic agents directly to or into tumor cells. The next chapter on growth factors will explore one of the most imaginative, new, approaches to cancer.

PARADIGM POINTERS

When a new discovery presents knowledge that challenges an accepted paradigm it is not always necessary to expunge the old one. A new paradigm that co-exits with the old one or incorporates some of the old one may take its place. There is no need to throw out the baby with the bath-water.

When the properties of radium became known, the element was heralded as a cancer cure. Later, when radiation was proven to be a cause of cancer, it did not entirely invalidate the use of radium in treating cancer. We now work with several paradigms that make allowances for both the causative and curative effects of physical agents.

PERSONAE

CORNELIUS PACKARD RHOADS, M.D.:
Wilhelm Roentgen announced the discovery of x-rays in 1895. Barely three years later, on June 20, 1898, Cornelius

Packard Rhoads came into the world in the small town of Springfield Massachusetts. After graduation from Bowdoin College he decided to follow his physician father's footsteps and attended Harvard University where he received his M.D. degree in 1924. While a surgical intern at Peter Bent Brigham Hospital Dr. Rhoads contracted pulmonary tuberculosis which necessitated spending a year as a patient at the Trudeau Sanitarium in Saranac Lake, New York. His first scientific paper, written with Dr. Fred Stewart, was concerned with tuberculin reaction. His illness fired up an interest in pathology which he pursued by becoming an Instructor in Pathology at Harvard University and Assistant Pathologist at Boston City Hospital. He held these appointments from 1926 through 1928 then joined the staff of the Rockefeller Institute for Medical Research where he acted an Associate 1928-1939, and Pathologist of the Hospital of the Rockefeller Institute from 1931 to 1939.

In January 1940 Dr. Rhoads accepted the directorship of Memorial Hospital and in 1945 was made Director of the Memorial Center for Cancer and Allied Diseases in New York. The program of research developed under Dr. Rhoads eventually became the Sloan-Kettering Institute for Cancer Research. Dr. Rhoads remained Director of the Institute as well as Professor of Pathology at Cornell University Medical College until his death in 1959.

Throughout his life Dr. Rhoads demonstrated keen insight and an uncanny intuition about how things worked. This was proven by his serendipitous association of lethal nitrogen mustard with life saving cancer chemotherapy. With the possible exception of Professor Umezawa, few people in the world of medicine have equaled Dr. Rhoads' accomplishments as researcher, scientist, physician, and administrator. He wrote well over 300 articles on cancer chemotherapy as well as review articles on trends in the diagnosis and control of cancer. He championed chemotherapy when it was still unpopular and was one of the first to recognize the contributions made by Rene Dubos at the Rockefeller Institute pointing the way toward the use of antibiotics for cancer therapy. He went so far as to make plans for a search for antitumor antibiotics, similar to those of Dr. Umezawa. His interests also extended to the medical uses of atomic energy. After World War II, in addition to his work at Memorial Hospital, Dr. Rhoads worked tirelessly to expand and improve the research programs of the American Cancer Society.

LEON O. JACOBSON, M.D.

Studying and advancing chemotherapy took up half of Dr. Jacobson's life. The other half was devoted to elucidating the biological effects of radioactive isotopes. Just after his graduation from the University of Chicago Medical School, in 1939, he became health officer for the university's Metallurgical Laboratory, a code name for the sector working on nuclear weaponry for the Manhattan Project. The valuable experience he picked up while protecting scientists from radiation was parlayed into establishing the foundations of nuclear medicine. Our current usage of radioactive substances to study normal and abnormal biological processes rests upon Dr. Jacobson's pioneering work.

HAMAO UMEZAWA, M.D.:

Hamao Umezawa's medical roots go back to his great grandfather who received his premedical education in Edo which, at that time, was the capitol of the central government of the Tokugawa Shogun, before the Meiji restoration; with the surrender of Edo Castle to Imperial forces in 1868, the Emperor made it his capitol and renamed it Tokyo 'the great eastern capitol'. The elder Umezawa studied medicine at Nagasaki in the Kyushu Prefecture, then his son-in-law, Hamao's grandfather, a graduate of Tokyo Medical School, took over the practice. Hamao's father, Junich, studied medicine at the University of Tokyo Medical School and took postgraduate training in biochemistry and internal medicine prior to being appointed director of the hospital in Obama.

Obama, where Hamao was born, is a small town about 300 miles west of Tokyo, surrounded by the cultivated coastal rice fields of Fukui Prefecture. Throughout childhood young Hamao spent many a day swimming and fishing in the clear brooks which fed the rice fields and the transparent river flowing past Obama to the sea of Japan. Those memories undoubtedly spurred his drive to find Kasugamycin so that mercury fungicides would not have to be used on rice.

Dr. Umezawa attained his medical and science doctorates while still in his 20's and during World War II was already developing antibacterial agents from soil fungi. During his illustrious career Professor Umezawa tested over 70 antimicrobial antibiotics, 30 antitumor antibiotics, and 45 enzyme inhibitors. He has clarified the enzymatic mechanism of drug resistance in bacteria and has demonstrated that drugs can be modified structurally so that resistance can be overcome without sacrificing efficiency. He gave the world bleomycin, kanamycin, and kasugamycin. Umezawa is well versed in the world of science literature. Every year his library receives thousands of books and journals, a two-foot stack every week. In 1979, at the age of 63, he became director of the Department of Antibiotics at Japan's National Institute of Health while retaining his Directorship of the Institute of Microbial Chemistry. This institute, incidentally, is the research base for 100 scientists in seven divisions: Microbiology, Ocean Microbiology, Cancer, Biochemistry, Analysis, and two sections devoted entirely to chemistry. At any given time it maintains 25 full time researchers, 20 students working toward Ph.D. degrees, plus outside workers from Universities and other organizations engaged in pharmaceutical and basic research.

At the time he received the Bunka Kunsho Medal, the highest science award bestowed by the Emperor of Japan, people asked how one man could accomplish so much. His peers jointly confirmed that Dr. Umezawa worked over 12 hours a day, he always gave sufficient thought to a new project before beginning laboratory investigation, he kept abreast of every research project, and he was ever on the alert to all kinds of clues.

An intimate glimpse into the ecumenical mind of this physician, biologist, chemist, and philosopher, comes with this anecdote. "There is a cultural phenomenon manifest in the Japanese people," Professor Umezawa once explained, "where a man who speaks very conclusively is not thought to be wise. A man who cannot reach a conclusion may make Japanese people think he is clever. This is bad for science." Unfazed by such notions Umezawa always spoke his mind and never let custom get in his way of doing things, even on a guess, hunch, or what he thought was right. Although science has preoccupied his entire life Dr. Umezawa wistfully admits, "I wish I could write beautifully and precisely. I admire the individual who can truly appreciate and understand the essence of great literature and the arts, of paintings and antiques. I enjoy television and watching sports, particularly baseball. And it is a basic truth that a man cannot have a good life without good friends. I have many friends. I thank all of them and hope to maintain and increase my friendships around the world." His greatest concern is about the food needs of the expanding world population. And above all, he feels that all nations must make sure that the world is supplied with qualified doctors and scientists who can teach young scientists-to-be.

chapter 11

CONCEPT OUT OF CHAOS

The Nobel Prize Committee's announcement that Dr. Rita Levi-Montalcini had been chosen to share the 1986 award in medicine for discovering nerve growth factor, stated, "this is a fascinating example of how a skilled observer can create a concept out of apparent chaos." The choice of the word *chaos* was strikingly appropriate since it underscored the major determinants of her life and times. Chaos taunted her as she sought a higher education. Chaos blocked her attempts to enter the medical profession. Chaos threatened her as a Jew in Fascist Italy. Chaos greeted her eyes as she peered through her microscope at entanglements of wildly growing nerves.

Now, exactly in the middle of the 20th century, in January 1950, chaos again confronted her. This time it was in St.Louis, Missouri. Temperatures had plummeted down to 20 degrees below zero and snow blanketed everything in sight, an enanthema for this lady who obsessively hated coldness. She had been born, raised and educated in Turin, Italy , and still pined for the temperate Piedmontese climate. She had come to the University of Washington four years earlier, in the fall of 1946, but still recoiled from its harsh winters.

Nonetheless, on this particular somber, gray, bone chilling, morning, there was work waiting to be done in her laboratory at the Zoology Institute's ivy covered Rebstock building, so she donned her parka and braved the frigid elements.

As she inched her way through the snow Dr. Rita Levi-Montalcini's stride seemed strengthened by the perversity of the weather—very much like her resolve to succeed in medicine was intensified by the prejudices she encountered as a woman in the medical profession. As late as 1930 when 21 year old Rita entered the Turin School of Medicine, she was constrained, both at home and in college, by the prevailing Bismarkian manifest destiny for women: *KUCHE, KIRCHE, KINDER*. Kitchen, Church, Children. Women had no business outside the home, especially not in the professions. But Rita rebelled. Together with her mother, the two women inveigled Adamo, father of the family, into allowing Rita to become a doctor. In later life she said, "My experience in childhood and adolescence of the subordinate role played by the female in the society run entirely by men had convinced me that I was not cut out to be a wife."

Often, to escape St. Louis' immobilizing cold winters and oppressively hot, humid summers, Rita would invoke halcyon memories of her family home on Corso Vittorio Emmanuele in Turin. She had lived there since her birth, in 1909, and could visualize the Roman living room embellished with plants, etchings, sculptures, and paintings by her twin sister, artist Paola. She could smell the buds of the horsechestnut trees as they burst into bloom in the spring. And she could hear her mother admonishing her to 'take a mouthful of sun' a she was shooed outdoors to play. Although she now lived in the university's faculty apartment house, in a city that boasted over 2 million people, nothing ever supplanted her own room, where, as a child, she could gaze upon the bronze statue of King Victor Emmanuel, with sword in hand, that maintained its protective vigil just outside her window.

Rita's father, Adamo Levi, had been a successful engineer. Though nominally Jewish, he was essentially indifferent to all religions. He epitomized Italy's dedicated middle class who served as teachers, artists, members of Parliament, generals in the army and civil servants. Families blended with the secular population through marriage, patriotism and a common culture.

Adamo, nicknamed "Damino the terrible" in deference to his volatile temper, rejected all ideas outside of domesticity and motherhood for his daughters. It was only in her stable and affectionate mother, Adele Montalcini, that Rita found support for a career in medicine as well as comfort and refuge throughout her childhood and the turbulence of Fascism. When Adamo died, in 1932, Rita acknowledged her indebtedness to her mother by changing her name to Levi-Montalcini.

As a member of the 1930 class at Turin Medical School, young Rita Levi-Montalcini fell under the spell of the man who was to shape her destiny, Giuseppe Levi (no relative). This bulky, red-haired, flamboyant Professor of Anatomy, was both feared and respected; feared because he brandished a bamboo cane to enforce discipline and decorum; respected for his intellect, knowledge, and uncanny ability to bring out the best in his students. He rubbed shoulders with great men of his time and regularly brought their theories and techniques back to his class room. On one occasion, at an international meeting of leading scientists, he met the illustrious Alexis Carrel from Rockefeller Institute who introduced him to the rudiments of culturing living tissues outside of living animals. Although Professor Levi never fully pursued this line of research himself he instigated an interest in tissue culturing techniques in others. Sensing that young Rita's talents could be put to better use in the research laboratory than at a patient's bed side, Professor Levi taught her the fundamentals of impregnating nerve tissues with sensitive silver stains and observing nerve growth in tissues separated from the whole organism. At this stage, neither one could suspect that Rita was on her way to becoming one of the world's foremost embryologists and neuroanatomists. Until his death at the age of 92, he would barge into her laboratory, whether in Italy, the United States, or Rio de Janeiro, and proffer advice, suggestions, criticism, or disbelief.

Rita Levi-Montalcini, through intelligence and tact, was able to quell the chaos instigated by being female in a predominantly male university but the inimical chaos provoked by being a Jew in Fascist Italy could not be similarly subdued. Mussolini had proscribed people with Jewish backgrounds from practicing medicine or holding positions in universities so she was forced into research, most of which she had to do in her room at home. When bombs began to fall on Turin's industries, after Italy's entry into World War II, the entire research operation was moved to the country.

While still under the tutelage of Professor Levi, Rita developed an insight into the nervous system by studying embryos. Primitive centers of development contained far fewer cells and neuronal circuits were much simpler than in mature organisms. Chick embryos were selected because they were easy to come by and they could be incubated at home. The family used to shudder at mealtime not knowing the source of the omelets. She ground down common sewing needles to make sharp microscalpels and spatulas. Except for a Zeiss microscope everything was home made, hand made, or adapted into use.

Scientific reports describing her early probing experiments were summarily rejected by major research journals. Non-Aryan science was enathematic to Fascism and Nazism. A Swiss journal ultimately did publish Dr. Rita Levi-Montalcini's studies on the chick nervous system. These, plus another one that demonstrated how the excision of embryonic tissues had a specific effect on the nerves designated to grow to missing parts, published in Belgium, brought her to the attention of Viktor Hamburger in the United States.

Viktor Hamburger, Chairman of the Department of Zoology, at Washington University had been immensely impressed by the precision and logic Dr. Rita Levi-Montalcini had displayed in her research papers. He wrote to Professor Levi, in 1946, asking if she could be released from her duties to work with him for a semester in St. Louis. Rite went. But the mysteries of how and why nerves grow as they do, the central concerns of both Viktor Hamburger and Rita Levi-Montalcini, could not be solved as fast as they had hoped. Rita's one semester sabbatical innocently stretched into its fourth year.

Once inside the Rebstock Building, Dr. Levi-Montalcini, a sprightly young woman with blue-green eyes and a beguiling smile, wearing a white lab coat with the sleeves rolled up because none were available to fit her 95 pounds and diminutive five foot three frame, went directly to an ordinary first floor office with an imprint on the door that read: Professor Viktor Hamburger, Chairman, Department of Zoology. Neither the office nor the sign betrayed any hint of the towering figure behind those dark umber doors. He towered in physical size, standing six and half feet tall, and towered in professorial presence and international stature. He had been the brilliant disciple of the 1935 Nobel Prize winner, biologist Hans Spemann and now, in his own right, he was regarded as the foremost authority in embryology and microbiology. Dr. Levi-Montalcini, currently a research fellow at the Institute, knocked on the door and waited for a call to enter. This was a needless formality for although Viktor was nominally her boss he was also her patriarch, mentor, colleague, working partner, and close personal friend. But Dr. Levi-Montalcini could not ignore the decorous rituals of academic deportment that had been instilled in her as an undergraduate in Italy. She felt constrained to follow old-world staid, traditional protocols, and wait.

"Come in! Come in!" Victor's voice boomed, knowing from the tenor of the knock and the time of day that it was Rita coming for their regular chat about her previous day's research. It was the highlight of the day for Victor because Rita, who spoke both French and Italian fluently, would detail her triumphs and defeats in a piquant French accented

English and enliven discussions with unbridled enthusiasm and brio.

Hamburger, with gray hair framing a high forehead and a congenial avuncular facade behind his thick horn-rimmed glasses characteristically displayed peripatetic energy and a piercing interest in everything and everybody. After the 'morning review' was finished, professor Hamburger perfunctorily grasped a stack of papers and casually said, "Rita, these are from Elmer Bueker, a former student of mine. He moved on to Jackson Memorial Institute in New York a couple of years ago. Like both of us, he had been working on the differentiation of nerve cells." Hamburger then extended a scientific paper with its attached letter to Dr. Rita Levi-Montalcini and continued, "This is most interesting. You must read it." The young woman in the over-size lab coat read the letter then delved into the substance of the scientific report which had been written by Bueker two years earlier, in 1948. From both she learnt that Bueker had come across a perplexing phenomenon. Nerves seemed to be growing wildly in some of his tumor cell experiments and he was wondering if Viktor or any one of his elite fellows had ever seen anything like it or could they explain it.

Bueker, in the same arena as Drs. Levi-Montalcini and Hamburger, had been investigating the mechanisms that govern the development and differentiation of motor and sensory nerve cells in chick embryos. In such studies it was customary to surgically remove or otherwise destroy the budding limb areas in embryos and then observe what happens at the interrupted sites. Sometimes limbs were grafted back again and junctions were studied. Bueker, however, instead of just removing and grafting limb buds, grafted fragments of cancer tissues, Rous Sarcoma and mouse tumors, onto chick embryos. He chose this line of research because it was well known by then that all kinds of mammalian tumors, from rodents to humans, grow nicely in embryos. This is so because immature life-forms have not as yet developed immune systems that cause tissue rejection. Of the several tumors Bueker assayed, one, the mouse sarcoma known as S.180, took hold and formed a tumor on the embryo as expected. In some of the embryos, however, the nerve fibers stemming from a clump of primordial sensory nerve cells had proliferated helter-skelter into the mass of tumor cells.

"Elmer believes that S.180 produced conditions favorable for nerve growth," Hamburger said. "You know, like fertilizer on a lawn, nerve cells migrated to richer pastures." Then, with the mischievous design of a sly fox he added, "Too bad he has abandoned this research."

Dr. Rita Levi-Montalcini swallowed the bait, hook, line, and sinker, and with Hamburger's consent and encouragement, interrupted her own limb grafting experiments and switched over to implanting tumors into chick embryos as Bueker had done. In her recollections of that momentous event, Dr. Rita Levi-Montalcini stated, "We wrote to Bueker who was flattered by the interest we took in his work and ten days later we received from Jackson Memorial Institute a box full of small albino mice that were carriers of S.180. I quickly took fragments of the tumors, the size of chick embryo limb buds, and grafted them to the sides of three day old embryos." Jackson Memorial Institute, incidentally, specializes in raising mice with malignant tumors. Tumor cells are perpetuated by means of serial transplants to successive generations of mice.

After one and two week intervals, when the graft sites on embryos were examined microscopically, Dr. Levi-Montalcini observed that the tumor cells were growing as well demarcated nodules. They were also interlaced with many blood vessels. Bits of tissue were excised and treated with special silver-containing stains which make nerve tissue brownish-black in color, therefore easy to examine under the microscope. Numerous bundles of nerve fibers could be seen emerging from the embryo and rampaging through the tumor nodules. Dr. Levi-Montalcini vividly remembers that event: "These fiber bundles passed between the cells like rivulets of water flowing steadily over a bed of stones."

There it was! The phenomenon as Bueker had seen it. Exactly as Bueker had described it. Levi-Montalcini had just invoked the foremost tenet and safeguard of scientific integrity. She duplicated experiments described by others and corroborated the data in an unprejudiced, dispassionate, manner.

But how could these observations be explained? Was this, as Viktor Hamburger quipped earlier, "like fertilizer on a lawn so nerve cells merely migrated to richer pastures?"

In order to find answers more tumor-bearing mice were ordered and the experiments were repeated. Again frenzied masses of nerves from the embryo invaded the grafts. Dr. Rita Levi-Montalcini then noticed that the second shipment of mice from Jackson Memorial Institute differed from the first. The order forms accompanying the two shipments stated that the first batch of mice were carriers of S.180 while the second group had been grafted with a tumor known as S.37. Both tumors originally came from mouse breast cancer cells. They had been serially transplanted, over and over again, so that they lost their original structure of mammary gland and now grew in a disordered and tumultuous fashion. Differences in tumor strains did not seem to matter. Both provoked massive nerve growth. After noticing that both S.180 and S.37 were equally effective Dr. Levi-Montalcini cautiously concluded that some element common to these tumor cells triggered the Bueker phenomenon. More importantly, she realized that this extraordinary growth of nerves was entirely different from the slow, uniform growth seen in her former limb grafting experiments.

To celebrate this achievement Dr. Rita Levi-Montalcini played a Bach cantata on her phonograph, something she always did whenever the light of triumph replaced the spectral shadows of doubt. But joy was abruptly terminated when dear, old, Turin professor Levi, on one of his many globe-girdling junkets, brusquely entered her laboratory, peered through the microscope and querulously proclaimed, "Rita, those are connective tissue fibers. What you are calling nerves are really connective tissues." Internecine arguments raced through her mind: Is he right? Am I right? Are we both wrong? In any event, good Professor Levi, now close to 80, was packed off to see the Grand Canyon and Dr. Levi-Montalcini used this window of time to strengthen her case.

During one of her many stints at her microscope Dr. Levi-Montalcini noticed that embryonic nerve fibers showed a tendency to grow into veins that were leaving the tumor. Aha! Two possibilities presented themselves: either the flourishing nerves had a predilection for veins or the tumors were secreting something into the veins that attracted the nerves. Let us see which it is!

She then proceeded to graft tumor cells to the chick chorio-allantoic membrane, that parchment-like sheet of tissue just under the shell that completely envelopes the developing embryo. It has no nerve fibers but is criss-crossed by a thick network of blood vessels belonging to the embryo itself. She made a small opening in the egg shell through which she deposited tumor cells. The hole was sealed with a removable piece of adhesive tape. Day after day Dr. Levi-Montalcini peered through her stereoscopic microscope to observe the developments. The tumor cells took hold and adapted to their new host, as expected, but now, for the first time, human eyes beheld the migration of nerves from the embryo to its normally nerve-less enveloping membrane. Many nerve fibers actually penetrated the veins in the membrane. This proved, irrefutably, that the tumors secreted something into the embryos circulation which not only attracted nerve cells but also fomented wild, rapid, growth. Since nerve tissue selectively sought out veins, which are leaving tumors, rather than arteries, which are entering tumors, Dr. Rita Levi-Montalcini rightly concluded that the tumor cells made and secreted a nerve stimulating substance.

The next step was to use tumor extracts on tissue cultures and see if they would have effects similar to those observed in chick embryos. But not knowing much about growing tissues in cultures outside the body Dr. Levi-Montalcini thought it best to look up her old friend, tissue culture expert, Hertha Meyer. Dr. Meyer had been chased out of Europe by the Nazis and was now working in Professor Carlos Chagas' laboratory at the Institute of Biophysics of the University of Rio de Janeiro.

At the end of summer in 1952, Dr. Rita implanted mouse tumor cells S. 180 and S. 37 into some mice and fitted them into a cardboard box which she carried in her overcoat pocket. She flew first to Italy to see her family then on to Brazil. Along the way she shared her apples with the traveling companions in her pocket.

Dr. Meyer quickly taught Dr. Levi-Montalcini the basics needed to culture tissues in flasks but the experiments that followed turned out to be both a boon and a bane. Disappointments came first. Tumor cells taken directly from her little mice failed to induce tumultuous growth in nerve cells growing in tissue culture media. But joy returned when after several transplants the tumor cells regained their magic. When tumor cells were added to cultures of living nerve cells the latter grew so fast that they seemed like an entirely new tissue. Beautifully latticed bunches of nerve fibers, for which Dr. Rita Levi-Montalcini coined the term fibrillar halos, appeared wherever she looked. When tumor cells were placed eccentrically in a culture dish, let us say at two o'clock, nerve cells orientated themselves toward two o'clock. Now she was sure that there was indeed something in tumor cells that had the ability to stimulate the growth of nerve cells. Tumor cells also made *neurotactants*, substances that attract developing nerve cells. This faculty of attracting developing nerve cells and forcing them to grow towards a specific site is known as the *neurotropic effect*.

Did all tumors have this ability to whip nerve tissue into frenzied growth? To find the answer to this question Dr. Levi-Montalcini performed two daring experiments. First she took tumors not related to mouse tumors S. 180 or S.37, transplanted them to chick embryos or tissue cultures and proved that they did not induce the formation of conglomerations of nerves or fibrillar halos; they did not affect nerve growth. The next experiment raised more questions than it answered, questions that remained unanswered until Stanley Cohen came on the scene a year later.

Thus far Dr. Levi-Montalcini had shown that there was something in mouse tumor cells S.180 and S.37 that could attract nerve cells and stimulate their growth. It did this in both chick embryos as well as in tissue cultures. Were there any other cells or tissues that could do the same thing? To find out, Dr. Levi-Montalcini grew fragments of normal mouse tissues, such as heart and skin, in tissue cultures, then tested them on pure cultures of nerve cells. To her dismay these 'normal' mouse cells produced growth substantially similar to that of mouse tumors S.180 and S.37, perhaps not as great but unmistakably present. This finding discombobulated all of her previous work and theories. First she proved that a nerve growth enhancing substance in tumor cells produced dramatic proliferations of nerve tissue in chick embryos. Then she reproduced this effect in tissue cultures, thereby showing that the chick embryo was just an innocent host and contributed nothing to the events that followed. Finally she demonstrated that only specific mouse tumor cells, S.180 and S.37, had this unique ability. But now normal mouse cells did the same thing. Since there was no ready explanation for this awkward finding it was called *the mouse effect* and conveniently tucked away into subliminal recesses of the mind. The Russian neurophysiologist Alexander Luria used to say, "We tend to obey the law of disregard of negative information"

Dr. Levi-Montalcini left the Carnival atmosphere of Brazil's Rio de Janeiro and returned to St.Louis in January, 1953, almost exactly three years after starting her pursuit of the Bueker phenomenon. She was still excited by her triumphs with tissue cultures but equally bewildered by 'negative information'. Viktor Hamburger was at the airport to greet her and after a warm hug he said, "I have a surprise for you.

We are getting a biochemist as a research associate. His name is Stanley Cohen. I suppose I should call him *Doctor Cohen* since he has a doctorate in biochemistry and finished a postdoctoral stint here at Washington University under the physical chemist Martin Kamen but he doesn't look like a doctor, doesn't dress like a doctor, plays the flute when he's thinking or dreaming, and takes his little dog, Smog, with him wherever he goes. I hope you can put him to good use."

In retrospect it is hard to say who put whom to good use. Stan, as he was called by everyone, knew everything about earthworms, which he had studied for several years, but knew nothing about neurophysiology. Rita knew neurophysiology but nothing about biochemistry. As a team however, their combined abilities far exceeded the simple sum of their individual talents. "I have often asked myself," Dr. Levi-Montalcini used to say, "what lucky star caused our paths to cross."

Dr. Cohen set up his biochemistry laboratory on the ground floor of Rebstock Hall while Dr. Rita Levi-Montalcini worked just above him on the second floor. This arrangement enhanced the closeness of Viktor, Stan and Rita and allowed them to meet almost every day to discuss new findings and developments.

Stan lived in an outlying prefabricated cottage. Early every morning and late every night until 1958 when he left for Vanderbilt University, in Tennessee, this motley dressed, pipe smoking man, limping slightly because he had had polio as a child, could be seen hiking across the university campus with scruffy little Smog. Once inside the laboratory Smog would generally lie down at Stan's feet while the scientist fidgeted with his test-tubes or worked at his desk. When Stan relaxed by playing haunting, melancholy, oriental strains on his flute, Smog invariably drifted off to sleep.

Dr. Levi-Montalcini compulsively brought into use everything she had learned under Hertha Meyer. She established a tissue culture center in her laboratory and then set out to confirm the observations she had made in Brazil. It was now apparent that certain tumor cells, when first transplanted into chick embryos and then excised from them and put into tissue cultures, produced something that attracted nerve fibers and made them grow profusely. In 1954, as a result of many references to it in scientific reports in the United States and abroad, the substance was officially dubbed nerve growth factor. It is usually referred to in its abbreviated form as NGF.

Before moving on to the next incredible series of events we should stop for a moment and look back over the enormous amount of work, patience and skill that was required to get this far. Remember, in 1953 the science of culturing tissues was in its infancy and the slightest variations in technique resulted in contaminated or dead cell cultures or failed experiments. And, as the Nobel Prize committee said, "The discovery of Nerve Growth Factor, NGF, in the beginnings of the 1950's is a fascinating example of how a skilled observer can create concept out of apparent chaos."

By chaos, the authors of this statement alluded to the vagueness and confusion of notions about the mechanisms of cell differentiation that had baffled scientists ever since the Dutch janitor, shopkeeper and lens-grinder, Anton Leeuwenhoek, assembled the first microscope in the mid-seventeenth century. At that time, scholars, with limited magnification but unlimited imagination, swore that they saw a homunculus, a tiny complete human being, compressed within every human sperm. After fertilization, they explained, it simply grew within the egg and nine months later a carbon-copy of the parents emerged. Improvements in microscopes and advances in science later proved that sperm merely deliver chromosomes and fertilized eggs just keep dividing into 2,4,8,16,32 cells, and so on. This then presented an enigma. Whereas the old homunculus theory neatly explained why a fertilized egg became a baby the new observations did not tell us how a cluster of cells knew enough to differentiate into one head, two eyes, and four limbs, and why toes and fingers stopped at ten. Attempts to find 'the thing' that controlled and directed the orderly spatial growth and specialization of cells proved exasperatingly futile. Thomas Hunt Morgan, the early 20th century Columbia University geneticist, came awfully close, coining terms like "head stuff" and "tail stuff" to denote biological substances that determined the head and tail polarities of amphibian embryos. A half century later, Levi-Montalcini's and Cohen's advanced sophisticated biotechnology carried Morgan's suppositions closer to fact.

Once the reality of NGF was established it was time to ask: What is this strange substance? A Hormone? An enzyme? Or something entirely new? The ball was lobbed into Stan's court. It was up to him to define the exact nature of NGF.

The tumors that grew in the form of little nodules in chick embryos contained infinitesimal amounts of this nerve stimulating factor. Rita would have to spend inordinate amounts of time extracting tumor tissues from dozens of embryos before Stan had enough material to work with. After one year of such intense labor the tumoral factor was identified. It proved to be a nucleoprotein, complexes of nucleic acids and proteins.

Enter happenstance. Stan, in the course of a conversation with Stanford's Nobel prize winning biochemist Arthur Kornberg, expressed his suspicion that the nucleic component in his test samples might be a contaminant. Dr. Kornberg suggested treating tumor tissue extracts with snake venom because it contained the enzyme *phosphodiesterase* which breaks down, therefore gets rid of, nucleic fractions. Following Dr. Kornberg's advice, Stan added snake venom to tumor tissue extracts and asked Rita to test them on some of the nerve cells growing in tissue cultures. Lo! and behold! The nerve cells grew better and faster than with standard extracts. The halos of nerve fibers were enormous and their density and symmetry without precedent. Excitedly she called Stan who looked through the microscope and in astonishment said, "Rita, I'm afraid we've just used up all the good luck we are entitled to."

There is an old adage in science that says, "when you find an answer to one question you can bet that it will give rise to two more." That is exactly what happened here. Drs. Cohen and Levi-Montalcini could not tell whether the snake venom had neutralized an inhibitor in their tumor extracts and this increased NGF's activity, or whether the venom itself contained a substance that had the same stimulating properties as nerve growth factor. Ten hours later they proved the second hypothesis. A tiny quantity of venom added to a nerve cell culture produced the same wild growth of fibers as had appeared in cultures treated with original tumor extract.

This breakthrough fulfilled a long-standing dream for both investigators because, for the first time, it handed them a large and reliable source of NGF, enough to make extensive and sophisticated tests. Stan subsequently purified a fraction that proved to be 3,000 times richer in NGF than extracts derived from tumors. He never talked about his competence and extraordinary intuition which always guided him with infallible precision in the right direction. With true modesty he admitted "If I manage to solve a problem, its only because I've really plugged away at it. I have to work hard, very hard, to find the solution." And with that resolve he also unraveled the molecular weight of NGF and most of its other physical and chemical characteristics.

Rita, meanwhile, conclusively proved that proliferative nerve fiber growth, whether induced by snake venom or mouse tumor extracts, was exactly the same. Then, with her new source of highly concentrated NGF, she clarified the realities and potentials of NGF, something that she had been unable to do with the tiny amounts heretofore at her disposal. She also devised techniques where hundreds of test samples could be added to cell cultures and in the space of a few hours she could tell whether the substance possessed nerve growth factor. She also laid to rest the supposition that NGF was a hormone. It was found in too many places, it stimulated different types of nerve cells, and responses were variable, all of these were not typical characteristic of hormones.

Instead of luxuriating in the warm glow of triumph the two scientists found themselves beset by a torment not unlike that faced by King Sisyphus, the ruler of Corinth, who was doomed to push a boulder up a hill but just short of the top the huge rock invariably rolled back down. "At first we believed that only cancerous cells made nerve growth factor. Then we found it in snake venom", the bewildered team reasoned. "Could it possibly be present in normal cells, even in organs and tissues of different species of animals?" The answer hinged on another stroke of good luck.

Stan reasoned that if snake venom contained nerve growth factor it might be present elsewhere in the animal kingdom. Since the venom of snakes comes from glands similar to the salivary glands in higher mammals, Stan decided to study the submandibular salivary glands of mice. He knew that there was some similarity between the attack functions of snake venom and the defensive functions of mouse saliva. The latter, in addition to digestive enzymes, contains a poison that the mouse depends on when attacking victims or defending itself from predators. He chose male mice because their salivary glands are much bigger than those in the female. Perhaps male mice, being more aggressive, have been endowed by nature with larger salivary glands.

That is why we find Stan in his laboratory in the winter of 1958 busily analyzing and testing mouse submandibular salivary glands for nerve growth factor. An extract of the gland was made and added in progressively higher dilutions to a culture of chick embryo nerve cells. "With a mixture of happiness and incredulity," as Rita later recalled, "the results far surpassed our greatest expectations." Nerve fibers had grown wilder, faster, and bigger than anything before, even when adult mouse gland material had been diluted 10,000 times to the equivalent of one gland in 50 liters of water.

This amazing result finally explained away the problem of the mysterious *mouse effect* that Rita had described to Viktor from Rio de Janeiro, six years earlier. All mouse tissues are rich in NGF and contain enough neurotropic material to enhance nerve growth. Interestingly, had Rita Levi-Montalcini used any tissues other than those of a mouse she would have missed the mouse effect. NGF is found almost exclusively in the mouse; even the rat, a close cousin, has virtually none. In her memoirs, Dr. Levi-Montalcini candidly assessed this episode: "It is a well known though often neglected rule that many apparently unsolvable problems, at one point or another, unexpectedly find their solution in future times."

From the winter of 1958 until the summer of 1959 Stan probed the physical and chemical characteristics of mouse submaxillary gland NGF and in that short time identified it as a protein whose molecular weight and structure was similar to the NGF in snake venom. Rita, too, with larger quantities of NGF from salivary glands to work with carried out numerous experiments to delineate its biological characteristics. She proved that it increased the size of nerve tissues in mice and newborn rats 10 fold after only a few days treatment and it caused a prolific growth of nerves in internal organs and skin tissues.

In December 1958 the budget at Washington University was slashed and Stan's position had to be terminated. But before leaving, between the fall of 1958 and June 1959 he carried out several investigations that still stand out as benchmarks in biochemistry. Knowing that snake venom caused nerve fiber proliferation Stan asked himself, "If snake venom anti-serum was mixed with venom, would the latter's NGF still work?" He then proved that indeed antiserum counteracted and inhibited the NGF in venom. But he did not stop there. He then tested antiserum to mouse salivary NGF on cultures of nerve tissues treated with tumor derived NGF. Again the effects of NGF were blocked. This proved that antiserum could block NGF regardless of its source.

Concurrently, Rita injected this antiserum into her chick embryos and was able to show that it could halt or retard the growth of nerve fibers even in an avian species. Subsequently it was shown that the inhibition or destruction of nerve buds in newborn rodents by antisera had no effect on the normal development of other systems; their fur, skin, body, size, eyes, ears, etc. grew to normal proportions.

Just before his departure from Washington University, Dr. Stanley Cohen made a momentous discovery, which, although serendipitous, could only have been made by an exceptionally astute scientist. He had injected embryonic and newborn mice with some mouse submaxillary gland extracts that had as yet not been completely purified. Upon examination a few days later he noted that the injected animals opened their eyes earlier than the controls. A lesser mind would easily have missed this subtle effect. He doggedly followed up this strange finding with several brilliant experiments. The initial ones proved that the premature separation of the eye lids, one week earlier than in untreated controls, was preceded by a rapid and frenzied growth of the skin-related portions of the eyelids. This must have been induced by some contaminant in his glandular extract. By means of other experiments and lots of biochemical wizardry Dr. Cohen established the fact that the contaminant was an altogether new growth factor, unrelated to NGF.

On a hot, muggy, July evening in 1958, Stan and Rita parted company. As Dr. Stanley Cohen limped over the horizon, pipe delinquently dangling from the corner of his mouth and Smog tramping at his side, Dr. Rita Levi-Montalcini reflected upon his legacy: "Stan, with his magical intuition and flute, played the part of the wizard, charming snakes at will and getting the miraculous fluid to flow forth from the minuscule mouths of mice."

Almost one decade to the day that Dr. Rita Levi-Montalcini discovered NGF, Stanley Cohen, in 1961, by now relocated at Vanderbilt University in Tennessee, reported to the world that he had isolated a protein from the salivary gland of a mouse that hastened the eruption of incisor teeth and the opening of eye lids in a newborn animal. He called this new entity EPIDERMAL GROWTH FACTOR, (EGF), a name it has retained even after it was shown to have powerful effects on connective tissues and being a key player in cancer. Besides adding a new growth factor, this discovery helped to unlock the secrets of their structure. Mouse salivary glands contains lots of epidermal growth factor so that now chemists could get enough to analyze. By 1972 the full amino acid sequence of mouse EGF had been worked out.

In order to fully appreciate Dr. Cohen's wizardry we must understand the laborious and extremely complex processes that were needed to isolate a rare factor from huge quantities of biological material. The job can be compared to sorting through a five-story mound of jelly beans in which 49.9% are red and 49.9% are yellow and separating out the 0.2% that happen to be orange. That is to say, the biochemist had to sort through the bewildering array of molecules normally found in tissue extracts, the reds and yellows, to pick out the few, orange, whose chemical structure resembled everything around them but nonetheless possessed growth factor properties. Growth factors ultimately proved to be peptides. If you visualize a protein as a long chain of molecules then a peptide would be a link in that chain. Although every peptide molecule has its own particular inventory of actions they all look and dress alike—separations and purifications are often next to impossible.

Despite her large coterie of friends and colleagues at Washington University and throughout the United States Rita Levi-Montalcini never lost her love for Italy. She resolved the conflict in loyalties by commuting between continents. In 1961 and 1962, between stints at Washington University, she established The Laboratory of Cell Biology in Rome and later enlarged this nucleus to handle the mechanisms of gene expression and immunology, making it Italy's second largest biological research facility. This move to Italy also allowed Rita to be near her mother who died shortly thereafter, in 1963. Two years later, on a January evening, Professor

Giuseppe Levi, at the age of 92, died. Dr. Rita Levi-Montalcini was at his bedside in the Turin hospital.

The loss of her mother and her intellectual father, Professor Levi, certainly dampened Dr. Levi-Montalcini's spirits for a time but the unsinkable researcher soon returned to the helm of The Laboratory of Cell Biology. It was under her aegis that Vincenzo Bocchini, "an able young Neapolitan chemist with imperturbable calm", as his friends depicted him, perfected the technique for purifying NGF from mouse salivary glands in an absolutely pure form, free from all contaminants. This was accomplished with almost no funding and with rudimentary facilities. Bocchini's purification methods, when expanded by Piero Angeletti, a young Italian doctor from Perugia who had worked at Washington University Medical School and took over biochemistry when Stanley Cohen left, handed Ruth Hogue Angeletti and Ralph Bradshaw the tools that they used to nail down the amino acid sequence of the NGF molecule.

Twelve years later this knowledge of NGF's primary structure led to discovery, in both man and animals, of the exact location on chromosomes of the genes that regulate the production of NGF in cells. It was now possible to synthesize human NGF in large quantities.

Ever since the turn of the century, when embryology became a legitimate branch of science, investigators accepted the fact that cell growth and differentiation were governed by protein molecules found predominantly in the embryo. It followed, naturally, that searches for these substances should be restricted to embryos. But the human embryo was, and still is, hard to come by, and the amounts of material obtained is barely enough to work with. Freedom from these constraints of infinitesimal supplies and embryonic experiments came from advanced genetic engineering and the development of recombinant DNA techniques. Given the means to transfer genetic material from one organism to another, scientists could program cells to produce growth factors in virtually pure form and in relatively larger quantities. Then, with advancements in tissue culture techniques, researchers could grow and maintain, albeit immortalize, cell lines derived directly from mammalian embryos. This not only proved to be a gold mine for growth factors but also provided a model, an alternative to unreliable embryos, where the workings of growth factors could be directly observed.

We now know that growth factors are elements of a complex biological language that provide the means of conveying information from one cell to another. They target and bind with certain unique receptors on the cell surface This then generates second messages which set off cascades of intracellular enzymes. The same signals are reused several times and may elicit very different biological responses in different cell types. Conversely, many distinctly different cell types respond to the same factor with fundamentally different consequences. It is this cross-talk between cells in the human embryo that tells them to make only one head, ten fingers, and so on. Our genes determine which and how much growth factor we make and provide a molecular basis for certain growth factor responses, a vital function that enables cells to respond to changes in their microenvironment.

Throughout our lives, cells die and have to be replaced. The lining of the intestinal tract, blood cells, and skin cells are shed and replaced constantly, and the cells in the lining of the nose require replacement every three weeks. Growth factors direct this traffic.

Wound healing, fracture repair, and the formation of tumor cells actually duplicate events that occur in the normal course of development of an embryo. Excessive growth stimulation could goad normal cells into a cancerous state. Indeed, malignant brain and skin cancers are made up of cells that have amplified copies of certain growth factor receptors and their respective genes.

Scientists have recently produced two types of antagonists, antibodies and receptors, that effectively block the effects of growth factors. Antibodies, hog-tie or engulf specific growth factors. Receptors, man-made chemical molecules 'look and feel' like those on cell surfaces, attract and bind growth factors; both prevent growth factors from fulfilling their missions. Receptors for nerve growth factor were identified at Stanford in 1979 and later, between 1986 and 1987, at Stanford and at Cornell Universities, two other receptors were cloned to organisms so that they could be manufactured through genetic engineering. The selective use of antagonists offers hope against cancer and provides a new alternative in the treatment of serious and blinding eye conditions.

Within the past decade molecular biologists have identified several large families of polypeptide growth factors, and, like mushrooms after a rain, structurally distinct new ones keep cropping up. Here are a few representative growth factors that may reshape the future of medicine.

- FIBROBLAST GROWTH FACTOR (FGF). Fibroblasts make the substances that hold cells together therefore FGF governs the making of the biological equivalents of glue, twine, and rubber bands, in organs and tissues. It is destined to find a role in the repair of wounds, promoting the healing of chronic sores, and providing a frame-work upon which tissues can be built. We already have proof that it can stimulate the repair of the cornea, the lens, and the retina of the eye. In combination with other growth factors, FGF prevented photoreceptor degeneration and blindness in an inherited type of retinal malformation in the rat and increased the rate of survival of photoreceptors after argon laser treatments.

- INSULIN-LIKE GROWTH FACTOR (IGF), has the ability to activate the formation of new blood vessels and hastens the repair and healing of wounds. Increased levels of IGF were found in the eyes of diabetic patients and this, it is believed, is the main cause of retinal detachment and loss of vision in diabetics. Researchers have developed receptor-blockers and antibodies that attack and neutralize excessive IGF, hopefully to head-off such catastrophes. IGF is being

used experimentally to restore sight by regenerating rods and cones, the photoreceptors in the retina.
- EPIDERMAL GROWTH FACTOR (EGF), accelerates healing and the repair of surgical wounds. It is present in tears and plays a role in replacing surface corneal cells, an important part of keeping the eye healthy. EGF and some members of the following TGF family show promise in assisting the healing of full thickness wounds and chronic ulcerations in the eyes of diabetics.
- TRANSFORMING GROWTH FACTORS (TGF), is a fairly large, unique, super-family of growth factors, which, because of size and broad range of functions, has been divided into alpha (a) and beta (b), and these further subdivided into 1 and 2. The TGF-betas are the master growth factors. They can trigger or inhibit cell division and can start or stop the differentiation of tissues. Their main function, however, is to lay down matrix, the web-like material to which cells become attached in order to form tissues and organs. They also control hormone secretion, immune function, and cellular differentiation. TGF beta can juggle genes so as to code for other polypeptide growth factors.
- NERVE GROWTH FACTOR (NGF), is the simplest of all the factors. It and its relatives, called neurotropic factors, keep brain and spinal cord neurons alive during development. Conversely, a reduction or absence of these factors may be the cause behind degenerative nerve diseases like Alzheimers Disease, Parkinson's Disease, and Lew Gehrig Disease (Amyotrophic Lateral Sclerosis, ALS). Dr. Eric Shooter (see Figure 11.1.), Professor of Neurobiology at Stanford University Medical Center, has shown that neurotrophic factors can rescue nerve cells affected by these three diseases and they may provide the means for a future cure. His research team has also demonstrated that rats who suffered from impaired short term memory due to brain damage regained this function after a one month infusion of nerve growth factor. Ciliary Neurotrophic Factor (CNTF), a relative of NGF, seems to ameliorate ALS and Alzheimer's Disease.

Figure 11.1 Eric Shooter, M.D. (Photo by Kathleen M. Podolsky.)

Right now we see 10,000 Americans paralyzed from spinal cord injuries every year. Once injured or damaged, cells of the mammalian central nervous system cannot regenerate themselves; electrical nerve signals cannot be transmitted, muscles stop working and there is a loss of all sensation. Hope for a cure surfaced about ten years ago when a group of researchers led by Albert Aguayo at McGill University in Montreal proved to the world that rat spinal cord axons, the long conducting part of a nerve cell, could be made to regrow. Now, NGF, by virtue of its ability to promote the healing and regeneration of nervous system tissues, promises to give us the tools necessary to finally treat such injuries.

There are about 60 other growth factors with names like brain derived growth factor, eye derived growth factor, and retina derived growth factor whose functions are still under investigation. The field has grown so fast that each family of growth factors has its own select enclave of specialists, with its own argot and its own group of sub-specialists. When growth factor people get together at international meetings they more or less understand what is being said but each group speaks its own particular language.

Growth factors still await government approval for human applications. Once this happens they will be impressed into use most quickly by neurologists, as noted above, and by ophthalmologists for eye conditions that are presently untreatable or result in blindness. Indeed, several medical institutions have reported successes, both in animals and humans, in the management of today's most frustrating eye conditions:

- CORNEAL WOUNDS: Several teams have already reported that epidermal growth factor, which is normally secreted in tears, promoted healing of persistent corneal erosions when applied as eye drops. Corneal grafts treated with EGF achieve a higher tensile strength several days earlier than untreated grafts.
- PROLIFERATIVE VITREORETINOPATHY (PVR): is really an "over-healing" disease. Following retinal reattachment surgery the healing process generates so much scar tissue that it disrupts the entire eye and ultimately causes blind-

ness. There is a three to four fold increase in TGF-beta in PVR eyes over normal eyes implying that scarring results from excessive growth factor production or increased sensitivity of the cells. Receptors and antibodies for TGF-beta are being tried to limit the scarring that follows an insult to the eye.
- RETINAL WOUND, RETINAL BREAK, MACULAR HOLE and RETINAL DETACHMENT: Growth factors have successfully glued back the retina to its base and receptors have prevented or limited unwanted scar tissue. IGF is being used experimentally to restore sight by regenerating rods and cones, the photoreceptors in the retina.
- SECONDARY OR COMPLICATED CATARACTS: TGF and the receptors of TGF and IGF, by virtue of their ability to interfere with scarring could prevent the cascade of events leading to lens opacification. TGF-b2, acting as an immune suppressant might, at best halt uveitis, at least, prevent secondary cataracts after uveitis.
- PROLIFERATIVE DIABETIC RETINOPATHY (PDR): TGF-beta has already patched up holes in the human retina and it has shown great promise as a chorioretinal glue, that is to say, it can reattach the retina to its base without any of the drawbacks connected to lasers. Fibroblast growth factor and insulin-like growth factors also have reduced scar tissues in the eye and promoted healing.
- AGE RELATED MACULAR DEGENERATION (AMD) also called SENILE MACULAR DEGENERATION, a major disease of the elderly, is the leading cause of new blindness in the United States. TGF-beta has the potential to control the natural course of AMD, stopping excessive capillary growth and scar tissue, while speeding up retinal repair. Studies to date indicate that the receptors of IGF and TGF-beta, might prove useful in preserving normal retinal and corneal tissues; they could also serve as adjuncts to laser surgery to limit or control excessive scarring.

Besides nerve regeneration and eye pathology, growth factors will also prove invaluable in bone conditions. Not only will they help to heal fractures and add or replace bone, they show great promise in artificial joint replacements. Studies are now going on where prosthetic joints are coated with growth factors so as to trigger the growth of natural bone around them after being implanted.

The sophisticated biotechnology at our command has given us a battery of pure growth factors and specific antibodies and receptors to make them do our bidding. As Dr. Michael B. Sporn, at the National Cancer Institute, so aptly stated, "one can conceive of totally new, and essentially unexplored therapeutic possibilities." After we learn everything about their potentials and limitations, figure out correct dosages and delivery pathways, and test them on humans in clinical situations, we will give the gift of sight to countless people, hand doctors decisive weapons against cancer and paralysis, and provide treatments for many other conditions that are currently beyond help or hope.

PARADIGM POINTERS

Big breakthroughs often hinged upon the lucky choice of a particular life-form for experimentation. THE MOUSE EFFECT would have been missed completely if, let's say, rats or guinea pigs had been used in Dr. Rita Levi-Montalcini's initial experiments. Dr. Gerhard Domagk might have missed discovering the bacterial-killing properties of sulfanilamide had he used a different strain of streptococcus or different test animals. Ehrlich and Heidelberger missed sulfanilamide and Fleming missed penicillin because of the use of inappropriate experimental subjects or procedures.

Perhaps the most compelling story in regard to choosing the right experimental model deals with Dr. Friedreich Stertürner (1783-1841), the German chemist who first separated the components of crude opium extracts. In 1803, after having isolated morphine in its pure form, he tested it on dogs and proved that, indeed, it had the same sedating effects that it had in humans. Had he tested this material in cats, a species in which morphine is a stimulant, it is more than likely he would have failed to recognize the value of his discovery.

PERSONAE

Dr. Levi-Montalcini started as a research associate at Washington University in 1947 and eleven years later was appointed full professor there. In 1977, at the age of 68, she retired from the University with the title of Emeritus Professor. Two years later she gave up the directorship of her brainchild, The Laboratory of Cell Biology, in Rome, but continued to work there, as a guest.

At this writing Dr. Levi-Montalcini is still active in Italy's scientific circles where she continues to teach, guide, and explore. She is president of the Italian Multiple Sclerosis Association and is a member of the Pontifical Academy of Sciences. Drs. Stanley Cohen and Rita Levi-Montalcini shared the 1986 Nobel Prize in medicine, she being only the fourth woman to ever become a Nobelist.

Dr. Stanley Cohen was born in Brooklyn, New York, in 1922. He majored in Biology and Chemistry at Brooklyn College, received his M.A. from Oberlin College, and his Ph.D. in biochemistry from the University of Michigan. He worked in pediatrics under Dr. Harry Gordon at the University of Colorado, then in 1952, went to Washington University to learn about radioisotope methodology under Martin Kamen. While the foregoing narrative tells much about his quintessential role in uncovering growth factors and unraveling their biochemical characteristics, it nonetheless stops short of acknowledging his many contributions to science. His most recent curriculum vitae lists 126 scientific papers and two full pages of honors, awards, and memberships in prestigious scientific bodies. Currently he is Distinguished Professor, Vanderbilt University School of Medicine, Nashville.

Viktor Hamburger came to St. Louis' Washington University in 1935, because, as a Jew, he could not return to Germany after the termination of his Rockefeller Foundation fellowship at the University of Chicago. He served as chairman of Washington University's Zoology department from 1941 to 1966. His international fame was established in 1951 when he published his seminal classification system for embryology. He made many false stabs at retiring but finally closed his laboratory in 1983, at which time he was the Edward Mallinckrodt Distinguished Professor Emeritus in Biology. He received numerous national and international honors for his contributions to experimental embryology, his most recent award, in 1989, was the National Medal of Science, which was given to him by President George Bush at a ceremony at the White House.

Eric Shooter M.D. is currently Professor of Neurobiology at Stanford University Medical Center, having served as chairman of the department of neurobiology from 1975 to 1987.

Just as Nikola Tesla carried electricity beyond the wildest dreams of Volta and Faraday, Dr. Eric Shooter carried NGF beyond all limits envisioned by Rita Levi-Montalcini and Stanley Cohen. In an interview with him in his laboratory at Stanford University Medical Center on August 24, 1995, he gave these personal insights into NGF and his line of research:

Dr. Shooter's career started out modestly in the late 1940's at Cambridge University where he analyzed the proteins in peanuts for his Ph.D. in biochemistry. He then moved to University College in London where he studied the mutations of hemoglobin, the oxygen carrying material in our red blood cells. Needing a better understanding of the emerging science of genetics he took a year-long sabbatical in 1961 which brought him to the United States and Stanford University.

Here he met Nobel Prize laureate Joshua Lederberg. "We started to talk about biochemistry and that is how I became interested in neurobiology." Dr. Shooter did not come to Stanford seeking a permanent change but the attraction of the genetics department Lederberg was building made it difficult to refuse an offer to stay. He returned, permanently, to Stanford in 1964.

Dr. Shooter plunged into biochemistry of the brain and this brought him face to face with the nerve growth factor. "I was introduced to NGF by Silvio Varon in 1964. He came to Stanford from Washington University where he collaborated with Rita Levi-Montalcini and Viktor Hamburger. He was familiar with the work of Stanley Cohen who had shown NGF to be a protein.

"Varon, Junichi Nomura and I set out to purify NGF from the mouse salivary gland but our bioassay procedures were tedious. Only 5% of the mouse salivary gland has nerve growth factor in it. Soon we realized we weren't getting anywhere; we had to get new analytical tools to assess the purity of NGF.

"Quite unexpectedly a young post-doc in Arthur Kornberg's lab came to our help. He was Tom Jovin now a scientist at Max Plank Institute in Goettingen. Tom set up a computer program for us to produce a new series of conditions for electrophoreses, a technique for detecting the many proteins in a given sample. It took us two years, all of 1966 and 1967, but that break allowed us to focus on the one protein that had the biological activity of NGF. Jovin permitted us to successfully isolate the NGF protein and open a whole new field of study. NGF has been purified, crystallized and its amino acid sequence has been determined."

A cell must have a receptor on its surface in order to attract NGF to it. The first such receptors for nerve growth factors were identified at Stanford in 1979. NGF receptors were cloned to other organisms at Stanford and at Cornell Universities between the years 1986 and 1987, and later, in 1991, additional receptors were cloned by Bristol-Meyers-Squibb pharmaceutical company and the National Cancer

Institute. Subsequently all of these places made antibodies against nerve growth factor. By carefully programming the use of NGF and receptors scientists hope to reverse neurodegenerative diseases.

"In 1991 we isolated and cloned brain derived neurotrophic factor, BDNF. We now have several neurotrophins which are proteins. They won't pass through the blood-brain barrier therefore they cannot be administered to humans as a pill. The plan is to find very small molecules that will go through the barrier that can imitate the neurotrophins and thus fool the body into believing that there are neurotrophins present. Alzheimers Disease, Parkinson's Disease and ALS are degenerative diseases with no known cures; the processes seem to be the same in all of them where the nerve cells just seem to age and degenerate. We have seen some favorable responses to combinations of nerve growth factors and neurotrophins. In ALS we have stopped degeneration of the muscles of the diaphragm and extended the life of these patients because they can now breathe.

"BDNF will rescue retinal ganglion cells and photoreceptor cells. It prevents blindness in laboratory animals and protects them from free radicals. The problem with the eyes, however, is getting these growth factors into the eye, because any method you use to get it into the retina or photoreceptor cells produces scars in the eyes.

Kaz Kagiwai, one of our Fellows, is now working with the Japanese newt which can generate a whole new eye. So far he has been able to lift out the retina or cut the optic nerve and follow the eye as it degenerates to the precursor cells, then it regenerates itself in six weeks. We have a whole eye with an optic nerve. Now if we can only adopt this to humans ....

"Spinal cord injury and nerve injury can be helped by growth factors. In some experiments where NGF and neurotrophins were injected into muscles, neurons grew to them. If you cut the sciatic nerve in a rat it will walk again in six weeks, probably not perfectly, but it will be able to walk again. Nerves go to the first muscles they see, that is, in peripheral nerve injury. NT3 will get the spinal cord to heal; 7% of the fibers will regrow back.

"The American Paralysis Association has just assigned a large block of money to a spinal cord injury consortium, a group of scientists who are going to focus on how to apply nerve growth factors to repairing spinal cord injuries. We already know that we can slow the rate of loss by 30% with what we have now. With future experiments we hope to allow for regeneration."

chapter 12

BING, BANG AND LEVIN AND THE HORSESHOE CRAB

The horseshoe crab is a living fossil. None of its original, primitive, components ever changed with time. The workings of their rudimentary systems are relatively easy to understand and discoveries at this basic level often provide valuable information about humans. The crab's elementary photoreceptors, for example, are equivalent to eyes in higher vertebrates, and from them, researchers are finding out how the complex human visual system operates. This is why scientists choose the horseshoe crab for experiments.

Recently, Dr. Robert Barlow, a neurobiologist from Syracuse University's Institute for Sensory Research explained why he is using light-amplifying video cameras to record every facet of the horseshoe crab's life cycle, both in and out of the water. "We're studying vision but the real goal is understanding how the brain works. We're really looking for the neuro basis of behavior." Amidst the tangled wires from portable television monitoring equipment he adds, "Now, near the end of May, the water temperature has crept up a few degrees and the new moon brings hundreds of these animals to the shallow waters off Mashnee Dike where they are engaged in their ancient mating ritual. That's when we do our work." By *work* he is referring to his gathering of data on how the eye sees things then transmits the information to the brain.

In 1977 Barlow and two of his graduate students severed the optic nerve in the crab's lateral eye and continued to measure the eye's sensitivity to light. The eye stopped making rhythmic changes and became a static organ. They then recorded the neural activity from the part of the severed nerve that was still connected to the brain. When they attached an electrode to the end of the nerve still attached to the eye and played the neural activity recorded from the brain, the eye converted to its highly sensitive night time state. The experiment showed that electrical impulses from the brain were responsible for the daily changes in the eye. Barlow and his team have worked out methods for recording impulses from a single optic nerve fiber. And from all of these studies they have learnt much about the human eye and its communications with the human brain.

This is but an isolated example of the many contributions that the horseshoe crab makes in improving the welfare of the human race. Its most fascinating contribution, however, comes from its royal blue copper-containing blood, more exactly, *hemolymph*. It has given us *Limulus Amebocyte Lysate*, commonly called LAL. This remarkable substance, discovered in 1964, can identify infinitesimal amounts of bacterial endotoxins, those powerful and often lethal poisons made by many germs. LAL can detect an endotoxin even when a teaspoon-full of it is dropped into the Great Lakes. To appreciate the magnitude and implications of this discovery we have to take into account the contributions of doctors Bing, Bang and Levin, who, at one time or another, used the horseshoe crab in their experiments.

Richard J. Bing MD (see Figure 12.1.), is an anachronism. He is just as comfortable with a stethoscope in his ears at an ailing patient's bedside as he is with strain-gauges and

Figure 12.1 Richard J. Bing, M.D.

sophisticated research equipment on a laboratory workbench. He is one of the last of that rare and rapidly disappearing breed of physician-scientists who find themselves equally disposed to carry on a full schedule of research while still practicing clinical medicine, administering an academic medical faculty, fighting for funding, teaching medical students, heading an international research society, and editing a prestigious medical journal. After 60 years of such intense activity he wistfully looks back to recount his many brilliant successes as well as his one 'near-miss'. The latter happened in 1936 when he was recruited by Alexis Carrel and Charles Lindbergh to work with them at Rockefeller Institute for Medical Research, now Rockefeller University, in New York City.

In its heyday, between World Wars I and II, Rockefeller Institute was the most preeminent research facility in the world. Sinclair Lewis, in his popular novel ARROWSMITH, used Rockefeller Institute, under the pseudonym of McGurk Institute, as the paradigm of a superbly efficient medical research compound staffed by extraordinary, talented and dedicated scientists. The atmosphere within Rockefeller's hallowed walls was always electric. Working more or less like atoms in a chain reaction, where one atom strikes another and imparts its energy to ultimately set off a massive explosion, its cadre of brilliant scientists bounced ideas and information off each other in such profusion as to ignite the imagination and inspire breakthroughs that astounded the rest of the world.

Among Rockefeller Institute's super-stars was Dr. Alexis Carrel, one of a handful of surgeons ever to receive a Nobel Prize. He was a skilled and daring craftsman who pioneered the techniques for sewing blood vessels together. He could cut out organs and then successfully sew them back into place. The first experimental coronary bypass operation was performed by Dr. Carrel as far back as 1912.

Carrel was fascinated by the prospect of keeping organs alive outside of the body, and to do this, he devised glass containers in which organs could be seen and kept germ-free while nutrients and oxygen were circulated through the tissues. Although he was an adept surgeon and a keen-thinking physiologist he realized that he did not have enough mechanical ability to assemble the tubes, pumps, valves, and gadgets, so necessary for his circulation apparatus. In his search for someone with superior technical and intellectual abilities he met and then teamed up with Charles A. Lindbergh, America's Lone Eagle. Lindbergh was extremely receptive to such an alliance because he had a relative who suffered from diseased heart valves and felt that working alongside Carrel might lead to a cure.

The match proved perfect. Carrel was cocky, strong-willed or obnoxious, depending on how you interpreted his directness, and his imperious manner made it easy for most people to dislike him. Lindbergh, a tall Midwesterner with a genial approach to people was immediately welcomed into the fold. He had the ability to get along with glass-blowers, bottle washers, and laboratory technicians. When Carrel exploded with one of his tantrums it was Lindbergh's charm and compassion that re-established harmony. Lindbergh's mind 'never slept'. He proposed methods for operating on a patient's bloodless heart and went so far as to suggest that a cardio-pulmonary bypass would be feasible. All this happened two decades before the 1946 introduction of open-heart surgery by Dr. Charles Gibbons.

In collaboration with Carrel, Lindbergh designed equipment and devices that circulated special blood-like fluids through tissues and organs that had been totally removed from experimental animals. By being kept alive, yet under easy observation, they could be studied without interference by extraneous factors. The fluid in the Carrel-Lindbergh apparatus contained salt, in the same concentration as that found in human blood, fortified with glucose and vitamins. It flowed from a reservoir into the artery of an organ, through the organ, then out through a vein, and finally back to the reservoir again. The unusual feature of this system was that the rate of pulsation and hydraulic (blood) pressure could be regulated, and sterility was easily monitored. Organs were 'kept alive' with this artificial circulation for weeks and studies on the isolated heart contributed greatly to our current understanding of rate and rhythm disorders, heart failure, transplants, cardiac nutrition, and physiology. When Carrel and Lindbergh published their results they handed other researchers the key to the interplay between living cells and the fluids that nurture them.

But the Carrel-Lindbergh circulation and perfusion system always failed. Sooner or later the tissues died. It was not the nutritional qualities in the fluid, waste products, or infections, that killed them; pure and simple they ran out of oxygen. This lack of oxygen-carrying capacity in the system also limited the Carrel-Lindbergh apparatus to very small organs. Both men were keenly aware of this shortcoming and were always on the lookout for ways and means of getting past this roadblock. During this period, both men happened to visit Carlsberg Biological Institute in Copenhagen, Denmark, where they met the eager, young, highly intelligent physician-scientist, Richard J. Bing, who was engaged in advanced research with an animal tumor virus in cell cultures. They intuitively sensed that here was talent that they and Rockefeller Institute could use. Dr. Bing, had been educated in Germany, picked up additional training at Bern University Hospital in Switzerland and had a good insight into complexities related to oxygen use by cells. Yes, he would come in handy.

Upon learning that Carrel had obtained a Rockefeller stipend for him, Bing quit Copenhagen, stopped off for a brief visit with Lindbergh at his home in Seven Oaks England, and then proceeded on to New York. He started to work on tissue and organ cultures in Rockefeller Institute in 1936. "Youth needs heroes and there were plenty of them at Rockefeller Institute," Dr. Bing recalls. Working around him in that 'golden age of science' were his heroes: Carl Landsteiner MD, the Nobel Laureate who discovered blood types, Peyton Rous, MD, the man who discovered the Rous Virus (chicken virus) and at age 80 received the Nobel prize, and Oswald

Avery, one the foremost biochemists of that time. Surrounded and inspired by such luminaries, Bing set out to find a way of oxygenating the Carrel-Lindbergh fluids so that bigger organs could be studied and tissues could be kept alive longer.

Human red blood cells contain hemoglobin, an efficient carrier of oxygen to the tissues, so Bing tried this first. It did not work. Once it left the body it turned into methemoglobin, a substance unable to carry oxygen. The relentless scientist had read somewhere that lower animals had unique oxygen carrying capacities and remembered that one of nature's oldest creatures, horseshoe crabs, *Limulidae,* used a beautiful copper-containing blue respiratory pigment to carry oxygen to its cells. Bing collected his horseshoe crabs, harvested the respiratory pigment, *hemocyanin,* and then tried it in the Lindbergh-Carrel artificial circulation apparatus. It too did not work. The horseshoe crab is a slow-moving crustacean that lives in cold ocean waters; it can get by on extremely low oxygen levels consistent with minimal activity. Hemocyanin works well in an animal surrounded by a naturally cold environment but it proved useless in warm blooded high oxygen users.

Dr. Bing never realized how close he had come to finding LAL, Limulus Amebocyte Lysate. His failure to grasp the significance of this material, even though it was in his hands and before his eyes, is not unduly remarkable because when Frederik Bang and Jack Levin did find it, it was quite by accident. Bing's greatness was in no way diminished by this oversight. He went on to become one of the foremost research physiologists of this century. By studying the effects of chemicals injected directly into the artery of the kidney, he found out how and why high blood pressure arises in humans. Then he proceeded to devise methods and find drugs that would control, even roll back, abnormally high blood pressures.

It was Bing who grasped the baton from Werner Forssmann and carried it beyond all expectations. Forssmann, in 1929, inserted a long flexible tube through an arm vein into his own heart, injected a dye, and recorded the event with X-rays.

"Weren't you scared?" people asked.

"I must confess that I was slightly nervous," he replied. Dr. Forssmann received the Nobel Prize in Medicine in 1956.

Following up on this lead, Bing and a few other top-notch scientists discovered ways of passing a catheter from blood vessels in an arm, leg, or neck, into the heart so that the heart itself, its valves, and chambers could be studied in detail. Indeed, he pioneered the way for the use of catheters to diagnose, with precision, birth-related heart defects in children. Bing also discovered that afflicted hearts use nourishment and oxygen differently than their normal, healthy counterparts.

Providence compensated Dr. Bing for 'missing the boat' with the horseshoe crab by handing him a lucky break in an entirely different endeavor, namely, in visualizing the coronary arteries in the heart. In a paper published with his associates in 1947 Dr. Bing wrote, "When the catheter is in the coronary sinus, it is seen curved upward toward the base of the heart ..... In the first five cases, intubation was fortuitous .... In the remaining four cases catheterization of the sinus was carried out deliberately." What Dr. Bing is modestly telling the world is that while he was experimenting with passing catheters into the chambers of the heart some of those catheters accidentally ended up in the blood vessels that nourish the heart, the coronary arteries. But that was only as far as 'luck' would take him. It was Bing's acute perceptiveness of these events and their implied potentials that led him to change directions, to concentrate on finding methods for getting catheters and test equipment directly and precisely to the blood vessels of the heart and thereby giving us the vital information about the heart's circulation and nutrition that is saving lives today.

The bizarre circumstances that led to the discovery of LAL bring to mind the motion picture **Annie Hall** in which Diane Keaton, watching the eruption of a herd of clowns from a tightly packed Volkswagen exclaims, "Gee. I wonder how they do that!" A slouched, bespectacled Woody Allen replies "I wonder why they do it." In the same context we might ask

Figure 12.2 Jack Levin, M.D. (Photo by Kathleen M. Podolsky.)

"Who, in his or her right mind, would spend inordinate parts of a lifetime experimenting with the ooze from a horseshoe crab?"

Jack Levin MD (see Figure 12.2), a research specialist with the University of California School of Medicine in San Francisco, did just that. His main interest, the mechanics of blood coagulation, directed him, as early as 1960, to the Marine Biological Laboratories, MBL, Woods Hole, Massachusetts where he would have ready access to the simplified physiology of the horseshoe crab. Although California is replete with university think-tanks and superior research facilities, Jack Levin journeyed clear across the country to MBL because, as he put it, "I wanted to be close to the source of the horseshoe crab but more importantly this is the place where serious scientists go to find the stimulation and encouragement so necessary for research."

MBL is to biology as Rockefeller University is to medicine. Both are human institutions possessed of a life of their own. MBL was chartered in 1888 but it actually started earlier, in 1871, when Woods Hole in Massachusetts was selected for a bureau of Fisheries station. It is located just off-shore where the gulf stream collides with northern currents, where all sorts of marine and estuarine life congregate. When word got out about the fecund waters and plethora of species at Woods Hole, all the big brains from Harvard University and Massachusetts Institute of Technology migrated down to it.

Among its many amenities is Stoney Beach, a hop, skip and a jump from the laboratories, where investigators bolt down their lunch while criss-crossing the sands with ordinates, abscissas, and curves to account for everything in nature. And when biologists confront and confound the pure physicists from the National Academy summer headquarters nearby, a cacophony of raised voices drowns out the crashing surf. Dr. Lewis Thomas likes to recount the electricity that usually accompanied the Friday evening lectures, MBL's weekly grand occasion when guest lecturers present their most stunning pieces of science: "As the audience flows out of the auditorium, there is the same jubilant descant, the great sound of crowded people explaining things to each other as fast as their minds will work. You can not make out individual words in the mass except that the recurrent phrase, "but look —" keeps bobbing above the surf of language."

It was at MBL that the invertebrate eye was so extensively studied. The giant axon of the Woods Hole squid became the standard for understanding neurobiology. Sea-urchin eggs provided the keys to understanding developmental and reproductive biology. And long before the rest of the world became aware of the pejorative changes to our atmosphere and ecology, MBL was busy cataloguing these assaults and exhorting man and industry to 'stop fouling the nest'. MBL's prominence in science continues to grow. New talents, attracted by notable investigators, make astounding discoveries, which attract more new talent—repeating the cycle to the point where spectacular discoveries are the rule rather than the exception. Today, MBL together with its neighbor, Woods Hole Oceanographic Institute, are recognized as the uniquely national centers for biology in this country, the equivalent of a National Biological Laboratory.

"If you ask around" Lewis Thomas tells us in his book THE LIVES OF A CELL, "you will find that any number of today's leading figures in biology and medicine were informally ushered into their careers by the summer course in physiology; a still greater number picked up this or that idea for their key experiments while spending time as summer visitors in the laboratories, and others simply came for a holiday and got enough good notions to keep their laboratories back home busy for a full year." At one time or another thirty Nobel Laureates worked at MBL.

Blue-blooded horseshoe crabs, Limulidae, so essential to this saga, belong to an ancient family of sea animals that have remained virtually unchanged for almost 300 million years. They are not related to any other creature in the sea and are believed to be the only living links to fossilized trilobites. Land spiders and the scorpions are their closest relatives. There are four species. *Limulus polyphemus*, the species that concerns us, is found along the eastern coast of North and Central America, from Maine to Yucatan. The three other species are distributed over the Indo-Pacific areas. Although the horseshoe crab is often called King Crab, it is not really a true crab. Its outer shell is not brittle like that of a crab but is flexible, rather tough, and hornlike.

In the spring and early summer, *Limulus polyphemus* migrates to the beaches along the eastern seaboard of the United States to mate and lay eggs. Males have specialized claws, claspers, on the end of the first large pair of legs. These look like a ball with a small hook on them and are used in the mating ritual. It is not uncommon to see female crabs dragging the smaller males behind them. Female crabs instinctively wait for a full moon so that they can lay their eggs in 'nests' or clusters well above the high-tide line of the beach. About 3,600 eggs are deposited in these nests dug in soft sand and the attached male fertilizes them while he is dragged over the nest. The eggs then incubate in the spawning beach's warm sand for about two weeks then, with the help of wave action, the young larvae break out of their membranous egg cases. These feeble swimmers soon settle down to the bottom of the shallows or the mud flats not far from where they were conceived. Both eggs and larvae of the horseshoe crab provide fish and birds with a rich source of proteins, making them a vital link in the complex food-chain of the coastal ecology.

Horseshoe crabs feed on young oysters, clams, mussels, and bottom-dwelling worms. Its penchant for shellfish makes it an enemy of clammers and oystermen. *Limulus polyphemus* navigates easily through the water to find prey. Many sensitive nerve sites on its body help to locate and identify prey. Food is grasped by the crab's pincer-like legs and then moved to the 'crusher legs' near its mouth. Everything is first crushed then swallowed. Shellfish are eaten completely and shells are later purged.

The horseshoe crab has two pairs of eyes that make it aware of its environment. One set, called lateral eyes, located behind and on the sides of the crab's head are composed of many transparent, lens-like facets, very much like an insect eye. The other smaller set of eyes, the median eyes, are used primarily for detecting light.

A crab must molt, shed its shell, sixteen or more times during a lifetime in order to grow. Crabs may be 25% larger with each shell change. The crab's shell is an eco-system unto itself. Mussels, oysters, oyster drills, assorted snails, worms, barnacles, protozoans and several species of algae call this their home and universe.

Even though they are inedible, horseshoe crabs have been exploited mercilessly. In the 1920's they were harvested from Delaware Bay by the tens of thousands and piled high on the beaches before being taken to factories where they were ground up into meal and sold as fertilizer. This industry thrived for over 25 years until chemical fertilizers took over the market. The shell of the horseshoe crab still has a commercial value. It is composed primarily of chitin, a white, somewhat flaky substance, similar to plant cellulose, that is used as a filter for cigarettes, a thickener for ice cream and a base for facial make-up. It can be made into a film that acts like a natural skin covering for wounds and surgical incisions. Some studies indicate that it actually speeds the healing of wounds.

The blood or hemolymph of the horseshoe crab contains only one type of circulating blood cell, an amebocyte, whose disc-like shape resembles a human platelet. Amebocytes contain hemocyanin, the blue, copper containing protein that carries oxygen to the tissues, just as hemoglobin does in humans. These cells are also full of cytoplasmic granules which are little more than packages of coagulation substances. They are mainly an ancient infection fighting tool. Should bacteria ever invade a horseshoe crab, through a wound or other source of infection, amebocytes would be sent to fight them. Granules released from the amebocyte's cytoplasm would effectively imprison germs in an encircling gel or clot and then they would be destroyed .

Now we can address the question posed earlier. Why would anyone want to meddle with the ooze from a horseshoe crab? Simply put, Dr. Jack Levin was interested in the mechanics of blood clotting so it was just as natural for him to study the primitive blood of the horseshoe crab as it was for neurobiologists to study its simple eyes and brain. In the early 1960's we find Dr. Levin gathering up his usual quota of horseshoe crabs, then, to reach their minute blood containing channels for blood samples, he masters the art of fine dissection. Once this was accomplished , blood was drawn into small test tubes which contained the same anticlotting substances that we use to prevent clotting in human blood held for transfusions or laboratory tests. But this did not work. As Dr. Levin put it, "Initial studies of blood coagulation of the horseshoe crab were hampered because samples of blood, which were initially liquid, became solidly clotted by the next morning, despite the presence of various anticoagulants."

The tricky, if not exasperating part of research is finding ways to overcome roadblocks. One course of action around this annoying tendency of the horseshoe crab's blood to clot would be to try and find some other chemicals, not necessarily those used in humans, that would prevent clotting and then go on with the studies. Another alternative, the one followed by Jack Levin, was to sit back and say, "Hey! If there are anticoagulants already in the tubes and blood is still clotting, something else must be doing it." Numerous possibilities flashed through the scientist's mind. Could it be the shaking of the tube? Could it be light? Could it be heat? Could it be temperature? Could it be the contact with the metal in the needle or the rubber in the stopper of the tubes? Or could there be trace elements like detergents on the insides of the tubes? All of these possibilities were examined and tested but in the final analysis Jack Levin concluded that the cause had to be bacterial contamination. He said, "I considered the possibility of bacterial contamination of the blood specimens, and when samples of blood were collected into sterile and absolutely clean containers the blood remained liquid. It was then possible to demonstrate that the coagulation system, or at least some of its components leaked from amebocytes, the only type of circulating blood in L. Polyphemus, and that plasma which contained cellular components could be jelled..." In other words Dr. Levin deduced correctly that the blood of a horseshoe crab clotted in the presence of bacteria. He set out to find out how this worked.

Dr. Levin teamed up with Dr. Frederik Bang, also at Woods Hole, who shared similar interests in *L. Polyphemus* and its exotic blue hemolymph. The two scientists separated the cells from the liquid portion of the hemolymph by spinning it at high speed in a centrifuge. The lighter plasma stayed on top and the heavier amebocytes compacted on the bottom of the centrifuge tubes. Next, they took the crab's plasma, completely free of all cells, mixed it with bacteria, and observed that nothing happened. They then placed a few of the amebocytes in the clear plasma, added some bacteria, and eureka! the mass turned into jelly. The clotting substance was present only in granules within amebocytes. In the final step the amebocytes were *lysed*, that is they were forced to swell until they exploded and extracts were made from the intracellular granules. When these were added to plasma containing bacteria, the mass clotted and turned into jelly. Inasmuch as the amebocytes were *lysed* to obtain this material it was christened *Limulus Amebocyte Lysate*, or LAL. Subsequent studies proved that bacterial endotoxins triggered the release of LAL and this caused clotting.

Bacteria make two kinds of toxins or poisons which are dangerous, if not lethal, to their hosts. Exotoxins are secreted by bacteria into their surroundings. Diphtheria germs, for example, live in the throat, but their exotoxin gets into the blood and damages the heart and kidneys. Endotoxins are poisons found within bacteria and are only released when the germs are broken down. Humans are exquisitely sensitive to them, so much so, that even the tiniest traces, as contaminants of injectables or medical devices can precipitate acute illnesses and extremely high fevers. Because they are fever-producers these endotoxins have been labeled as *pyrogens*, from the Greek "pyros", meaning fire. Manufacturers of drugs and medical equipment go to great lengths to test for the presence of pyrogens. Products must be destroyed unless all pyrogens can be eliminated. In the past, rabbits were injected with material under test and observed for signs of fever or illness. Often they were sacrificed for post mortem examinations. The LAL test has changed all that. Besides being more sensitive, more economical and easier to use, it enables us to spare animals from the torments of testing. Anything that must be endotoxin-free, such as large and small volume injectables, vaccines, biological products, intravenous fluids, medical devices, and pharmaceutical preparations, presently undergo LAL testing. The test is also used to detect the presence of endotoxins in short-lived radioactive isotopes that are used for diagnostic purposes. LAL is useful for field-testing water and milk supplies. Research laboratories routinely use the LAL test to exclude endotoxins from biologically active reagents; contamination could confound results and generate misleading data. Since its discovery, LAL has found more new uses than ever imagined. Most recently it has been introduced into the routine testing of medical equipment and devices before they enter the market and in detecting contaminants in blood derived products such as plasma protein fractions. Thanks to the LAL test, we no longer need the term pyrogen; it has no real meaning. Henceforth all biological materials will be labeled toxin free.

Doctors Levin and Bang published their first report on LAL in 1964. They then proceeded to show that the reaction was enzymatic in nature and it was initiated by endotoxins, particularly those made by gram negative bacteria (see GRAM, Chapter 8). This finding became increasingly important as new antibiotics came on the scene. Because antibiotics killed bacteria rapidly, greater quantities of endotoxins were released and patients often succumb 'cured of their infection'—a situation not unlike killing a poisonous snake after being bitten by it; the snake may be gone but its poison is still capable of killing.

One of the most promising new uses for the LAL test will be in the detection of endotoxins in body fluids. A great deal of intense research is now going on to see if the test will discover early meningitis and infections in urine, urethral secretions, corneal tissues of the eye, and in blood. In the near future we may be able to scrap all of our complicated, expensive, tests that pin-point infections. Instead we will add a drop of LAL or a related compound to a specimen and identify an infection in seconds.

PARADIGM POINTERS

Paradigms are helpful inasmuch as they act like maps in a strange country—they tell us where we are. But they can also act as deterrent to discovery by fixing our focus on the map rather than on the landscape.

For instance a person gazing at a window can focus on the reflection in the glass or the vista far beyond, both images being miles apart. In much the same way, scientists, not infrequently, focus on some distant objective and miss the great discovery in front of them. Dr. Bing did not see LAL because his mind was concentrating on oxygen mechanics rather than on immunity. And, as noted in Chapter 1, both Gelmo and Heidelberger who worked with sulfanilamide for two decades before Domagk, failed to recognize its germ killing properties. Abel, Rowntree and Turner discovered everything necessary for the artificial kidney as early as 1914 but this went unnoticed until 1943 when Kolff applied this knowledge to man. And Waksman, even though he did discover streptomycin in 1944, missed it while experimenting with *Streptomyces griseus* throughout 1916, 1919, and again in 1932 and the early 1940s.

POST SCRIPT

1. The way scientists are funded often acts as a serious drawback to innovative research. Here is a case in point. When

Dr. Jack Levin, as a young physician, became interested in hematology and the way that blood cells were manufactured, he received grants from the Atomic Energy Commission. The agency at that time, was primarily interested in the effects of atomic radiation on bleeding; it needed to know how to treat people exposed to radiation. As soon as Dr. Levin's work took him away from bleeding into more basic research the funds were cut off.

2. The greatest hang-up for researchers is the stringent training and indoctrination that goes hand in hand with attaining advanced degrees. It forces people to stay on conventional paths, which, in essence, redefines what is already known. Scientists who respect the boundaries of conformity, albeit an orderly process, tend not to see the incongruity before their eyes whereas the exceptional, unorthodox risk-taker, sees the score differently and recognizes something entirely new. Young undergraduates and medical students who do not have to unlearn the nonsense that sometimes passes for gospel are free to make chance discoveries. They can see what older, structured, scientists have been taught not to see.

James Gleick, in his thought-provoking book CHAOS, gives us a good example of this. He cites an experiment where subjects were shown playing cards with red spades and clubs and black hearts and diamonds. When viewed rapidly and regularly no abnormalities were detected. When the cards were displayed slowly or irregularly, the subjects encountered all sorts of confusion. Scientists, being no more than human, learn to see the conventional, the routine, the normal, and in so doing, often miss the big breakthrough at their finger tips.

PERSONAE

Richard Bing, MD, taught cardiology and physiology to physicians and medical students and carried out research at

Columbia University's College of Physicians and Surgeons, Johns Hopkins Hospital, the University of Alabama School of Medicine, and Washington University in St. Louis. He also served as head of the department of medicine at Wayne State University in Detroit.

He is the director of experimental cardiology and scientific development at Huntington Medical Research Institutes, Pasadena, California, and Professor Emeritus of Medicine at the University of Southern California School of Medicine. Dr. Bing has received many honors for his lifelong contributions to the field of cardiology, including the institution of the Richard Bing Award for the Best Young Investigator in the Field of Heart Research awarded by the International Society for Heart Research, of which Dr. Bing was the first president and now an honorary life president. He acts as editor-in-chief of the society's publication, The Journal of Molecular and Cellular Cardiology.

Frederik Bang, MD, came from a distinguished Danish family that nurtured many physicians. During his long and distinguished career he worked at the Rockefeller Institute and during World War II he served with the US Army in the department of Epidemic Intelligence. Here he was part of the group that carried on surveillance of possible germ warfare. As a specialist in vector-borne diseases he investigated areas of diarrhea and other infections that might affect military and civilian populations.

After World War II Dr. Bang joined the faculty at Johns Hopkins University in the Department of Pathophysiology. This brought him to laboratories around the world. He worked in the French Roscoff Laboratory in Brittany, the Johns Hopkins branch laboratory in Calcutta, as well as at Woods Hole. Dr. Bang was an expert in finding marine models for human illnesses. He showed other scientists that many marine animals had physiology and pathophysiology similar to humans but were not so complex; it was easier to unravel the nature of diseases, infections, and immunity in them than in humans.

Dr. Bang learnt about Dr. Levin's interest in hematology and was instrumental in bringing him to Woods Hole in 1963. Thereafter, because of common interests, they teamed up together as collaborators. As a matter of fact Dr. Bang had already completed some path-finding experiments which would later help in discovering LAL. When he happened to notice that one of his horseshoe crabs became ill after being injected with the bacteria from the local sea water he isolated the germ, identified it as a *marine vibrio*, and succeeded in culturing it. When this organism was then injected back into another crab he noticed that it triggered a cascade of coagulation within that crab. The mechanics behind this phenomenon were investigated together with Dr. Jack Levin. Dr. Frederik Bang died in 1984.

Jack Levin, MD, was born and raised in New Jersey. He was graduated magna cum laude from Yale University in 1953 and magna cum laude from Yale University Medical School in 1957. He actually he got into research while still a medical student at Yale. For three summers he did endocrinology studies with *Relaxin* a substance that releases the symphysis pubis in lower animals at the time of delivery. After serving as a lieutenant in the US Public Health Service he accepted a two year fellowship at Johns Hopkins, then went to National Institutes of Health where Dr. Nathaniel Berlin introduced him to hematology, the field he still works in.

Dr. Jack Levin worked all summer at Woods Hole but spent the winters at Johns Hopkins teaching and doing his research. He was primarily interested in blood platelets and was looking for a simplified marine model with which to do his studies. That is when he and Dr. Bang came together. In 1968 they published their scientific paper that announced the discovery of LAL.

When asked if he ever regretted not patenting LAL he replied, "We (Levin and Bang) never did get patent rights and although there are some regrets there is also some compensating values. Although I am not rich I am free to do as I please and do my research wherever and in which directions I care to follow. I am not tied down to a company.

"I now work 7 days a week. I teach hematology courses at the University of California in San Francisco for medical students and graduate students. In addition to that I do my research and do a little clinical material and consulting at the Veterans Administration Hospital in San Francisco." Dr. Levin is Professor of Laboratory Medicine and Professor of Medicine at UCSF.

chapter 13

MIGHTY LIKE A ROSE

In the Public Broadcasting television series CONNECTIONS, historian James Burke explains how seemingly isolated and disparate events in the chronicles of man are actually connected. One episode, for example, relates how England became master of the seas and merchant of the world. Colonies were acquired, they bought England's textiles, this paid for more ships, which then established more colonies, and so on. A similar chain of events precedes and follows the discoveries about our white blood cells and their peculiar immunity-related surface molecules. If made into a television special, the trail from mysterious blood transfusion reactions, to heart transplants, to daring research, to the rescue of a celebrated operatic tenor, would provide a story as astonishing as anything seen on CONNECTIONS.

The opening shot of such a television drama would pan the frenetic yet orderly Tissue Typing Laboratory at Stanford University Hospital Medical Center then focus on its peripatetic director, Dr. Rose O. Payne (see Figure 13.1.). She is a slender silver haired woman in a starched white lab coat with a name tag pinned to her left lapel. Her plastic rimmed eyeglasses furtively sneak down her nose moments after she

Figure 13.1 Rose Payne, M.D. (Photo by Kathleen M. Podolsky.)

unconsciously pushes them up. And crows feet appear at the temples when she smiles. It is a no-nonsense smile, a smile that is a metaphor for "I'm pleased to meet you and I'll help you all I can, but don't make small talk or stay too long; the work I don't do today will be stacked up on tomorrow's load."

The camera now zooms in on a surgically garbed assistant carrying blood and tissue samples from an accident victim. They have just been donated to the Medical Center's transplant program. Dr. Norman Shumway and his team of surgeons anxiously await typing and matching before proceeding with a heart transplant. "Dr. Payne. Dr. Payne. What's going on here?" the narrator asks. In a measured cadence, an offshoot of her many years of teaching, she methodically explains, "We do all the tissue typing for transplantations here. We type donor organs and the recipients' cells and tell the surgeons how close the two match up. The closer the match the less chance for rejection." Then, as if reminiscing about those tender early years before she made her momentous discoveries, she pensively adds, "We do the screening for antibodies in patients to explain transfusion reactions and we also do typing for patients who require special blood products such as platelets."

To prepare the audience for the story of Dr. Payne's long and arduous struggle to reach the top of her profession there would be a flashback to 1884 when 15 year old Leland Stanford Junior dies from typhoid fever. He was the only son of California's wealthiest and most prominent citizen, Leland Stanford, Governor of California during the Civil War, president of the Central Pacific Railroad, and a U.S. Senator until his death in 1893. As a memorial to their son, the distraught parents, Jane and Leland Stanford, created a university and donated over 20 million dollars plus their prized 8,800 acre Palo Alto Stock farm, the present site of the campus. It is officially named the Leland Stanford Junior University, but today, everybody just calls it Stanford. The cornerstone was laid in 1887 and four years later its doors were opened for business. When a promising young woman bacteriologist by the name of Rose Payne needed a job in 1948 Stanford University was already established and able to provide a niche for her. And when she needed resources and support for some highly speculative research, the University that had gained a reputation for nurturing Nobel prize winners, now had the prestige and prudence to provide them.

For the time remaining in our CONNECTIONS-like drama the television screen shows sequences of Rose Payne's spectacular breakthroughs: how they enabled transplant surgery to forge ahead and how they made bone marrow transplants a reality. In the final scenario opera tenor Jose Carreras has just been told that he has acute leukemia. But, in the happy ending, he is plucked from the jaws of death by a bone marrow transplant, a direct result of the discoveries made by the modest, diminutive woman we met at the beginning of the drama.

Like any other showpiece this one touches only the highlights. The real and most intriguing part of the story lies in the details on how Dr. Payne became the foremost authority on tissue typing and how, together with a handful of individuals on the international level, she was credited for discovering the human leukocyte antigens. These molecular combinations, usually called HLA, are unique to each individual. By means of HLA typing every person can be immunologically fingerprinted. HLA types are inherited, half from the father and half from the mother. Blood relatives in general have closely related types. Study of HLA types is useful in paternity testing and in other situations where there is a dispute as to whether two people are related. It was effectively used to prove that Indians on the North American continent came from Asia by way of the bygone land-bridge over the Bearing Sea; both populations have closely related HLA types.

Antigens are substances that have the ability to send out signals telling the body that something is amiss. They are usually introduced by bacteria, toxins, viruses, and foreign blood cells. Unlike poisons that kill by destroying vital nerves and enzymes, antigens alert our immunity-makers to get into action. This entails two main countermeasures. First, antibodies are generated; they effectively trap or neutralize antigens, more or less like sticky fly-paper immobilizes flies. Second, an army of special white blood cells are dispatched to gobble up or otherwise destroy intruders.

The term *antibodies* was invented in 1890 by the man who conquered diphtheria, German bacteriologist Emile Von Behring, when he discovered that blood serum of immunized animals contained substances that killed germs. American and British scientistrs, Gerald Maurice Edelman, and Rodney R. Porter, shared the 1972 Nobel Prize in Medicine for determining, in 1969, that a typical antibody contains over 1,000 amino acids. Sometimes antibodies get a little confused, as described in Chapter 2, A LESSON FROM THE WIZARD OF OZ, and attack the body that made them, causing autoimmune disease.

White cells, leukocytes, are the least plentiful of the formed elements in blood, but at the same time they are the largest blood cells. Even the smallest white cell, the *lymphocyte*, is much larger than a red cell. One classification breaks them into two groups, depending upon whether or not they contain granules when stained. *Polymorphonuclear leukocytes*, the most numerous of the granular types, comprise 67 to 75 percent of the white cells in humans. They engulf and digest bacteria. Those that do not have granules, 20 to 25 percent of the white cells, are lymphocytes, the makers of antibodies. Lymphocytes made in the thymus gland are called T-lymphocytes; the T stands for Thymus; those originating in bone marrow are called B-lymphocytes.

The number of white cells in the circulation changes throughout the day and varies with certain diseases. White cells are able to move about freely hence they are found in tissues as well as in the bloodstream. Leukemia is characterized by an overabundance of leukocytes, the excess resulting either from overproduction in a deranged bone marrow or slower-than-usual destruction of cells elsewhere. Allergies and infections increase the number of some white cells.

In 1896 Paul Ehrlich drew a picture of a lymphocyte that looked like a soft rubber ball but resembled a T-lymphocyte as we now know it today. Ehrlich also suggested that this cell might have the capacity to recognize diphtheria and nothing else. Early investigators surmised that as the antigens of germs struck the surface of white cells they would leave characteristic indentations which would act as a template capable of recognizing that particular antigen if it ever reappeared. Specific chemicals that would attack specific antigens would reside within these templates. This theory held sway until about 25 years ago. Today we know that the thymus gland programs responses to antigens. It marshals four groups of T-cells: Inducers provide our first line of defense; two groups of information producing T-cells actually get rid of antigens; and the fourth immunoregulatory group regulates the production and distribution of immune materials.

These immune materials circulate in the blood plasma as soluable proteins called globulins. They are labeled alpha, beta, and gamma; the last, gamma globulin is the best antigen neutralizer therefore the best disease fighter. Newborns would be terribly vulnerable to infections if not for the gamma globulin borrowed from their mothers. Some immune substances are donated through the placenta before birth, some, after birth, through breast milk. A full-term baby at birth has adult levels of immune gamma globulins to prevent and fight off invaders. This stays with them for about three months, long enough for them to start making their own. When a baby is born prematurely it receives a short-supply of the mother's immune globulins and faces more risks. We now have the ability to "top up" such babies with intravenous infusions of concentrated commercially made immunity conferring materials. Since babies are apt to pick up germs from their mucous membranes and from the intestinal tract, nature loaded mother's breast milk with immunity agents. No biological fluid has a higher concentration of gamma globulins than colostrum, the early milk secreted by the breast in the first few days after delivery. Protective immunity-factors remain high in breast milk for the first six months of life then begin to taper off. The concentration of immune factors in breast milk is so high that some countries, especially those in Scandinavia, have created human breast milk banks and use it as eye drops, ear drops and nose drops.

Human leukocyte antigens are present in all tissues as well as in white blood cells, and, since they play such an important role in recognizing 'self', that is, they can tell if something is really part of us or a foreign intruder, they are collectively known as histocompatibility antigens. Genes within all chromosomes control the production of diverse antigens called HLA-A, HLA-B, HLA-C, HLA-D, and HLA-DR. Each histocompatibility antigen is composed of two parts, a constant region, which is the same for all individuals, and a variable region, designated by a number, which differs from person to person. All antigens in the HLA system have a letter indicating the series it belongs to, and a number corresponding to the form within the series, for example, HLA-A3, HLA-B13, HLA-C5. The possibilities for different combinations is astronomical, so that, apart from identical twins, each person has his or her unique, inherited, label, or tissue type.

Histocompatibility antigens within the series HLA-A, B, and C are present on virtually all living cells in the body. They are essential for the function of special lymphocytes known as killer T cells whose job it is to identify and eradicate abnormal cells that might be infected with a virus or tumor inducing virus. Histocompatibility antigens within the HLA-D series are present on the surfaces of various other defense-related cells and are relegated to combating infection and tumors.

When an organ is transplanted from one person to another, the histocompatibility antigens in the donor organ are generally recognized as foreign by the recipient and are attacked. This leads to rejection. However, if a donor can be found whose HLA types are very similar to those of the recipient, often a blood relative or ideally an identical twin, the chances of rejection are minimized.

Certain HLA types show up quite regularly in patients with particular diseases. For example, multiple sclerosis is associated with HLA-A3, celiac sprue with HLA-B8, and ankylosing spondylitis with HLA-B27. There is reason to believe that susceptibility to these diseases is influenced by the HLA types, presumably as a result of their immunologic actions. The associations are of interest because they give us the means of identifying individuals at risk and help to confirm the diagnosis of certain diseases.

Before Dr. Payne came on the scene none of this was known. Most of our knowledge about the HLA system resulted from her pioneering work, which, unlike discoveries that burst forth suddenly and quickly from luck or serendipity, covered most of a liftime.

Born August 5, 1909 in Lake Bay, Washington, Rose Payne spent most of her young years attending to the niggling chores of rural farm life. Her father was a lean, rustic, farmer and her mother, an èmigrè who had fled from the pogroms in Russia. There was nothing in Rose Payne's background to suggest that she possessed an aptitude for science. When she entered the University of Washington she had no idea of what she wanted to do for the rest of her life. Then something happened, something that we in the western world call luck or coincidence, but more appropriately, what the Hindus attribute to Karma, the cosmic force that decides a person's destiny.

"I grew up in an era where you didn't expect much. It was in the depths of the depression. When I was a sophomore," Dr. Payne recalls, "I still had no profession or occupation in mind. All I knew was that I had to earn a living. I was talking with a sociology professor one day about the possibility of becoming a social worker. I had dabbled in some science courses and he said, 'Why don't you become a technologist?' I began taking all kinds of science courses. That's how I found bacteriology; it was fascinating." Rose Payne went on to get her Bachelor of Science degree, then her Masters, and finally her Ph.D. in Bacteriology, all at the University of Washington.

She also served as Research Fellow at Woods Hole Oceanographic Institution in Massachusetts and as a Teaching Assistant in Bacteriology at the University of Washington. She then went on to become Assistant Professor of Bacteriology at Oklahoma Agricultural and Mechanical College but returned to her alma mater from 1938 to 1939 to serve as Research Fellow, at McDermott Foundation for the Study of Tuberculosis. The next four years were spent as Lecturer in Bacteriology at Seattle College. During World War II when teaching opportunities dwindled she made ends meet as a case worker for the welfare department.

In 1948 Dr. Payne accepted a job as Research Assistant at Stanford University Hospital in San Francisco and returned to science. It wasn't an easy choice. It meant leaving Washington where she had a lifetime of friends, family, and halcyon memories. But an inexorable *Karma* forced the final decision. "I was looking for a job," she recalls. "It was that simple. I had been out of science for a while so I accepted what I could get. I was a glorified technician and I helped to teach a class in laboratory procedures. The medical school was a gentlemen's school for the well to do and the very poor. It was not a research institution. There were very few women on the staff. I thought it would provide a way for me to function economically."

Dr. Payne's career in science began in an old, dilapidated building in Stanford's original School of Medicine. It had been a private school, Cooper Medical College, incorporated in 1882, the successor to earlier schools which had been established in San Francisco as part of the University of the Pacific. In 1908 the properties of Cooper Medical College were transferred to Stanford University.

Yet, despite close quarters and dismal facilities it was here in 1950 that a most extraordinary event took place, the one that ultimately led to personal fame and paved the way for tissue typing and a new era of transplants. "I was doing research on autoimmune disease," Payne recalls, "and was looking into a theory that presupposed that anemia was induced by an antibody. I happened to pick up a professional journal and read a paper by the French scientist, J. Dausset, who described a new white blood cell antibody. "It related to the work I was doing and I thought it would be an additional test for antibodies that caused anemia. I decided to try it and see."

I DECIDED TO TRY IT AND SEE. These are the magic words. Here is the clue to the workings within the mind of a scientist. While chance may have been a factor in bringing a particular scientific report and Rose Payne together, it was her astute perceptiveness, sharpened by years of study and a stubborn refusal to acquiesce to the easy way out that compelled her to try and see. She knew that she would have to make observations without bias and to let nature dictate the answers. Her commission was to repeat experiments, no matter how painstaking, and to publish everything that supported or contradicted her data. But this try and see process turned out to be a major undertaking, one that entailed endless solitary hours in the laboratory, shut off from friends, entertainment, and carefree pleasures. It demanded study, study, study, and probing into unknown and uncharted areas of physiology and genetics. Where to start?

Blood transfusions originated with the famous British architect, Sir Christopher Wren. Although extremely busy in 1668, rebuilding London after its great fire, he was asked to design a device that could be used in experiments where blood was to be transferred in animals. He perfected a hypodermic needle and syringe, very much like the one in use today. Samuel Peyps recorded in his diaries that blood had been successfully passed from one animal to another at the home of a doctor in London. The doctor is quoted as saying "I feel that this is the beginning of a new era that will see the mending of bad blood by the borrowing from a better body."

Human blood transfusions became practical and popular only after Carl Landsteiner demonstrated, in the 1920's, that there were four blood groups and later, in 1937, when the first blood bank was established. Yet, despite better blood typing methods and marked improvements in transfusion techniques over the years, physicians continued to be puzzled by the chills and high fevers that accompanied many blood transfusions. Although patient's and donor's serums and red blood cells had been typed and correctly matched these horrible reactions still appeared all too often. That of course meant that the culprit was not in the red cells but in some other element, probably the white cells. Alien white cells, when transfused to a person with a different white cell type, triggered the manufacture of antibodies by the recipient in order to get rid of these interlopers. "It was obvious the antibody was a result of transfusions and that it had caused the reaction," Dr. Payne explained. "There must be different white blood cell types and we set out to hunt for those types." The "we" refers to Dr. Herbert Perkins who for the next 15 years collaborated with Rose Payne in elucidating the causes of transfusion reactions.

"It was suggested there might be white blood cell types similar to the red cell types, A, B, ABO, and Rh, as early as 1924," says Payne, "but techniques for demonstrating these were not available until the late 1950s." With no little trepidation, Payne decided to look for them. Initially she and her collaborator devised and carried out a number of brilliant experiments that showed that indeed all white blood cells were not alike. The next step was to figure out a means of separating and labeling them and here she resorted to the tools she knew best, antibodies.

With masterful insight, honed by training, education and experience, Dr. Payne started to hack away at one of nature's most complex secrets by analyzing the records of desperately sick patients who, by virtue of the protracted nature of their diseases, had received many blood transfusions. She knew that such patients must have developed many antibodies after repeated exposures to transfused blood. She also correctly surmised that women who had given birth to many children and subsequently suffered transfusion reactions probably had spouses and offspring whose white cell types differed from hers. When these 'different' cells entered the mother's circulation her body manufactured antibodies against them. It was a simple matter to mix a husband's or child's white cells with the mother's serum and watch the white cells clump together, the sine qua non of antigen-antibody interaction. Later, as Rose Payne put it, "the common denominator was that all of the patients had multiple transfusions. When I studied their charts I also found unexplained transfusion reactions." It was now time to find the antibodies responsible for these events.

Shortly after Dr. Dausset announced that he had discovered an antibody that attacked white blood cells, in 1954, Dr. Payne repeated his experiments and confirmed his findings. Her studies clearly established the fact that leukocyte or white blood cell antibodies were produced by the body in response to foreign white blood cells introduced at the time of transfusions. The fever and chills characteristic of many blood transfusion reactions were due to the interaction of these antibodies with the transfused white blood cells: *leukocyte antibodies were the cause of transfusion reactions*. This finding was immediately put to use at Stanford where incompatible white blood cells were removed from the units of blood prior to transfusions. Adverse transfusion reactions disappeared. Dr. William P. Creger, associate professor of medicine and head of the Division of Hematology at Stanford Medical School remarked, "The point is that because the work was done at this Center some of the patients here benefited immediately. They received better care and were saved some very distressing complications."

Early investigators tried to separate and type white blood cells by using antibodies formed in patients who had received multiple blood transfusions but they all failed. Nature obscured her clues by creating vast numbers of antigens amidst vast numbers of cells. It was like trying to match up two identical fish in the ocean—too many fish, too big an ocean.

"Generally, you are working on something with an entirely different objective," Dr. Payne once explained, "and you make an observation that leads you off in another direction. You have to be prepared to think in other terms." Embarking on another tack, Dr. Payne began examining the blood of pregnant women who had never received any blood transfusions. She reasoned that just as the mother's immune system would react against any alien substances transmitted to her by a fetus, it should react against any incompatible white blood cell types that might find their way from fetus to mother. Her hypothesis was right on the mark. The sera of pregnant women contained antibodies against a number of white blood cell types and this now gave her a way to label them. Soon, other types were pinned down in this manner. Researchers around the world adopted these discoveries, giving a tremendous boost to international collaborative projects dedicated to cell and tissue typing.

In the midst of her research, just shortly after she had been promoted from Research Assistant to Research Associate at Stanford University School of Medicine, another crisis threatened to stop her in her tracks. The rug was about to be pulled from underfoot. The Trustees of the University had decided to move the School of Medicine to the University campus in Palo Alto, some 50 miles away from San Francisco. Dr. Payne had the option of moving with it or finding a another job. It was a difficult choice. Men equate reallocations with job advancements and new career opportunities; they mitigate inconvenience with dreams of higher salaries. But for Rose Payne it not only posed the logistical problems of moving a laboratory, fragile research equipment, and tons of papers, notes, and files, but it also meant uprooting a husband whose job was in San Francisco, leaving friends and colleagues, and exchanging a scintillating city that was understood and loved for some nebulous, unknown site in hicksville. The trauma from her first big move from Washington to San Francisco was ameliorated by vistas of the Golden Gate, the Berkeley hills, the Bay and the ocean, not unlike those surrounding her native Washington. Palo Alto promised none of these.

After much soul searching Payne just knew she could not part with the institution so essential for her research so she stuck to her job, relocated with the medical school, and commuted back and forth every day between Palo Alto and San Francisco. She doggedly pursued her research at the new site and ultimately handed the world an understanding of the HLA system—the benchmark for much of the genetic, transplant, and immunological knowledge and technology enjoyed today. In her laboratory, in 1984, 25 years after that fateful move, she recounted, "When I began work in the 1950's nobody knew there were that many white blood cell types." Then, pointing to a poster on the wall of her office that contained a list of 80 HLA types, she added "I found this one and that one and that one......" What was left unsaid, however, was that her discoveries enabled Stanford to take the lead as the most imposing heart transplant program in the nation and later enabled Dr. Michael De Bakey to do the same at Baylor University in Houston.

Much of the phenomenal growth of kidney transplantations can be attributed to Dr. Rose Payne. When they were first being tried at the University of California Medical Center in San Francisco, Payne was called upon to direct the tissue typing. "When we began to be concerned about the role of our findings in kidney transplants we retrospectively typed the kidney transplant recipients and their donors and the research just continued to expand," she explains. As things now stand it is incumbent upon transplanters to get as perfect an HLA match as possible between donor and recipient in order to minimize rejections. "For many years," Dr. Payne said, "there had been controversy about the degree of matching required and that still goes on. But there are good statistics to show the better the match, the better the course of the patient. With immunosuppressant drugs, you can suppress the effects of mismatch but the course of the patient may not be as good."

The final word on the HLA system is far from being entered into the books. We know that lymphocytes play a major role in transplantation reactions and that they collect in large numbers at the site of a graft which is being rejected.

How and why they carry out their nefarious deeds is still being studied.

As mentioned earlier, specific white blood cell types appear as markers for susceptibility to a form of arthritis known as ankylosing spondylitis, as well as for psoriasis, juvenile onset diabetes and other disorders. One disease Payne and her colleagues investigated was thyroiditis, an inflammation of the thyroid gland, which is unduly common in the Japanese population. In Caucasians the disease is associated with white blood cell antigen B8, however, this antigen is rarely found in Japanese people. Instead, as Rose Payne proved, thyroiditis in Japan is associated with another antigen, BW35. Researchers are now trying to determine the relation between the gene coding for the two antigens.

All the honors and accolades that Dr. Rose Payne accumulated throughout a lifetime dedicated to science did not come in a logical and facile manner. She escaped neither the mundane chores of everyday subsistence nor prejudices and discrimination encountered by women generally, and especially by women in science. As a housewife there was shopping to do, meals to prepare, laundry and dirty dishes to be washed. When Dr. Payne brought home brief cases crammed with work, Earl, her husband, who also put in a hard day's work knew that 'Chinese take-out' would be fare for the day. On weekends it was always an agonizing decision whether to grapple with experiments and data or escape to Golden Gate Park, the De Young Museum, or Davies Symphony Hall.

Dr. Payne handled the problems of bias and discrimination against women in science with the same aplomb as that used in blending the demands of science with the requirements of a personal life. She neither surrendered nor took up jousting with windmills. In 1955, when she was elevated in rank to Research Associate at Stanford University School of Medicine Dr. Payne took it in stride but was painfully aware that no improvement in laboratory facilities or increased funding went with the promotion. As she recalls, "One of the main difficulties is that we didn't have much space. As far as rank was concerned, I eventually was named a senior research associate, which at the time was considered the equivalent of an associate professor, without advantages of even a sabbatical."

Rose Payne resigned herself to a secondary platform in the hierarchy of science, accepting a feminine subsphere within the larger male sphere. Although this strategy may have created research opportunities it doomed her to women's work—work that was poorly remunerated, low in status and generally regarded as intellectually inferior. But she wanted to do what she was doing. Her research was more important than the issues of male and female and she chose to continue her career in a 'woman-in-her-place world'. Because she refused to 'make waves' and possibly jeopardize her research she was not named professor of medicine at Stanford until 1972, three years before she retired. "It really didn't make that much difference and when I was made a professor, my life didn't change that much. I already had an international reputation and had done what I wanted to do."

Others have fought the battle differently. In June 1991, one of the nation's first female neurosurgeons, Frances K.Conley, MD, audaciously resigned from her tenured faculty position of 23 years at Stanford University School of Medicine because the male acting-chairman of her department, who allegedly failed to treat women as equals, received the permanent appointment as chairman. Dr. Conley withdrew her resignation after the gender insensitivity issues in the 'good old boys' world of medicine and science had been exposed to the media and Stanford officials agreed to rectify the matter. The chairman of the neurosurgery department subsequently stepped aside.

When Dr. Payne began her work in the early 1950's there were few women in science. It was without a doubt, a man's world. But quietly, without any ruckus, she used her knowledge and skill to rise to the top of her field. Although she has received some of the most prestigious awards, national and international, people who know about her accomplishments and her hardworking, persistent, and diligent efforts to explain the HLA maze think that Rose Payne certainly should have shared the Nobel Prize with Dausset who pioneered the same area of discovery and research. Some say because of being a women she never received the real recognition she deserved.

Payne's peers and associates have had candid as well as differing views of Payne's handling of the women's rights issues. Bernice Hemphill from the Irwin Memorial Blood Bank remembers those days well. "Payne will stand up to the best", Hemphill said. "She is totally focused on her work; a no nonsense lady with the courage of her convictions. She knows how to get what she wants, often against great odds." Her long time associate, Dr. Herbert A. Perkins, research director of the Irwin Memorial Blood Bank in San Francisco, adds, "She is a hard pusher for things she thinks are right, but not for herself."

Reflecting upon our current scene Payne says, "Many more doors are open today. I think a women should try to have everything, a family, and a husband who understands her as well as her work. Men in science still find it difficult to integrate women into the picture. It is still the "old boy's club" and they are more comfortable talking turkey to each other; they can't talk that way to a women. It's a cultural problem that is not easily overcome." But things are improving. When Payne joined the staff there were very few women students. Today women comprise almost half of each freshman medical school class at Stanford University. Also, we now have The Association For Women In Science (AWIS), a 20 year old organization with 36 chapters nation-wide whose members serve as advisors to college women and graduate students in science. They also proffer advice on how to behave at a job interview, how to deal with sexual harassment and how to balance career and family.

Dr. Payne's contributions to science, in addition to handing the medical world the prerequisites so necessary for transplanting hearts and organs from one human to another, also led the way to making bone marrow transplants practical and successful. HLA typing of donors and recipients of bone marrow assures histocompatibility which means that the transplants will take hold and generate new blood components. Bone marrow transplants are currently the best means at our disposal for combating immune deficiency diseases and leukemia. Dr. Richard Hong, professor of pediatrics and immunology at the University of Wisconsin Medical School and director of the bone marrow transplant program at the university's Children's Hospital has observed, "Successful bone marrow transplants rely on matching HLA types. When a transplant is performed quickly following an early diagnosis the results can be remarkable."

It was through chance and a brilliant series of experiments several decades ago that we actually obtained the clues as to how our bone marrow functions. Toward the end of the 1960's the United States Department of Agriculture tried to see if chickens could be made to grow faster. To do this scientists injected testosterone, the male sex hormone, into embryonic chicks. When the chicks hatched they all died within a few days of birth from overwhelming bacterial infections. Examinations subsequently proved that the birds failed to develop the *Bursa of Fabricus,* an organ unique to fowl, where their lymphocytes and antibodies are manufactured. The equivalent of the Bursa of Fabricus in human beings was found to be the bone marrow; this is where our B-lymphocytes come from.

Armed with this new understanding about bone marrow's immune funtion and Dr. Rose Payne's HLA typing technology, doctors were now ready to attack their most irrepressible foe, leukemia. A case in point is this extraordinary account of Jose Carreras' bout with the disease.

In early 1987, as he rounded the curve of his 39th year of life, the brilliant Spanish operatic tenor began limping because of pain in his legs. This was soon followed by constant fatigue, and he had to be heavily bandaged to limit joint pains. In July, after extensive tests and examinations, doctors confirmed the ominous diagnosis of leukemia. The celebrated conductor Herbert von Karajian, who was recording the Verdi REQUIEM with Carreras at that time remarked, "he has this terrible illness, but he is full of hope."

His days were numbered. The final curtain was about to descend on a career that started in 1970, at age 22, when he was discovered by diva Montserrat Caballè in his native Barcelona. The voice that enraptured audiences in opera houses all over the world was about to be stilled. It would join the glorious echoes of Enrico Caruso, Beniamino Gigli and Jussi Bjoerling. His chances of recovery ranged from that of a 'snowball in hell' to an optimistic 1 in 10.

But a kind fate intervened. Thanks to the widespread impact of Dr. Payne's research, medical centers around the world adopted her HLA typing techniques. One of them, The Fred Hutchinson Cancer Research Center in Seattle, Washington, developed special expertise with bone marrow transplants for the treatment of leukemia, offering hope and reprieve to those with the affliction. That is where Carreras was sent.

Leukemia is essentially cancer of the white blood cells. They must be destroyed with chemicals, radiation, or both. Such drastic measures destroy the normal bone marrow as well as the cancerous cells. Heretofore this was a major obstacle, but now, thanks to HLA typing, new healthy marrow could be transplanted to the treated patient. And that is what Jose Carreras had to endure. Initially he underwent months of chemotherapy during which time his cancerous white cells and his bone marrow were destroyed. He also had to take agonizing radiation treatments, which, despite dizziness and nausea, required him to stand in a small cubicle at least 3 times a day. He was told that treatment would take about 20 minutes but having no way of keeping track of time he found the sessions almost unendurable. "So I decided to sing to myself—Sotto Voce, of course. Since I knew how long each aria from certain operas would take I knew approximately how long I had been in the cubicle." As soon as all the cancerous cells were eliminated, four months after his leukemia was diagnosed, the life-saving bone marrow transplant was performed. He stayed in the Hospital 100 days and upon his release he realized that he had lost 40 pounds. But he was alive and the future looked good.

Just a few months after his release from the hospital he gave a free open air concert for 150,000 fans in his native Barcelona. It was, as the tenor nostalgically recalls "a psychosis of affection, something only the Catalan people could summon." Since then he has given dozens of concerts many of which were benefits for the Jose Carreras International Leukemia Foundation, which is based in Seattle and Carreras' home town Barcelona. The foundation helps raise funds by staging benefit concerts.

At his first big recital at Royal Opera House in London he received a tumultuous standing ovation both before and after the concert. The usually reserved British crowd cheered for 30 minutes, tossing flowers onto the stage until Carreras was ankle deep in carnations, roses, daffodils and lilies.

In July, 1990, Carreras shared the stage with Placido Domingo and Luciano Pavarotti at the outdoor auditorium in Rome, the Baths of Caracalla. This unique concert was telecast throughout the civilized world to broad acclaim—which, in a sense, was also a tribute to surgeons and staff at the Fred Hutchinson Cancer Research Center and to quiet, non-assuming, gentle genius, Rose Payne.

Dr. Payne was appointed professor of Medicine at Stanford University School of Medicine in 1972 and became professor emeritus in 1975. Her laboratory remains an important site for the training of individuals who will be qualified to promote the field of immunogenetics.

Advancing age has inexorably reduced her long stints in the laboratory but Dr. Payne still manages to find time to act as consultant on the serum samples from all over the United States and the rest of the world that are sent to her laboratory at Stanford for analyses, an invaluable resource for other investigators with complex clinical or research problems. And she makes time to work with post-graduate fellows, passing on her great body of knowledge to young and eager researchers. One of these, Dr. Rajendra Desai, in 1963, proved that two kinds of white cells, as well as platelets, can actually pass through the placenta, from mother to baby. This might explain the transmission of certain diseases to the fetus during pregnancy, especially if infectious agents are located within platelets or white cells. Also, if these 'foreign' cells settled in the child, then multiplied, they could give rise to tumors or provoke autoimmune diseases which are conditions that arise when the body angrily reacts to some of its own tissues.

Pride in her proteges has always been one of Dr. Payne's redeeming pleasures. "It's nice to think you worked in an area that has opened up all kinds of new perspectives and is still doing so," she likes to tell visitors. Not content to rest on her laurels, Dr. Payne is now engaged in further identification of the antigens of a population of B lymphocytes, important players in the scheme of immunity.

Bernice Hemphill, the erstwhile director of the Irwin Memorial Blood Bank at San Francisco, summed things up nicely when she said, "Payne was a pioneer in studies so necessary for transplants. She was a great humanist, totally involved in caring for her patients as part of her research." But Payne, herself, tends to soft-pedal her monumental contributions to science, while wistfully reflecting upon her early trials and the missed potentials of education. "It is unfortunate that people have to earn a living. We don't go to school completely for an education, but to get some kind of a skill. It would be nice if we could afford to go to school and worry about the skill later. But I've always enjoyed what I've done." Yes! Rose Payne has always enjoyed what she has done, and done it well.

PARADIGM POINTERS

Paradigms do not always arise from explosive, intuitive or lucky discoveries in the laboratory. All too often they come from the painstaking, diligent, assemblage of knowledge that is already available but is so scattered that each fragment makes little sense by itself. A good illustation of this is seen in the work of Maude Abbott, M.D., who spent a good portion of her life studying and cataloging specimens of congenital heart disease. Only after she had assembled the world's most comprehensive atlas of such conditions did the many facets of the disorders present an intelligible picture. This paved the way for Dr. Robert E. Gross, in Boston, Drs. Alfred Blalock and Helen Taussig in Baltimore and Dr. Clarence Crafoord in Stockholm, to diagnose these heart defects in living babies and then perform the surgery that saved them.

POST SCRIPT

Women have had a long, rancorous struggle in science, and although things are improving, justice and equality have persistently eluded them. They have had to fight substantial built-in cultural and political biases, and even when they did succeed, they ended up in patterns of segregated employment and under-recognition. Madam Marie Curie, for instance, remained an unpaid assistant in her husband's laboratory, even after sharing the Nobel Prize in physics in 1903, and only after his death in an accident did the Faculte des Sciences bestow his chair to her. This, in effect, permitted her to teach at the Sorbonne—the first woman to do so.

One of the most invidious examples of discrimination shows up in the life and times of Maude Abbott, MD. She completed her undergraduate education at Montreal's McGill University but was denied entrance to its Medical School. A professor of surgery announced that he would resign if women were allowed to take the medical course. Undismayed, Dr. Abbott obtained her medical degree from the University of Bishops College, in 1894, the only woman in her class, then went on to study abroad in England and

Europe. We are indebted to Sir William Osler who asked Dr. Maude Abbott to undertake the preservation and study of the Osler collection, a virtual museum of pathologic specimens that Sir William had put together while at McGill University. From this came the first comprehensive understanding of heart defects seen at birth.

Mary Shelley's mother, Mary Wollstonecraft, issued the first feminist manifesto in 1792, in her book **A Vindication of the Rights of Women**, wherein she asked that women receive an education. Yet today, two centuries later, studies show that gender differences in mathematics begin to appear in the middle school level and in science as early as the elementary school level. In High School girls take fewer years of mathematics and science than boys and are not as likely to enroll in advanced courses.

Pejorative patterns take hold early in our society. When parents give their little girls dolls and their little boys Leggo Blocks they begin a long process of socialization that steers girls away from science. While girls learn nurturing their brothers learn manual dexterity and sharpen their ability to think creatively, skills that will help them succeed in science. Parents teach girls to be passive and tend not to give them puzzles and toys that stimulate abstract or complex thought. Teachers in elementary schools seldom push girls into mathematics. How often have you heard a mother say, "Your father knows math, I don't know numbers." While businesses are spending millions to attract women into undergraduate and Ph.D. science programs the message that still goes out is "you're needed, but there probably won't be a job for you when you finish your studies."

Margaret Rossiter the author of **Women Scientists In America: Struggles and Strategies to 1940**, stated, "they (women) were the shadows. It is very tricky to focus on these minor people. They lived in a sort of segregated world and were never connected to the main stream of science." The author also shows that although women held 13% of all doctorates in the science fields for which the National

Research Council provided post-doctoral support, 1920 to 1938 fewer than 1/2 of these women received post-doctoral fellowships. A Rockefeller Foundation official once admitted that an extra burden of proof of their qualifications was required for women to receive fellowships.

Although we have come a long way in eliminating discrimination against women we have not entirely stamped out its subliminal endurance. Even the Wall Street Journal has on occasion exhibited undercurrents of bias against women in science. In 1958 the paper criticized a federal program that promised to give 1/3 of its scholarships to women: "It is inevitable that some government money will go to train scientists who experiment only with different household detergents." Other newspapers, too, often reflected similar insensitivitys. In 1963 newspaper headlines commended 57 year old Maria Goeppert-Mayer as the "San Diego Mother" who won a Nobel Prize in Physics. Women entering the world of science inevitably face different types of sex discrimination during different phases of their careers. When they first start out men do not take them seriously; most suspect that women will drop out after a few years to start a family. And if a woman does succeed she is criticized as being 'unfeminine' or too aggressive for a team leader.

If science is to remain independent there should be no place for gender in any discourse. And if scientific minds are truly disembodied it is irrelevant whether a scientist is male or female. Yet, as we look at Rose Payne's progress through the academic hierarchy we see that this was not so.

PERSONAE

Dr. Rose Payne co-authored 87 scientific papers which have appeared in the most prestigious national and international publications. Her awards include the John Elliot Memorial Award (1964) and the Karl Landsteiner Award (1977), both from the American Association of Blood Banks. Dr. Payne is a member of numerous professional societies, including the Transplantation Society, the American Society of Hematology, and she is councilor of the International Histocompatibility Workshops. She is a member of the World Health Organization Expert Advisory Panel on Immunology, a member of the committee on organ transplantation and tissue typing for the American Association of Blood Banks, and she is on the editorial boards of several highly respected scientific journals.

chapter 14

A BACKYARD FULL OF DIAMONDS

Peaks of popularity and valleys of abandonment best describe the roller-coaster endurance of aspirin. It was first made by the French chemist Charles Gerhart in 1853 then put on the shelf as a curiosity. Almost a half century later it was reinvented by Felix Hoffmann in Germany's Bayer pharmaceutical company. It became the most successful drug in the world, even outselling antibiotics. When newer pain-killers and non-steroid anti-inflammatory drugs, NSAIDs, came on the market, aspirin went into a decline. But when aspirin's ability to interrupt the process that leads to heart attacks and strokes became evident, the drug once more became king of the drug empire. The United States alone produces an estimated 35 million pounds of aspirin a year and Americans consume 50 billion aspirin and aspirin-containing tablets annually.

Aspirin, acetylsalicylic acid, chemically links an acetyl vinegar-like molecule to relatively weak salicylic acid. People began to learn about aspirin around the begining of the 20th century when it was commercially produced on a large scale but knowledge about its main component, salicylic acid, goes back to the dawn of civilization. In their continual search for

375

food, ancient peoples ate leaves, berries, roots, grasses and tree bark, and through accident or trial and error, discovered that many of these had the power to ease their ailments. Cinchona bark quieted down malaria, foxglove ameliorated the swollen ankles that accompanied weak hearts, and willows and meadowsweet reduced pain and fever. Tribal witch doctors probably acted as absolute custodians of these secrets but later on, much of this healing information, especially that pertaining to the powers of willows and salicylate containing plants, found its way into folklore, home remedies and finally into the hands of physicians. The Ebers papyrus, which is approximately 3,500 years old, prescribes a decoction of myrtle leaves for rheumatic and womb pains. Pliny, in his medical encyclopedia written 2,000 years ago, advised the use of extracts from the leaves and bark of willow trees to ease pains and to reduce fevers. The fathers of Western Medicine, Hippocrates and Galen, agreed with this advice. The curative effects of salicylate containing plants were recognized by the Chinese, the North American Indians as well as the South African Hottentots.

Throughout the middle ages, extracts from the bark and leaves of willow trees provided medicines for dysentery, wounds, plasters and external emollients. When willow branches were needed for basket-making, herbalists substituted meadowsweet, *Spiraea Ulmaria*, from their herb gardens. The first almost-scientific report on willow bark as a pain reliever was made on June 2, 1763 by Reverend Edmund Stone. He reported to England's Royal Society that he had accidentally tasted powdered willow bark and found it extremely bitter. It reminded him of cinchona bark which was already in use for malaria. As a staunch advocate of the **doctrine of signatures,** which stipulates that cures are found in the locations where the disease prevails, he argued, "Because the willow delights in a moist and wet soil, where agues (rheumatic pains) chiefly abound I gathered willow bark and dried it, pulverized it and used it on 50 patients....I have no other motives for publishing this valuable specific, than that it may have a fair and full trial in all its variety of circumstances and situations and that the world may reap the benefits occurring from it."

The bitterness that Reverend Stone tasted was due to salicin, a close relative of salicylic acid, a component of common white willow. Salicin was first isolated in pure form in 1829 by the French pharmacist Leroux, who then demonstrated that the pure compound was effective in reducing fevers. Nine years later the Italian chemist Piria broke salicin down chemically and thus stumbled on salicylic acid. In 1853 chemist Charles Gerhart added an acetic molecule, like that found in vinegar, to salicylic acid and came up with acetylsalicylic acid. It lay forgotten as a useless chemical compound and might even have disappeared into oblivion were it not for serendipity during the search for germ killers and the chemical wizardry of Adolph Wilhelm Hermann Kolbe and Felix Hoffman.

From time immemorial it was known that extracts from salicylate containing plants would sour milk but preserve meats. After Pasteur established the connection between bacteria and disease, scientists reasoned, rightly, that the active substance in these plant extracts, salicylic acid, had germ killing properties and went to their laboratories to prove it. Sure enough, when salicylic acid was added to germs growing in test tubes and culture plates it killed them. In Leipzig, Professor Hermann Kolbe followed up on this lead and during the period from 1873 to 1874 he proved that salicylic acid gradually broke down to carbolic acid, the foremost disinfectant of that era.

Roughly six years earlier, in 1867, Dr. Joseph Lister, Professor of Surgery at Glasgow University, had introduced carbolic acid into medical usage as a disinfectant and antiseptic. He fell upon this discovery after learning that carbolic acid was being used in Carlysle township to disinfect sewage. He tested it in his hospital for two years before proclaiming its virtues to the medical world.

After Lister's pronouncement, carbolic acid became the world's most popular germ killer. One day chemist Kolbe asked his good friend, the eminent surgeon Dr. Karl Thiersch, "Where do you use carbolic acid?"

"Everywhere," replied Thiersch. "We wash down the operating rooms, we soak our instruments in it, we wash our hands in it, we do everything with it. You know it is the best antiseptic the world has ever known."

"If it is that good," asked Kolbe "why don't you give it to patients to kill germs in the body and cure disease that way?"

"I'm glad you are not my doctor!" Thiersch replied. "Carbolic acid or, as you chemists call it, phenol, is also quite poisonous when taken internally. Just like it poisons germs it poisons people."

"Suppose we could give the patient a non-poisonous substance that changed gradually to carbolic acid? Wouldn't that just kill the germs but spare the patient?" Then, without giving his friend a chance to reply, Kolbe went on, "I have just such a substance, salicylic acid."

The chemist went on to explain to his medical friend how salicylic acid in the test tube slowly broke down into carbolic acid and this might be an effective way of killing germs within the body. The logic convinced Thiersch, but being a cautious scientist, he decided to try it externally at first. He liberally doused the stump of a recent amputee with salicylic acid and beamed with excitement several days later when the bandages were removed and the wound appeared clean and healing. The foul odor of infection, which had characterized amputations until then, was absent.

As news of this success spread to medical institutions in Leipzig and the rest of the world, physicians and surgeons routinely began swabbing wounds with salicylic acid. The reports coming back to Kolbe and Thiersch praised salicylic acid glowingly. Sensing a potential market, Professor Kolbe assigned his best and brightest student, Fredrick von Heyden, to start up a factory in Dresden to supply the world with this new miracle antiseptic. The enterprise flourished. Demand always exceeded the supply. Salicylic acid was formulated as liquids, pills, and powders. Cuts and wounds were doused with it and it was prescribed internally as treatment for tuberculosis, rheumatism, dysentery, and plagues. Food manufacturers used it as a preservative.

But this euphoria was dampened when reports began to cast doubts on some of the claims made for salicylic acid. When it was given to typhoid fever patients their high temperatures went away but they died nonetheless. The recovery rate from diseases like typhoid fever and tuberculosis was the same with or without salicylic acid. At about this time, the Swiss physician, Dr. Carl Emil Buss, at the small Swiss Canton Hospital of St. Gallen, noted that although salicylic acid did not save patients with infectious diseases, it did, however, reduce fevers and patients felt better. Further analysis of patient records showed that there was a drop in temperature with every dose of salicylic acid and that regular doses would prevent temperatures from coming back. After Dr. Buss' findings were reported in medical journals, von Heyden's factory in Dresden could not make enough salicylic acid to keep up with demand.

Dr. Buss was able to back up his observations because clinical thermometers for medical use had been introduced in the 1870's. They allowed doctors for the first time to actually document the rise and fall of a patient's fever and to correlate this with the course of an illness. Concurrently, temperature charts became an integral part of every clinical and hospital setting. When the salicylates and acetylsalicylate came on the market, those precipitously descending lines on the graphs of fevered or pyrexial patients showed how effective the new medicines were. Consequently they became known as *antipyrexics*, against fever, and *analgesics*, against pain.

Then, to compound good fortune, Dr. Franz Stricker, a physician in the German army, discovered that salicylic acid worked wonders on rheumatism. He compiled a number of glowing reports showing that it reduced the size of swollen joints, it stopped the redness and pain in the joints and even had a soothing effect on the nerves and muscles. By now the world knew that salicylic acid could be relied upon to treat rheumatic conditions and to reduce temperatures.

Salicylic acid was enormously profitable. This attracted many would-be competitors one of which was the pharmaceutical manufacturing house of Fredrick Bayer. The company embarked on a search for a derivative of salicylic acid, or even an improvement, with the aim of bypassing patent infringements. Arthur Eichengrun, head of research, and Dr. Heinrich Dreser, chief chemist at the Bayer Laboratories, assigned the task to a young chemist, Felix Hoffman, who, incidentally, did most of his work at the experimental laboratories at Elberfeld, the same place where Dr. Gerhard Domagk discovered sulfanilamide.

Hoffman had personal reasons for seeking a more acceptable form of salicylic acid. His father had been taking it for arthritis and found that besides being bitter, it caused stomach pain, bleeding and vomiting. The treatment was worse than the disease.

There are no exact accounts of what led Hoffman to aspirin. Some colleagues maintained that Hoffman liked oil and vinegar on his salads and knew that vinegar, acetic acid, did not upset the stomach, so he decided to make salicylic acid more like acetic acid. Scientists familiar with Hoffman's professionalism believe he searched the literature of chemistry, read Charles Gerhart's 1853 report on acetylsalicylic acid, and recognized a potential in the novel compound. In any event Hoffman hooked an acetyl molecule to salicylic acid and came up with acetylsalicylic acid. When he brought it to the head chemist for evaluation, Dreser employed his favorite form of testing living tissue reactions to chemicals. He reached into a small aquarium that he kept in his office, netted two tiny fish that had transparent tails, and placed one tail in salicylic acid and the other in acetylsalicylic acid. The former turned cloudy, the latter remained clear.

In 1899 the esteemed Dr. Witthauer prepared such a favorable report that Bayer then decided to release aspirin for general use. But it is amusing to read what Witthauer had to say about new drugs in the late 19th century: "Nowadays a certain amount of courage is required to recommend a new medicine. They are thrown on the market almost every day, and one needs a marvelous memory if one wants to remember all the new names. Many pop up, are praised and recommended by a few authors and especially by the manufacturers, and after a short time one hears no more of them." Now, almost a century later, this still rings true.

ASPIRIN, as a commercial, easy to remember name for acetylsalicylic acid, came from the fertile mind of Bayer's Heinrich Dreser. There are two versions of how that came about: One has it that Dreser named his new drug after St. Aspirinius, the Neapolitan patron saint who warded off headaches. The other, a more prosaic explanation, purports that the name is derived from the genus *Spiraea*, salicylate containing plants. By adding acetyl to Spiraea, Dreser concocted aspirea, which then became aspirin. Bayer succeed in copyrighting ASPIRIN and then advertised it so extensively that the term completely supplanted acetylsalicylic acid. When the Bayer copyright expired in 1917, and many companies began to manufacture generic aspirin, the Bayer Company sued in an attempt to block the use of the name ASPIRIN. It lost because by then the term aspirin had become part of the world vernacular. To offset this defeat the company resorted to GENUINE BAYER ASPIRIN as its slogan. When introduced into the United States after World War I, aspirin supplanted salicylic acid and became the most widely used drug in the world. The US Pharmacopeia adopted **aspirin** as the official term.

But the scientists who engineered the original breakthroughs profited the least. Charles Gerhart died before the true potential of aspirin was recognized and poor Hoffman was cheated by a cruel injustice. Hoffman's contract with Bayer stipulated that he would receive royalties from the sale of all patentable products that he invented. Dreser on the other hand had an agreement with Bayer whereby he would receive royalties on any product that he introduced. The German patent office refused to grant a patent for aspirin because they did not consider it a sufficiently novel process. Since there was no patent, Hoffman received only a modest salary, but Dreser received substantial royalties and retired as a rich man.

Heinrich Dreser was no slouch when it came to promoting his company, its products, or himself. The same year that Hoffman produced aspirin, 1898, Dreser told the Congress of German Naturalists and Physicians that he had created a new, non-addicting form of morphine by adding acetic acid to it—in much the same manner as acetic acid was added to salicylic acid. Its chemical name was di-acetyl-morphine. "But that's too complicated a name," Dreser said. "Since it is such a heroic drug we will call it heroin." Within a year, Jean Jarrige, a young medical student at the University of Paris proved that heroin was a good deal more addicting than morphine. Bayer then went back to hawking aspirin, emphasizing that it was a much better analgesic for pain and headaches than salicylic acid. Besides, aspirin was not addicting.

Interestingly, the chemists at the Bayer Company found another drug, phenacetin, which duplicated many of the actions of aspirin. In the courtyard of the Friedrich Bayer Dye Factory a 50-ton pile of a yellowish powdery waste product had been built up. It would be costly to move, and besides, it would certainly discolor any area where it happened to be dumped. So there it stood until someone could decide what to do with it. The renowned chemist Carl Duisberg, whom we met in Chapter 1, inquired about the pile and was told that it was para-amino-phenol, a useless by-product of every dye factory. Duisberg said, "We must haul it away or change it into something we can sell." (This manifesto became the guiding philosophy, first, of German industry, and later, of all industries, world-over.) Chemists went to work on the pile and converted it into phenacetin, a drug that proved to be as effective as the salicylates in treating rheumatism and fevers. Subsequently, Bayer sold as much phenacetin as aspirin and then combined it with aspirin for a double-barreled effect. Phenacetin, with and without aspirin, sold very well and, incidentally, got rid of the waste pile. Phenacetin was later found to have adverse effects on the urinary system. Its use has declined since 1982, when the U.S. Food and Drug Administration proposed removing phenacetin from analgesics.

Under special circumstances and when administered in high doses, aspirin too is believed to have untoward effects on the kidneys. The death of the multimillionaire Howard Hughes has been attributed to the shut-down of his kidneys as a result of taking large doses of aspirin over long periods of time. Additional unpleasant side-effects of aspirin are nausea, vomiting, ringing in the ears and rarely, deafness itself. It sometimes produces headaches rather than relieving them. People who are allergic to aspirin may develop hives, asthma-like wheezing, or a stuffy nose similar to hay fever. Highly allergic individuals may go into shock within three hours after taking medication.

In 1963 the Australian pathologist R. Douglas K. Reye, M.D., collected a number of cases where children who had received aspirin while recovering from viral infections, notably influenza and chickenpox, began to vomit and later showed signs of brain and liver involvement. This became known as Reye's Syndrome and in 1986 the British Committee on Safety of Medicine recommended that aspirin not be given to children under the age of 12 except on a physician's advice. The number of such cases has been declining. Whether this is due to better parent education, changes in the viruses, or less aspirin, remains to be determined.

But it is aspirin's propensity to induce hemorrhages in people with blood abnormalities that has created the most consternation among physicians. Indeed, this property may have profoundly affected the course of history. The young Tsarevich of All of the Russias, Alexis, suffered from hemophilia in which defective blood clotting leads to devastating hemorrhages. In 1915, at age eleven, when his affliction was particularly aggressive, he undoubtedly received substantial doses of aspirin for pain that arose from the bleeding into his joints. Since aspirin itself induces bleeding, a vicious cycle was established. After the boy's father, Tsar Nicholas II, had been persuaded to allow the monk Grigorii Efimovich Rasputin to manage his son's condition by mystical and spiritual means, physicians were dismissed, the administration of aspirin was stopped, and the boy seemed miraculously better. Rasputin's prestige rose to such heights that no one was able to throttle his malignant influence over the ruling house and the nation.

That adversity is just opportunity in disguise is aptly illustrated by the way aspirin rose to even greater glory after being humbled by its uncomfortable side-effects. It was aspirin's propensity to promote bleeding that turned out to be a double-edged sword—one side leading to doubt and worry about intractable hemorrhaging, the other to marvelous new and extended uses for preventing strokes and repeat heart attacks.

In order to understand this paradox we must learn how and why aspirin works, and that calls for the introduction of Sweden's Ulf von Euler and Bengt Samuelsson, and England's John Robert Vane—the scientists who discovered prostaglandins and figured out their relationship with aspirin. They earned Nobel Prizes in Medicine for this work.

In 1935 physiologist Ulf von Euler identified a family of hormone-like chemicals in the prostate glands of mammals which he labeled prostaglandins. Then, throughout the 1960s and 70's, Bengt Samuelsson of Stockholm's Karolinska Institute unraveled their actual composition, structure, and functions. He proved that there are more than a dozen biologically important forms of prostaglandins, that they are found in almost all tissues in the human body, and they play important roles in digestion, reproduction, circulation and in the immune system. In 1971 the British pharmacologist, Dr. Vane, with colleagues of London's Royal College of Surgeons proved that many of aspirin's actions stem from its ability to block the production of certain prostaglandins. The group demonstrated that any assault on the body triggers the production of prostaglandins which, as part of our normal defense mechanisms, give rise to inflammation, pain, and fever. Aspirin prevents the formation of prostaglandins, this interrupts the cascade of annoying symptoms, and one feels better.

Prostaglandins cover a wide spectrum of actions. Normally they protect the stomach, which explains why aspirin often upsets it. They initiate uterine excitability and contractions and are used medically to induce labor or to terminate pregnancy. Excessive amounts of prostaglandins cause headaches, fever, menstrual pain and cramps, inflammation and altered immune responses. Too little disrupts blood clotting, a cause of excessive, uncontrolled, bleeding. But, as we have seen, a particularly undesirable side-effect of aspirin, one that goes back almost to the time of its discovery, is its tendency to cause bleeding identical to that of prostaglandins deficiency. Could there be a connection?

Prostaglandins encourage platelets in blood to aggregate, to stick together, so they can effectively trap other elements in the blood to plug a hole in a blood vessel when it is torn or injured. It is a vital function in normal blood clotting but can be troublesome when it becomes part of the process that clogs blood vessels, causes heart attacks, and sends blood clots to the brain. Two groups, one headed by John Vane and the other by doctors J. Brian Smith and Anthony L. Willis, both working at Royal College of Surgeons Institute of Basic Medical Sciences in London, discovered that small amounts of aspirin block the action of enzymes in blood platelets that are needed to manufacture prostaglandins. They also found that two agents control platelet aggregation. *Prostacyclin,* a true prostaglandin, made both in platelets and in the lining of blood vessels, inhibits aggregation while *Thromboxane,* a prostaglandin derivative, stimulates it. Aspirin blocks the enzyme *Cyclo-oxygenase,* which is critical in the manufacture of both Thromboxane and Prostacyclin. Platelets are pretty much like red blood cells in that the moment they get into circulation they cannot generate any more enzymes like Cyclo-oxygenase. The lining of blood vessels however, can do so continuously. Although aspirin blocks Thromboxane and Prostacyclin from platelet sources, Prostacyclin production recovers quickly because the blood vessel linings keep up the supply. The end result is that platelets are prevented from aggregating. This means that they are less likely to block blood vessels or interrupt the blood flow to the heart or brain.

Think of platelet generated thromboxane and prostacyclin as a lion and tiger respectively and blood vessel lining-generated prostacyclin as a cat. They are all confined in a cage where aspirin acts like the bars, that is until the cat discovers that it can squeeze through the bars and does so. It can now interrupt platelet aggregation. This then provides the rationale for the use of aspirin in preventing second heart attacks and strokes.

To test this premise, at least six large-scale clinical trials were launched between 1974 and 1983. They proved that aspirin definitely reduced the rate of second heart attacks. In another United States study that covered 22,071 participants who were given one adult aspirin every other day and observed over a period of 60.2 months, 44% evidenced a reduction in heart attack risk compared with the group taking a placebo. In parallel studies, aspirin reduced the number of interruptions of blood circulation to the brain and strokes.

On the basis of initial promising reports the Food and Drug Administration, in 1985, approved the labeling of aspirin for treating patients with prior heart attacks and unstable or uncontrolled heart pain. Aspirin is an over-the-counter drug whose advertising is regulated not by the FDA but by the Federal Trade Commission, but promotion for specific uses requires FDA blessings. Having gained this approval, aspirin manufacturers coined a whimsical catchphrase: An aspirin a day may put a heart attack at bay.

It is entirely possible that aspirin's ability to ward off heart attacks could have been recognized as early as 1953. In that year Lawrence L. Craven, M.D., a general practitioner in

Glendale California, observed that none of the 1,465 overweight, sedentary men between 45 and 65, to whom he had administered one or two tablets of aspirin daily, had heart attacks or heart failure. When he reported this finding in the Mississippi Valley Medical Journal, he candidly admitted that " Such a finding is contrary to statistical expectations as well as to the consistent experience of 36 years in general practice." Unfortunately, Dr. Craven's paper was never taken seriously. It was partially discounted because he used no controls and did not make allowances for a placebo effect. The discoveries of the relationship between aspirin and prostaglandins, two decades later, vindicated him.

In 1953, the same year that Dr. Craven published his observations, a group at Harvard University reported in the October issue of the New England Journal of Medicine that when they tallied causes of death among aspirin-treated rheumatoid arthritis patients they found less coronary artery disease than would statistically be expected. Statisticians saw the relationship long before it appeared in the laboratory.

Russell H. Conwell, a well-known lecturer on the science and chemistry circuit during the 1920's and 30's, knew that aspirin was the most popular medicine in the world and was fascinated by the many novel uses for it that kept cropping up. The title for his lecture on aspirin was ACRES OF DIAMONDS IN YOUR OWN BACKYARD, an obvious intimation to the valuable, often unnoticed, and yet to be discovered, potentialities of aspirin.

Since that time, and since the discovery that aspirin inhibits prostaglandin biosynthesis, there has been an explosion of research which has resulted in new aspirin-like drugs and an extension of the uses of aspirin itself. Today, just short of being a century and a half old, aspirin is headed for a whole new career in therapeutics and preventive medicine—proving how prescient Conwell really was.

Here is a short summary of some of the practical and theoretical uses for aspirin:

1. It alleviates pain, fever, inflammation, and menstrual pains
2. It can reduce the risks of second heart attacks and strokes.
3. In an incidental observation, aspirin was found to reduce redness of the skin resulting from exposure to ultraviolet light. When taken early it can prevent sunburn.
4. Aspirin may raise or lower blood sugar levels, depending on a variety of circumstances. In 1876 Ebstein recognized the blood-sugar lowering effect of salicylates in diabetics and diabetes was treated with aspirin until the advent of insulin. New trials with prostaglandins in diabetes are still in progress.
5. Low doses of salicylates promote the build-up of proteins within the body; large doses have the opposite effect.
6. High cholesterol levels in the blood have been lowered by high doses of aspirin, 5 grams a day. This produces unfavorable side effects and is not recommended.
7. Aspirin imitates the actions of the adrenal and thyroid gland by increasing oxygen consumption, increasing the metabolic rate, decreasing cholesterol, and stimulating the production of cortisone.
8. Aspirin may precipitate asthma in some people; in others, it relieves the condition.
9. Changes in the electroencephalograms of humans taking aspirin resemble those seen with the administration of sedatives and tranquilizers, suggesting possible sedative and anti-anxiety effects.
10. Aspirin increases the excretion of the crystals responsible for gout.
11. Aspirin stimulates the immune system, thereby minimizing graft-versus-host reactions, interrupting shock-like allergic reactions, and helping to fight infections.
12. Aspirin has all the effects of salicylates plus many more derived from the acetyl part of the molecule.
13. It has been observed that patients who were taking aspirin for arthritis, over a period of eight to ten years, seemed to have less cataracts. Aspirin was also found to

either prevent or relieve other ophthalmic conditions such as corneal abrasions and optic nerve problems. These observations are now under investigation.
14. Scientists at Boston University School of Medicine, in a study of more than 1,300 patients with colon or rectal cancer, concluded that aspirin's anti-prostaglandins effect may play a role in stopping or slowing down the development of these cancers.

The discovery of aspirin's interactions with prostaglandins, its approval as a preventative of second heart attacks and strokes and sky-rocketing sales triggered massive searches for drugs that might duplicate or improve upon aspirin. This led to the manufacture of ibuprofen, fenamates, indomethacin, piroxicam and naproxen, which are classified as non-steroidal, anti-inflammatory drugs, NSAID. Despite having different chemical structures these new drugs have all the advantages and disadvantages of aspirin. They alleviate swelling, redness and pain of inflammation, reduce fever, and cure a headache; they also upset the stomach, delay the birth process and may even irritate the kidneys. When a chemically diverse group of drugs shares not only the same therapeutic quality but also the same side effects it is fairly certain that the action of those drugs are based on one common biochemical intervention—inhibiting the enzyme that regulates prostaglandins.

Experiments with aspirin indirectly gave us a whole new generation of drugs that could relieve problems associated with hyperacidity in the stomach. Once we learnt that aspirin caused stomach pain and that prostaglandins counteracted this effect—that is, they stopped hyperacidity in the stomach and thereby relieved such pains—pharmaceutical companies began to develop and manufacture prostaglandin analogues for this purpose. These will become the antacids of the future.

The final word on aspirin and aspirin-like drugs has not been written. Now, a century and a half after its discovery, aspirin appears to be on the threshold of a whole new career.

PARADIGM POINTERS

It took 36 years, the lifetime of a working scientist, to find a working paradigm for prostaglandins and aspirin's effects on them. Remember, Von Euler identified prostaglandins in 1935 and Vane made his discoveries in 1971.

The reason for this lies in the paradox that prostaglandins were to be found everywhere and nowhere. In 'classic' hormone research investigators can remove the gland or organ that makes a specific hormone and study the effects. This approach could not be used with prostaglandins, since they are made throughout the body rather than in one easy to study place; they could be everywhere. On the other hand, prostaglandins are not stored up but are produced on demand, which makes them nearly impossible to find.

Try to imagine that you are on shore trying to count cars crossing a long bridge. This bridge has turnstiles at both ends that allow only one car to enter when one leaves. By looking at this bridge sporadically you would hardly ever see a car and would swear that the bridge is always empty. This is essentially what happens with prostaglandins. No prostaglandins are released into the body until those already present disappear—exit through the turnstile—hence, almost impossible to find. It is easy to understand why it took so long to develop this leg of knowledge.

POSTSCRIPT

Our ancestors discovered many useful remedies in plants, minerals, insects, marine life and parts of animals; the search keeps going on. As recently as September 1995 Japanese investigators reported that something in the Chinese herbal medicine *Sho-saiko* controlled certain cancers in the liver, and prior to that, in the mid 1990's, the anti-cancer drug taxol was extracted from the Pacific yew tree. Advocates of natural remedies and modalities of healing based solely on nature have, for a long time, pointed to such accomplishments and urged that more time and money should be committed to analyzing the myriad herbal and folk remedies on hand and continue the search for new medications in the land and sea about us. In 1992 congress succumbed to these agitprops and ordered the National Institutes of Health to establish the Office of Alternative Medicine.

Before this can be called progress we ought to consider its invidious consequences. With few exceptions, most therapies derived from arcane natural sources lack conclusive proof of effectiveness. Some may actually do harm, as proven by the number of herbal poisonings reported every year. And where parts of animals are used medicinally, such as rhinoceros horn as an aphrodisiac, are we not encouraging the decimation of our endangered species?

Of the more than 500,000 natural-product extracts that have been tested for biotherapeutic activity, only one in 40,000 has ever shown any promise. And in those instances where nature handed us useful remedies they seldom, if ever, came to us in pure form. It was formal science that discovered the essential principles and then purified them. Once identified, active substances would then be manufactured or synthesized on a commercial basis, which usually brought the cost down and made the product universally available. Moreover, such products generally proved equal to or more effective than the natural product, and variants were developed that gave better performance as well as new and extended uses.

It is only human nature that makes us all look for bargains, in material goods as well as in health. Promises of a quick cure will always have a universal appeal, but when they are linked to 'natural healing' they automatically push people toward the occult or betray them by fostering hopes based on fallacy and emotionally charged illusion. When attempts are made to substantiate the effectiveness of natural panaceas we encounter such strong beliefs that it is impossible to separate fact from fancy or the real from the imaginary. Lavoisier, in the late 1700's recognized this when he stated, "Since the principle of life in animals is a force which is ever active, which is constantly endeavoring to overcome obstacles, and since nature when left to its own devices cures many diseases by itself, it follows that when a remedy is applied, it is infinitely difficult to determine what effects are due to nature and what to the remedy. The result presents itself to the wise man merely as a greater or lesser probability, and that probability can be converted into certainty only by a large number of facts of the same kind."

Obviously, it is not enough to accept anecdotal evidence of a remedy's effectiveness; strictly controlled scientific trials through double-blind studies provide the best safeguards for society. Simply having controls is not the absolute answer because subjects often experience the placebo effect, enjoying a miraculous cure even though no medication was administered. In the double-blind study there are two identical groups of patients, one is given medication under test and the other, the control group, receives the placebo; neither the patient nor the physician knows who receives the real medication, hence the term double-blind. Although not perfect, it is the best way to evaluate drugs and procedures.

Sometimes there is no sharp line between the useful and the useless; one may merge into the other. For example, Dr. Jonathan Moore's Essence of Life, a popular concoction in the late 1800s, professed to cure coughs, colds, asthma, influenza, whooping cough, consumption, indigestion, and dysentery. It contained morphine derivatives that eased coughs but simultaneously caused constipation. When the claims were branded as fraudulent, the product was marketed as a cure for dysentery; in this perspective it worked extremely well and thereby regained respectability. Only later, when its addictiveness became apparent, was it quashed.

It is not known what sort of a degree Dr. Moore possessed but even now people with prestigious university credentials, ostensibly members of the scientific community, regularly champion unproven therapies, often personal whims. Bookstores are loaded with tracts by authors with MD's and Ph.D's after their names, who, motivated by conviction or monetary gain, advocate unorthodox diets, antioxidants, super-vitamins, trace minerals, and remedies ranging from extracts of cactus to sundry herbs. Some of these items may have a tiny segment of truth, but deliberate obscurantism often misleads the public about their absolute value. Makers of spurious treatment devices and occult paraphernalia add more confusion to the public mind.

Notwithstanding science's insistence upon safeguards against charlatanism, we are currently seeing a gradual encroachment of non-science into the health-care sector. In a headlong flight toward Twentieth Century shamanism the city of Seattle, Washington, set up a government-subsidized naturopathic health clinic. The decision to follow through on this project was made, as the New York Times reported, after "rhapsodizing about garlic pills and the healing power of ginkgo-tree extracts." Also, the state of Washington has mandated that health insurers, as of January 1, 1996, can no longer deny payment for licensed natural health care. Medical insurers will be paying for massage, herbalism, acupuncture, and a host of other fringe-type therapies. And, as noted above, the National Institutes of Health now has an Office of Alternative Medicine. With a budget of $14 million, they will look at natural remedies, which in itself is not reprehensible, except that it funnels money away from bonafide research into cancer, detached retinas, and the many crippling conditions which are still untreatable.

There are two possible explanations for this drift toward non-science. The first is that we are witnessing a normal historical cycle similar to the strong reactionary backlash that followed the Age of Reason. The second is that society has been intimidated by the overwhelming advances of science over the past fifty years; people fear what they cannot understand. In either case we have nothing to worry about. In the first instance, time will take us through a full cycle, the regressionary period will wane and rationality will return. In the second instance, better comprehension will obliterate cynicism and fear. Just as the current generation of children accept space travel and computers, they will take in stride the complex technology that they grow up with and thereby escape the doubts and fears of their parents. In the meantime we can only look upon our present plight with forbearance and a bit of mirth—as depicted in the **New Yorker**. The caption under a cartoon showing a doctor collecting payment from a patient, reads: Medical science can do nothing for this illness but fortunately for you I am a quack.

SOURCES AND REFERENCES GENERAL

Aaseng, N. The disease fighters, the nobel prize in medicine. Minneapolis: Lerner Publications Company,1987.

Asimov I. Asimov's biographical encyclopedia of science and technology. New York: Doubleday, 1982.

Barber B. Resistance by scientists to scientific discovery. Science, 1961.

Burnham J. How superstition won and science lost: popularizing science and health in the United States. New Brunswick, NJ: Rutgers University Press, 1987.

Chandler C. Famous men of medicine. New York: Dodd Mead, 1950.

Clark RW. Einstein: the life and times. New York: World Publishing Company, 1971.

Dibner B. The new rays of professor roentgen. Norwalk: Burndy Library, 1967.

Dubos RJ. Louis Pasteur free lance of science. Boston: Little, Brown and Company, 1950.

Galton L. Medtech: the lay persons guide to today's medical miracles. New York: Harper and Row, 1985.

Gibson J. Great doctors and medical scientists. London: Macmillan, 1967.

Gibson W. Young endeavor: contributions to science by medical students of the past four centuries. Springfield, Illinois: Charles C. Thomas, 1958.

Gleick J. Chaos. New York: Penguin Books, 1987.

Kuhn TS. The structure of scientific revolutions, Chicago: The University of Chicago Press, 1962, 1970.

LaFollette M. Making science our own: public images of science. Chicago: University of Chicago Press, 1990.

Lehrer S. Explorers of the body. New York: Doubleday & Company, Inc., 1979.

Lord Cohen of Birkenhead. The fruits of error and false assumption. Nuffield Lecture. Proc R. Soc Med July 1976; 60:673–682.

Lyons AS. Medicine, an illustrated history. New York: Harry N. Abrams, 1978.
Marti-Ibanez F. A prelude to medical history, New York: M D Magazine, 1961.
Morton L. A medical bibliography: an annotated check-list of texts illustratig the history of medicine, 4th ed. London, England: Gower, 1983.
Nuland SB. Doctors. New York: Alfred A. Knopf, 1988.
Pflaum R. Grand obsession: Madame Curie and her world. New York: Doubleday, 1989.
Rapport S, Wright H. Great adventures in medicine. New York: Dial Press, 1952.
Riedman S, Gustafson ET. Portraits of nobel laureates in medicine and physiology. London: Abelard-Schuman, 1963.
Robbin I, Nisenson S. Giants of medicine. New York: Grosset & Dunlap, 1962.
Siegerist HE. Great doctors. London: George Allen & Unwin, 1933.
Singer C. (updated by Underwood EA). A short history of medicine. New York: Oxford University Press, 1962.
Thomas L. The youngest science. New York: Bantam, Dell, Doubleday, Publishing Group, 1984.
Thomas L. The lives of a cell: notes of a biology watcher. New York: Viking Press, 1974.
Trefil J. Reading the mind of GOD. New York: Charles Scribner's Sons, 1989.
Weatherall M. In search of a cure. New York: Oxford University Press, 1990.
Wear A, ed. Medicine in society: historical essays. New York: Cambridge University Press, 1992.
Weisskopf V. The joy of insight: passions of a physicist. New York: Alfred P. Sloan Foundation Series, Basic Books/Harper Collins Publishers, 1991.
Zinsser H. Rats, Lice, and history. Boston: The Atlantic Monthly Press by Little, Brown and Co., 1947.

SOURCES AND REFERENCES FOR CHAPTER 1 RIGHT DEED FOR THE WRONG REASON

Calvery HO, Klumpp TG. The toxicity for human beings of diethylene glycol with sulfanilimide. South Med J 1939; 32:1105–1109.

Chain EB. The Trueman Wood lecture. London: Royal Society of Arts, June 19, 1963.

Colebrook L, Kenny M. Treatment of puerperal infection and experimental infections in mice, with Prontosil. Lancet 1936; I:1279.

Domagk G. Chemotherapie der tuberkulose mit den thiosemikarbazonen. Stuttgart, Germany: Thieme, 1950; MDNM (c-1662 DNLM) MNUM (c9665 MnU-B).

Marquardt M. Paul Ehrlich. New York: Henry Schuman, 1951.

Mietzch F. Chemie und wirtschaftliche bedeutung der sulfanamide. Westdeutscher Verlas, Kohn: 1954; MDNM (c-9662 DNLM).Morton L. A medical bibliography: an annotated check-list of texts illustrating the history of medicine, 4th ed. London:Gower, 1983.

Multidrug resistant tuberculosis poses challenge, editorial. JAMA Feb.12, 1992; 267:6:786.

Otten H. Domagk and the development of sulphonamides. (Communication) J Antimicrobial Chemotherapy; 1986; 17:689–90. (Dr. Hinrich Otten was Domagk's youngest associate and one of his last collaborators; he was still at Bayer AG Pharma Research Centre in Wuppertal, when he wrote this report.)

Petersdorf RG. Brief History of Sulfonamides. JAMA, March 16, 1984; 251:11:1475–6.

Ravitch MM. The danger of stating an opinion of the hazard of going on record. Surg Gyn & OB 1980; 151:810–186.

Silverman M. Magic in a bottle. New York: The Macmillan Company, 1941.

Snider DE, Simone PM, Dooley SW, Bloch AB. Multi-drug-resistant tuberculosis. Scientific American Science and Medicine May/June 1994; 1:2:16–25.

Snow E. Red china today. New York: Vintage Books, 1971.
Tuberculosis among homeless shelter residents. JAMA January 22/29 1992; 267:4.
Interview with : Bob Parks and Raymond Teichman, Archivists of the Franklin D. Roosevelt Library at Hyde Park, New York, April 1983.

SOURCES AND REFERENCES FOR CHAPTER 2 A LESSON FROM THE WIZARD OF OZ

Ann Arbor News Note 1924. Courtesy of the Editor in 1974.
Chesney AM, Clawson TA, Webster B. Endemic goiter in rabbits: incidence and characteristics. Bull Johns Hopkins Hospital 1928; 3:262–277.
David Marine and Carl Lenhart: Thyroid Pioneers. The Bulletin of the Cleveland Medical Library; XXV:3, July 1979.
DeMaeyer EN, Lowenstein FW, Thilly CA. The control of endemic goiter. Geneva, Switzerland; Published by World Health Organization, 1979.
Hennesey WB. Goiter prophylaxis in New Guinea with intramuscular injections of iodized oil. Medical Journal of Australia, 1964: 1:505
Langer P, Greer MA. Antithyroid and naturally occuring goitrogens. New York: S. Kargaer (Basel, Switzerland), 1977.
MacKenzie CG, MacKenzie JB, McCollum EV. Effects of sulfonamides and thioureas on the thyroid gland and basal metabolism. Endocrinology 1943; 32:185–209.
Marine D, Baumann EJ, Cipria A. Studies on simple goiter produced by cabbage and other vegetables. Proc Soc Exp Biol and Med 1929; 26:822.
Marine D, Kimball OP. Prevention of simple goiter in man. Arch Int Med 1920; 25:661.
Marine D, Lenhart CH. Occurrence of goiter in fish. Bull Johns Hopkins Hospital 1910; 21:95.

Marine D. Etiology and prevention of simple goiter. Medicine 1924; 3:463.
Merke F. Ed history and iconography of endemic goiter and cretinism. Boston: MTP Press a division of Kluwer, 1984.
Pitt-Rivers R, Trotter WR. The thyroid gland. Washington,DC:Butterworth Inc., 1964.
Recent Progress in diagnosis and treatment of hypothyroid conditions. In: Bastenie PA, Bonnyns M, Vanhaelst L. eds. Proceedings of the XVTH International Congress of Therapeutics, Brussels, September 5–9, 1979 Published by Excerpta Medica, Amsterdam 1980.
Richter CP, Clisby KH. Graying of hair produced by phenylthiocarbamide. Proc Soc Exp Biol Med 1941; 48:684–687.
Rushdie S. Errata:on unreliable narration in Midnight's Children. In: Imaginary Homelands: Essays and Criticism, 1981–1991. Granta Books (in association with Viking Penguin), New York
Stanbury JB, Hetzel BS. Endemic goiter and endemic cretinism. New York: John Wiley and Sons, Inc., 980.

SOURCES AND REFERENCES FOR CHAPTER 3 ELEGY FOR A PLANET

Allen JR. Response of the nonhuman primate to PCB exposure. Fed Pro 1975; 34:1675–1679.
Allen JR, Norback DH. Pathobiological responses of primates to polychlorinated biphenyl exposure. Natl Conf Polychlorinated Biphenyls (November 1975, Chicago, Illionis), EPS-560/6–75–004, Washington, D.C: Environmental Protection Agency 1976; 43–49.
Blum A, Ames BN. Flame-retardant additives as possible cancer hazards. Science 1977; 195:17–23.Carter LJ. Michigan's PBB incident: chemical mix-up leads to disaster. Science 1976; 192:240–243.Centers for disease contol. MMWR 1978; 27:14:115.

Data for background information relative to PBB's in breast milk. Michigan Department of Public Health August 1976.Department of Health, Education and Welfare (DHEW): Subcommittee on Health Effects of PCBs and PBBs. Series of articles appearing in Environ Health Persp 1978; 24:146–198.

Dunckel AE. An updating on the polybrominated biphenyl disaster in Michigan. J Am Vet Med Assoc 1975; 167:838.

Harada M. Intrauterine poisoning: clinical and epidemiological studies and significance of the problem. Bull Inst Constitutional Med (Supp) 1976; 1.

Humphrey HEB, Hayner NS. Polybrominated biphenyls: an agricultural incident and its consequences, an epidemiological investigation of human exposure. Presented at the Ninth Annual Conference on Trace Substances in Environmental Health, Columbia, Missouri: June 1975.

Jackson TF, Halbert FL. A toxic syndrome associated with the feeding of polybrominated biphenyl-contaminated protein concentrate to dairy cattle. J Am Vet Med Assoc 1974; 165:437–439.

Jones RS. Physics for the rest of us. Chicago: Contemporary Books, 1992.

Kimbrough RD, Burse VW, Liddle JA. Toxicity of brominated biphenyls. Letter to editor. Lancet 1977; 2:602–603.

Kuratsune M, Yoshimur T, Matsuzaka J, Yamaguchi A. Epidemiologic study on Yusho, a poisoning caused by ingestion of rice oil contaminated with a commercial brand of polychlorinated biphenyls. Environ Health Perspect 1972; 1:119.

Miller RW. Pollutants in breast milk. J Ped 1977; 90:510.

Selikoff IJ, Anderson, HA, Wold MJ. Investigation of health effect of PBB exposure among Michigan chemical company workers. Presented at a Workshop on Scientific Aspects of Polybrominated Biphenyls, East Lansing, Michigan; 24–25 October 1977.

Selikoff IJ. Summary report on human health effects of exposure to polybrominated biphenyls.Presented at a Workshop on Scientific Aspects of Polybrominated Biphenyls, East Lansing,Michigan; 24 – 25 October 1977.

Stadtfeld CK. Cheap chemicals and dumb luck. Audubon 1976; Jan: 110–118.

Sturman JA, Rassin DK, Gaull GE. Taurine in developing rat brain: transfer of (35S) taurine to pups via the milk. Ped Res 1977; 11:28.Rassin DK, Sturman JA, Gaul GE. Taurine in milk: species variation. (abstr.) Ped Res 1977; 11:449.

Wilcox KR. Mothers' milk as a chemical transport medium. Presented at the Toxicology Forum, Washington, D.C., 21 Feb 1978.

Winters RW. Infant nutrition. (Address) Society of Ped Res; May 1976.

SOURCES AND REFERENCES FOR CHAPTER 4 AFFLICTIONS OF AFFLUENCE

Center for Disease Control: Legionnaires' disease: diagnosis and management. Ann Int Med March 1978; 88:363–65.

Fraser DW, McDade JE. Legionellosis. Scientific American Oct 1979; 241:82–99.

Glick TH, Gregg MB, Berman B, et.al. Pontiac fever: an epidemic of unknown etiology in a health department: clinical and epidemiologic aspects. Am J Epidem Jan 1978; 107:149–60.

Hudson RP. Lessons from legionnaires disease. Ann Int Med 1978; 90:704–7.

SOURCES AND REFERENCES FOR CHAPTER 5 TOO MUCH OF A GOOD THING

Ashton N, Henkind P. Experimental occlusion of retinal arterioles (using graded glass ballotini). Br J Ophthalmol 1965; 49:225.

Ashton N, Ward B, Serpell G. Role of oxygen in genesis of retrolental fibroplasia: preliminary report. Br J Ophthalmol 1953; 37:513.
Ashton N, Ward B, Serpell G. Effect of oxygen on developing retinal vessels with particular reference to the problem of retrolental fibrioplasia. Br J Ophthalmol 1954; 38:397.
Ashton N. The story of blindness in premature babies. Notes from the ordinary Meeting of Monday 28 March 1988.California State Department of Public Health: Statement on oxygen administration with reference to retrolental fibroplasia, 1955.
Campbell K. Intensive oxygen therapy as a possible cause of retrolental fibroplasia: a clinical approach. Med J Aust 1951; 2:48.
Committee on Fetus and Newborn: Hospital care of newborn infants: full term and premature, reved. Evanston, Illinois: American Academy of Pediatrics, 1957.
Comroe JH, Botelho S. The unreliability of cyanosis in the recognition of arterial anoxemia. Am J. Med Sci 1957; 214:1.
Crosse VM, Evans PJ. Prevention of retrolental fibroplasia. Arch Ophthalmol 1952; 48:83.
Crosse VM. Retrolental fibroplasia. Proc R. Soc Med 1950; 43:232.
Gordon HH. Oxygen administration and retrolental fibroplasia. Pediatrics 1954; 14:543.
Johnson L, Schaffer D, Bogg TR Jr. The premature infant, vitamin E deficience and retrolental fibroplasia. Am J Clin Nutr 1974; 27:1158–73.
Kinsey VE, Arnold HJ, Kalina RE, et al. PaO2 levels and retrolental fibroplasia: a report of the cooperative study. Pediatrics 1977; 60:655–68.
Kinsey VE. Retrolental fibroplasia: cooperative study of retrolental fibroplasia and the use of oxygen. Arch Ophthalmol 1956; 56:481–543.
Lucey J. Detailed study on retrolental fibroplasia. J Ped 1984; 73.
Owens WC, Owens EU. Retrolental fibroplasia in premature infants. II. Studies on the prophylaxis of the disease: the use

of alpha tocopheryl acetate. Am J Ophthalmol 1949; 32:1631–7.

Owens WC, Ownes EU. Retrolental fibroplasia in premature infants. Am J. Ophthalmol 1959; 32:1.

Patz A. Symposium on retrolental fibroplasia: summary. Ophthalmology 1979; 86:1871–3.

Patz A. The role of oxygen in retrolental fibroplasia. (E. Mead Johnson Award Address) Pediatrics 1957; 19:504.

Reese AB. Retrolental fibroplasia. Am J. Ophthalmol 1951; 34:763.

Terry TL. Extreme prematurity and fibroblastic overgrowth of persistent vascular sheath behind each crystalline lens. I. Preliminary report. Am J Ophthalmol 1943; 25:203.

Terry TL. Ocular maldevelopment in extremely premature infants: retrolental fibroplasia: VI. General consideration. JAMA 1945; 128:582.

Warkany J, Schraffenberger E. Congenital malformations of eyes induced in rats by maternal vitamin A. deficiency. Proc Soc Exp Biol Med 1944; 57:49.

Zacharias L. Retrolental fibroplasia: a survey. Am J. Opthalmol 1952; 35:1426.

I am indebted to the following experts for their assistance in preparing this chapter:

Norman P. Blair, MD, Professor of Opthalmology, University of Illionis, Eye and Ear Infirmary, Chicago, Illinois.

Susan Bressler, MD, Associate Professor of Ophthalmology, Wilmer Ophthalmology Institute, Johns Hopkins University, School of Medicine, Baltimore, MD.

M. Christina Leske, MD, MPH. Professor of Preventive Medicine and Ophthalmology. Chair: Dept of Preventive Medicine. State University of New York, Health Sciences Center, Stony Brook, New York.

Johanna M. Seddon, MD, Associate Professor of Ophthalmology and Epidemiology, Harvard University, Boston, MA.

Yale Solomon, MD, Associate Professor of Opthalmology, Stonybrook School of Medicine and emeritus Chief of Opthalmology, South Side Hospital, Bayshore, New York.

SOURCES AND REFERENCES FOR CHAPTER 6 FORESIGHT, NEARSIGHT, AND INSIGHT

Curtin, BJ. The myopias: basic science and clinical management. Philadelphia: Harper and Row, 1985.
Duke-Elder S, Abrams D. System of ophthalmology. Vol 5. Ophthalmic optics and refraction. St. Louis: The CV Mosby Co., 1970.JAMA (Editorial) Feb 23, 1990; 236:8.
Podolsky ML. Phototherapeutic keratectomy: from serendipity to high-tech. Ophthalmology World News Oct 1995; 1:10,17,18,30.
Waring GO. Refractive keratotomy. St. Louis, Missouri: Mosby-Year Book In., 1992.

I am indebted to the following experts for their assistance in preparing this chapter:
Andrea DeVeau and Tom Kennedy at the American Academy of Ophthalmology, San Francisco, CA.
Lisa Kelley, MD, Assistant Professor of Ophthalmology, Stanford University Medical Center.
Gerald B. Kara, MD, Former Director of Ophthalmology at the New York Eye and Ear Infirmary.
Marguerite B. McDonald, MD, Refractive Surgery Center of the South, Eye Ear Nose and Throat Hospital, New Orleans, LA.
James D. Reynolds, MD, University of Arkansas School of Medicine.
Ivan Schwab, MD, Professor of Ophthalmology, UC Davis, Davis CA.

SOURCES AND REFERENCES FOR CHAPTER 7 CAPTAIN OF MEN OF DEATH

Austrian R. Life with the pneumococcus: Notes from the bedside laboratory, and library. Philadelphia, University of Penn Press, 1985.

Austrian R, Bennett IL. Pneumococcal infections. 7th ed. M.M. Wintrobe, eds. In: Harrison's principles of internal medicine: New York, McGraw-Hill Book Co., 1974:771.
Osler W. The principles and practice of medicine: 4th ed. New York, D. Appleton, 1901:108.
Pasteur, L. (with MM Chamberland and Roux). Sur une maladie nouvelle, provoquee par la salive d'un enfant mort de la rage. C R Acad Sci (D) Paris 1881; 92:159–165.
Sabin AB. Immediate pneumococcus typing directly from sputum by the Neufeld reaction. JAMA 1933; 100:1584–1586.
Sternberg GM. A fatal form of septicemia in the rabbit produced by subcutaneous injection of human saliva. Natl Board of Health Bull 1881; 3:87–108.
Thomas L. The youngest science. New York: The Viking Press, 1983.

SOURCES AND REFERENCES FOR CHAPTER 8 OXFORD INCIDENT

Agricultural Research Service of the United States Department of Agriculture.
Chain EB. in Lyght CE (ed): Reflections on research and the future of medicine. New York: McGraw-Hill Book Co., 1967.
Chain EB. Thirty years of penicillin therapy. Proceedings of the Royal Society, London; 1971; 179, 293.
Florey HW. Antibiotics. London: Oxford University Press, 1949.
Jaffe IA. D-penicillamine. Bull Rheum Dis 1977–78; 28:948–52.
MacFarlane G. Howard Florey: The making of a great scientist. New York: Oxford University Press, 1979.
Mayberry DH. Penicillin chronology northern regional research center. Science and Education Adm, US Dept of Agri 1980.

Richards AN. Production of penicillin in the United States (1941–1946). Nature, February 1, 1964; 201:441.
Tyndall J. Essays on the floating-matter of the air in relation to putrefaction and infection. New York: D Appleton and Co, 1882.
Willson D. In search of penicillin. New York: Alfred A. Knopf, 1976.
Unpublished penicillin files at the Northern Regional Research Center, United States Department of Agriculture, Agricultural Research, North Central Region, 2000 West Pioneer Parkway, Peoria, Ill 61615.
Weinstein L. in Goodman LS, Gilman A (eds): The pharmacological basis of therapeutics. New York: Macmillan Publishing Co Inc, 1975.

SOURCES AND REFERENCES FOR CHAPTER 9 ORPHANS AND THEIR RELATIVES

AAP News, (American Academy of Pediatrics), October 1995; 13
Carter R. Breakthrough: the saga of Jonas Salk. New York: Trident press, 1966.
Dubos RJ. The professor, the institute and DNA. New York: Rockefeller University Press, 1976.
Fenner FJ, et al. The biology of animal viruses. New York: Academic Press, 1968.
Fisher PJ. The polio story. London: Heinemann, 1967.
Goodfield J. Quest for the killers. Cambridge MA: Birkhuser Boston, Inc, 1985.
Journal of the American Medical Association, short report, Jan 22/29 1992; 267:4:473
Smith JS. Patenting the sun: polio and the salk vaccine. New York: William Morrow, 1990.
Watson JD. The double helix. New York: New American Library, 1968.

Time Magazine, October 30, 1995; 83.

SOURCES AND REFERENCES FOR CHAPTER 10 JANUS AND THE CANCER CONNECTION

Rosenberg B, Van Camp L, Crigas T. Inhibition of cell division in escherichia coli by electrolysis products from a platinum electrode. Nature 1965; 205:698–699.
Thomas L. The youngest science. New York: Bantam, Dell, Doubleday, Publishing Group, 1984. Chapter 18: 181–183

SOURCES AND REFERENCES FOR CHAPTER 11 CONCEPT OUT OF CHAOS

Agardh E, et al. The prevalence of retinopathy and associated medical risk factors in type 1 (insulin dependent) diabetes mellitus. J. Int Med 1989; 226:47–52.
Am Diabetes Assn. Vital Statistics, 1991.Blueford P. Opthalmic indications under study at Celtrix. Personal communication April 1992.
Borisuth N, et al. Quantification of TGF-Beta, EGF, and bFGF in vitreous of patients with proliferative diabetic retinopathy (abstract). Invest Ophthal Vis Sci 1991;32: 1028
Bressler NM, et al. Age-related macular degeneration. Surv Ophth 1988; 32:375.
Campbell MT, McAvoy JW. Onset of fibre differentiation in cultured rat lens epithelium under the influence of neural retina-conditioned medium. Exp Eye Res 1984; 39:83–94.
Charles S, Flinn CE. The natural history of diabetic extramacular traction retinal detachment. Arch Ophthalmol 1981; 99:66–68.

Cohen S. Isolation of a mouse submandibular gland protein accelerating incisor eruption and eye lid opening in the new born animal. J Biol Chem 1961; 237:1555.

Connor TB, et al. Correlation of fibrosis in transforming growth factor-beta type 2 levels in the eye. J Clin Invest 1989; 83:1661–1666.

Dills DG, et al. Association of elevated IGF–1 levels with increased retinopathy in late onset diabetes. Diabetes 1991; 40:1725–30.

Dwyer MS, et al. Incidence of diabetic retinopathy and blindness; a population based study in Rochester, Minnesota. Diabetes Care 1985; 8 (no.4):316–322.

Frank RN. On the pathogenesis of diabetic retinopathy. Opthal May 1991; 98(No.5):586–93.

Fredj-Reygrobellet, et al. Acidic FGF and other growth factors in preretinal membranes from patients with diabetic retinopathy and proliferative vitreoretinopathy. Ophthalmic Res 1991; 23:154–161.

Gillis J, McIntyre L. Growth factors and their promising future. J Am Optom Assn 1989; 60:442–5.

Haik GM, Jr, Terrell, Lee W III, Haik, GM Sr. Macular degeneration: the major cause of severe vision loss in persons fifty-five years or older. J Mississippi State Med Assoc July1989; 30, #7:207–210.

Hilton GF, McLean EB, Chuang EL. Retinal detachment, Ed 5. San Francisco: American Academy of Ophthalmology Monographs Program, 1989.

Hjortdal J, Ehlers N. Exogenous, ocular, and systemic factors associated with keratitic ulceration. Acta Ophth 1989: 67:169–173.

Holsgrefe, JG. Memorandum to the Lash Group, June 20, 1991.

Hyman L. Epidemiology of eye disease in the elderly. Eye 1987; 1:330–341.

Kehrl JH, Taylor AS, Delsing GA, Roberts AB, Sporn MB, Fauci AS. Further studies of the role of TGF-Beta in human B cell function. J Immunol 1989; 143:1868–1874.

Kirschner SE, Brazzel RK, Stern m, et al. The use of EGF for treatment of persistent corneal erosions. J Am Vet Med Assoc, in press.

Kirschner SE. Persistent corneal ulcers: what to do when ulcers won't heal. Vet Clinics of NA: Small Animal Pract 1990; 20:3.

Kronke M, Leonard WJ, Depper JM, Arya SK, Wong-Staal F, Gallo RC, Waldmann TA, Greene WC. Cyclosporin A inhibits T cell growth factor gene expression at the level of mRNA transcription. Proc Natl Aca Sci USA 1984; 81:5214–5218.

Leschey KH, Hackett SF, Singer JH, Campochiaro PA. Growth factor responsiveness of human retinal pigment epithelial cells. Investigative Ophthalmology & Visual Science 1990; 31:839–46.

Levi-Montalcini, R. In praise of imperfection: my life and work. New York: Basic Books, Inc, 1988.

Limb GA, Little BC, Meager A, Ogilvie JA, Wolstencroft RA, Franks WA, Chignell AH, Dumonde DC. Cytokines in proliferative vitreoretinopathy. Eye 1991; 5, 686–693.

McAvoy JW, Chamberlain CG. Growth factors in the eye. in Progress in growth factor research. Great Britain: Pergamon Press, 1990.

Moss SE, et al. The incidence of vision loss in a diabetic population. Ophthal 1988; 93:1340–1348.

Nader D, Schwartz B. Cataract surgery in the United States, 1968–1976. Ophthalmology Jan 1980: 87:10–15.

Nishida T, Ohashi Y, Awata T, et al. Fibronectin. a new therapy for corneal trophic ulcer. Arch Ophthalmol 1983; 101:1046–1048.

Ohashi Y, Motokura M, Kinoshita Y, et al. Presence of epidermal growth factor in human tears. Invest Ophthal Vis Sci 1989; 30:1879–1882.

Paavan-Lansteon D (Ed). Manual of ocular diagnosis and therapy. Little Brown and Co, 1985.

Parikh PK. The sicca (Sjogren's) syndrome. Ent J 1987; 66:36–43.

Pena R, Jerdan J, Glaser B. 1992 ARVO Abstract Pesach S, et al. The prevalence of diabetic retinopathy: effect of sex, age, duration of disease, and mode of therapy. Diabetes Care 1983; 6 (No.2) :149–51.

Pizzarello LD. The dimensions of the problem of eye disease among the elderly. Ophthalmology 1987; 94:1191–1195.

Rand LI. Retinopathy: what to look for. Clin Diabetes 1983; 1:14–18.

Savage CR, Cohen S. Proliferation of corneal epithelium induced by epidermal growth factor. Exp Eye Res 1973; 15:361–366.

Schepens CL. Retinal detachment and allied diseases. Philadelphia: WB Saunders Co, 1983.

Smiddy WE, et al. Transforming growth factor beta: a biological choriorentinal glue. Arch Ophthalmol April 1989; 107:577–580.

Sporn M, et al. Peptide growth factors and their receptors,in Sporn M, Roberts A (eds): Handbook of Experimental Pharmacology. Berlin: Springer-Verlag, 1990.

Statistical Abstract of the United States 1986, 106th Edition. US Department of Commerce. Bureau of The Census. Surgical operations on the Eye in the United States in Short Stay Hospitals 19671–1983; also longevity.

Straatsma BR, Foos RY, Feman SS. Degenerative diseases of the peripheral retina. in Duane TD (ed): Clinical ophthalomogy. Philadelphia: Harper and Roe Publishers, Inc., 1983.

Talal N. The autoimmun diseases (Sjogren's Syndrome). Academy Press, 1985.

Wang H, Berman M, Law M. Latent and active plasminogen ativator in corneal ulceration. Invet Ophthal Vis Sci 985; 26:511–524.

Weidemann P, Weller N. The pathophysiology of proliferative ritreoretinopathy. Acta Ophth 1988; 66:supple 189.

Weidemann P. Growth factors in retinal diseases: proliferative vitreoretinopathy, proliferative diabetic retinopathy, and retinal degeneration. Surv Ophthalmol arch-April 1992; 36 (5):373–384.

SOURCES AND REFERENCES FOR CHAPTER 12 BING, BANG, LEVIN AND THE HORSESHOE CRAB

Bing RJ. An investigators journey in cardiology. JAMA 1991; 67:969–992.

Bing,RJ. ed. Cardiology:the evolution of the science and the art. Switzerland: Harwood Academic Publishers, 1992.

Cohen E, Bang FB, Levin J, et al. eds. Biomedical applications of the Horseshoe Crab (Limulidae). New York: Alan R. Liss, 1979.

Levin J, Bang FB. The role of endotoxin in the extracellular coagulation of Limulus blood. Bull Johns Hopkins Hosp 964; 115:265–274.

Levin J, Bang FB. A description of cellular coagulation in the Limulus. Bull Johns Hopkins Hosp 1964; 115:337–345.

Levin J, Bang FB. Clottable protein in Limulus: its localization and kinetics of its coagulation by endotoxin. Throm Diath Haemorrh 1968; 19:186–197.

Levin J, Buller, HR, Ten Cate JW, van Deventer SJH, Sturk A, eds. Bacterial Endotoxins. pathophysiological effects, clinical significance, and pharmacological control. New York: Alan R. Liss, 1988.

Levin J, Poore TE, Zauber NP, Oser Rs. Detection of endotoxin in the blood of patients with sepsis due to gram-negative bacteria. New Eng J Med 1970; 283:1313–1316.

Levin J. Blood coagulation in the horseshoe crab (Limulus polyphemus): A model for mammalian coagulation and hemostasis. In "Animal Models of Thrombosis and Hemorrhagic Diseases". (Proceedings of a Symposium of the National Academy of Sciences) DHEW Publication No. (HIH) 76–982:87–96, Washington DC. US Department of Health, Education and Welfare.

Levin J. The limulus amebocyte lysate test: perspectives and problems. Progress in Clinical and Biological Research Vol. 231/ . New York: Alan R. Liss, Inc., 1987.

Lewis A. The Horseshoe Crab—a reminder of Delaware's past. University of Delaware Sea Grant Marine Advisory Service Publication 1983; No. 6, Rev.
Liles G. Late nite thoughts of a horseshoe crab. MD Sept 1990; 66–76.
Mullen RJ. Neumann biotechnologies, Inc. Personal Communications, 1990.
Shuster CN. Distribution of the American Horseshoe crab, limulus polyphemus. Progress in Clinical and Biological Research, Vol 29. New York: Alan R. Liss, Inc., 1979.
Smith DC. Horseshoe crabs—from prehistory to modern medicine. Clemson University Extension Service Publication, No. 16–a, March 1976.
Thomas L. The lives of a cell. New York: The Viking Press, 1974.
Watson SW, Levin J, Novitsky TJ. Eds, Endotoxins and their detection with the limulus amebocyte lysate test. New York: Alan R. Liss, 1982.
Watson SW, Levin J, Novitsky TJ. Detection of Bacterial Endotoxins with the limulus amebocyte lysate test. New York: Alan R. Liss, 1987.
Weissman G. Editorial. MD Sept 1990.
Zuckerman KS, Quesenberry PJ, Levin J, Sullivan R. Contamination of erythropoietin by endotoxin: in vivo and in vitro effects on murine erythropoiesis. Blood 1979; 54:156–158.
Personal Interview Dr. Jack Levin, VA Hospital San Francisco, April 1992.

SOURCES AND REFERENCES FOR CHAPTER 13 MIGHTY LIKE A ROSE

Ames K. A little bit of wonder. NEWSWEEK; May 1 1989; 74.
Cohen S. Bumps on your blood cells - key to catching a disease. San Jose: Mercury-News March 11, 1978.

Conley FK. And ladies of the club. JAMA, Feb 5, 1992; 267–740.
Early transplantation of bone marrow. University of Wisconsin Hospital and Clinics; Winter 1990.
Goben R. Peninsula portrait. Palo Alto: Times; Octo 1, 1965.
Press H. Stanford University medical center news bureau. Stanford Medicine Fall 1987.
Roca O. An opera star soars past bout with cancer. INSIGHT; July 2, 1990:58–59.
Stanford (Alumni) Magazine. Rose Payne Curriculum Vitae and press clippings.
Stanford Medicine. Fall issue 1987.
Personal Interview with Dr. Rose Payne at Stanford University Blood Bank, March 1992.

SOURCES AND REFERENCES FOR CHAPTER 14 BACKYARD FULL OF DIAMONDS

Burlingham R. The odessey of modern drug research. Kalamazoo: Upjohn Company, 1951.
Fields WS, Lemark NA, Frankowski RF. Controlled trial of aspirin in cerebral ischemia. Stroke 1977: 8:301
Garnett HJ. A randomized trial of aspirin and sulfinpyrazone in threatened stroke. the Canadian cooperative study group. N Engl J Med 1978; 229:53.
Lasagna L, McMahon FG. New perspectives on aspirin therapy. Proceedings of a symposium co-sponsored by the aspirin foundation Inc, and Tulane university medical center. Am J Med June 14, 1983; 74:6A.
New and emerging uses for aspirin. Course delivered by The George Washington University School of Medicine and Health Sciences. December 2, 1986.
Opthalmology World News (OWN), Volumn 1, #10, October 1995; 24–25.

Regular aspirin intake and actue myocardial infarction, Boston collaborative drug surveillance group. BR Med J 1974; 1:440.

Silverman M. Magic in a bottle. New York: The MacMillan Company, 1941.

Vane JR, Botting RM. (eds). Aspirin and other salicylates. London: Chapman and Hall, 1992.

INDEX

Abbott, Maude, 370, 371
Abraham, E. P., 215
acetylsalicylic acid, 375–81
actinomycetes, 286
acupuncture, 393
Acyclovir, 258, 279
Adelaide University, 195
adrenal, 388
Age of Reason, 393
age related macular degeneration, 320
Aguayo, Albert, 319
AIDS, VI, 28, 40, 106, 171, 267
Albert Alexander, 201
Allen, J. R., 85
Allen, Woody, 133
allopurinol, 278
Alzheimers Disease, 317, 325
Amantadine, 258
amebocyte, 339
American Academy of Ophthalmology, 145, 147, 148
American Academy of Pediatrics, 123, 255
amino acids, 353
aminothiazole, 45
analgesics, 379

Anderson, M. J., 228
Angeletti, Piero, 314
Animal and Plant Health Inspection Service, 77
ankylosing spondylitis, 356, 363
antagonists, 315
Anthrax, 229
antibiosis, 193–213
antibodies, 69, 160–70, 217–89, 315–25, 351–62
antibody tests, 95, 97, 106
anticoagulants, 340
antigens, 167, 352
antioxidant, 110, 128, 129, 393
antipyrexics, 379
antiserum, 311
Antrypol, 3
Aranda and Sweet, 125
Argon-laser, 146
Arrowsmith, 195. *See* Lewis, Sinclair
arsenic, 11, 14, 20, 95, 181, 210
Arsphenamine Salvarsan, 20
arthritis, 163, 217, 278, 363, 380–88
Ashton, Norman, 110, 119, 122

415

Asian Flu, 258
Aspergillus terreus, 219
aspirin, 177, 375–91
asthma, 191, 383, 388, 392
astigmatism, 138–144
Astwood, E. B., 45
atabrine, 9
Atomic Energy Commission, 344
Atoxyl, 11
Auld, A.Z., 169, 170
aureomycin, 288
Australian aborigines, 257
Australian antigen, 258
Austrian, Robert, 3, 23, 165–76, 243
autoimmune disease, 353, 357, 369
Avery, Oswald, 169, 170, 175, 333
azathioprine, 278
AZT, 258

Babes, V., 193
Bach, 302
bacillus brevis, 216
bacteriocidal, 25, 38
bacteriophages, 229
bacteriostatic, 25, 205
Baine, William B., 106
Bakwin, Harry, 112
Balkan Grippe, 243
Ball, J., 136
Baltimore, David, 234, 260
Bang, Frederik, 328–48
Barcelona, 221, 367, 368
Barlow, Robert, 327, 328
Barr, Mason, 81
Barr, Y. M., 259
Barton, Benjamin, 55
Baum, L. Frank
  The Wizard Of Oz, 56
Baumann, Eugen, 50
Bausch and Lomb, 143
Bayer pharmaceutical
  company, 375, 382
Bayer, Fredrick, 375, 380–83
Behring, Emile Von, 168
Beijerinck, Martinus Willem, 230, 232, 235

Bellevue-Stratford Hotel, 90–104
Beloff, Anne, 221
Berlin, Nathaniel, 346
Bern University Hospital, 332
Bernard, Claude, 20, 42, 47, 240
Bernheim, 28, 29
Beta carotene, 129
beta-lactamase, 213
Beth Israel Hospital, 203, 206
Beyer, Karl H., 35
Bing, Richard J., 328–45
birth defects, 150
Bishop, Michael, 260
Bjoerling, Jussi, 367
Blalock, Alfred, 370
Blattner, Russell J., 227
bleomycin, 286, 288, 292
blindness, 135
Bliss, Eleanor, 22
blood clotting, 339, 383, 385
blood sugar, 388
Blumberg, Baruch, 257, 258
B-lymphocytes, 234, 353, 366
Bocchini, Vincenzo, 314
bone marrow, 219, 277, 289, 352–67
bone marrow transfusions, 277
bone marrow transplant, 352, 366, 368
Borel, Jean-Francois, 218–19
Boston Children's Hospital, 242, 244, 262
Boston University School of
  Medicine, 389
Boussingault, Jean-Baptiste, 49, 50, 67
Bovet, 21
Bowman's layer, 140
Bozeman, F. Marilyn, 103
Bradshaw, Ralph, 314
brain, 46, 56, 163–74, 203, 209, 217, 226, 244, 254, 260, 262–66, 325–28, 340, 383–86
brain derived growth factor, 319
Brassicace Cruciferae mustard
  plant family, 46
breast milk, 80–85, 354
Brenner, Donald J., 100

INDEX 417

Bristol-Meyers-Squibb, 324
British Committee on Safety of Medicine, 383
British Medical Research Council, 199
Brodie, Maurice, 250, 254
bromine, 54, 73, 74
Brooke, Blyth, 214
Brotzu, Giuseppe, 213–216
Brunhilde strain, 247
Bubonic Plague, 229
Bueker, Elmer, 300–305
Bunka Kunsho Medal, 292
Bunsen, Robert Wilhelm, 78
Burke, James, 349
Burkitt, Denis, 259
Burkitt's Lymphoma, 259
Burnet, Macfarlane, 244, 262
Burroughs Wellcome, 278, 279
Bursa of Fabricus, 366
Bush, George, 43
Bush, Vannevar, 207, 211
Buss, Carl Emil, 379
Bust, John Brown, 233

Caballé, Monserrat, 367
Caesar, 243
Calne, Roy, 219
Cambridge University, 323
Campbell, Kate, 118
cancer, 46, 51, 60, 74, 86, 172–183, 221–35, 253, 259–69, 271–321, 367, 389, 390, 393
cantaloupe, 209
capsid, 236
carbolic acid, 377, 378
carbonless copying paper, 83
cardio-pulmonary bypass, 331
Carlsberg Biological Institute, 332
Caroline Institute, 192
Carrel, Alexis, 238, 297, 330–332
Carrel-Lindbergh apparatus, 331, 332
Carreras, Jose, 352–368
Carrier, William, 107
Carson, Rachel, 77, 88
Caruso, Enrico, 367
cataracts, 320, 388

cats, 230, 234, 241, 248
Caventou, Joseph, 8
CDC
  Centers for Disease Control, 81, 90–107, 254–256
celiac sprue, 356
centrifuge, 341
Cephalosporins, 213–215
cereal boxes, 83
Chain, Ernst B., 39, 149, 177, 189–199, 202–23
Chamberland, 222, 232
Chandler, Francis W., 100
Chappell, 112
Charrin, C. R., 193
Chatin, Gaspard Adolphe, 50, 69
chemotherapy, 3, 11, 15, 25, 38, 39, 181, 273, 276, 283–291, 367
Cheng Zihua, 24
Chesney, A. M., 46
chick embryo, 238–244, 298–311
chickenpox, 245, 247, 258, 261
China, 47, 55, 135
Chinese, 134, 177, 363, 376, 390
chitin, 339
chloracne, 81, 85, 86
chloramphenicol, 171, 288
chlorine, 48, 54, 73, 74
chloroquine, 10
chlorthiazide, 35
cholera, 160, 185, 222–243
cholesterol, 219, 388
chorio-allantoic membrane, 303
chromosomes, 213
chrysoidine pyridium, 15
Churchill, Winston, 179, 180, 182, 204, 268
Ciliary Neurotrophic Factor, CNTF, 317
cinchona, 8, 376
cisplatin, 280–284
Claude Bernard Hospital, 20
Clement, Nicholas, 48
Cleveland, Grover, 57
Clutterbuck, P. W., 194
coal-tar products, 274
Coghill, Robert D., 205, 211, 212

Cohen, Stanley, 149, 305–19
Coho Salmon, 70
Coindet, Jean-Francois, 49
Colebrook, Leonard, 21, 22
Coleridge, Samuel Taylor, 12
colon cancer, 389
colostrum, 354
Columbia University, 170–175, 308, 345
Columbia-Presbyterian Medical Center, 203, 206
competitive inhibition, 26, 29, 40
compound 606
Salvarsan, 11, 20
Conley, Frances K., 364
Connaught Laboratories, 251
Conner, John T., 207
Conwell, Russel H., 387
Coolidge, Calvin, 1, 2
Coolidge, Calvin Jr. 1, 38
Copernicus, 192
cornea, 133–49
corneal abrasions, 389
corneal tissues, 134, 320, 343
corneal wounds, 319
coronary arteries, 334
coronary sinus, 334
Courtois, Bernard, 47, 48
Cowpox, 233
Cox, Haerald Rea, 241
coxsackievirus, 226
Craddock, Stuart, 185–89
Crafoord, Clarence, 370
cramps, 385
Craven, Lawrence L., 386, 387
Creger, William P., 360
cretinism. *See* Goiter
Creutzfeldt-Jakob disease (CJD), 263–266
Crick, Francis, 39, 222
Crookes tube, 273
Crookes, William, 273, 274
Crosse, Mary, 118
Curie, Pierre and Marie, 221, 370
Cyclo-oxygenase, 385
cyclosporin, 218, 219
Cylindrocarpon Lucidum, 218
cytomegalovirus, 226, 227, 247

cytoplasmic granules, 339

d'Herelle, Felix H., 229
Daraprim, 279
Dausset, J., 357, 360, 365
David, Hugo, 33, 43
Davies, John, 51
Davy, Sir Humprey, 48
Dawson, M. H., 203
DDT, 77
de Duve, Christian, 183
de Victor, Maude, 88
deGarilhe, Privat, 258
Delaware Bay, 339
Delbruck, Max, 262
Dengue-2, 261
deoxyribonucleic acid, DNA, 236
Department of Agriculture Laboratories, 208
DeRudder, J., 258
Desai, Rajendra, 369
Desormes, Charles Bernard, 48
detached retinas, 393
di-acetyl-morphine, 382
Dioxin, 88
diphtheria, 160, 168, 177, 185, 229, 257, 353, 354
diphtheria antitoxin, 25, 168
diplococcus, 155, 156
DKB, 288
DNA, 155, 228, 236, 257, 278, 281
doctrine of signatures, 376
Domagk, Gerhard Johannes, 3, 6–44, 149, 223, 321, 343, 380
Domagk, Gertrud, 6, 36
Domagk, Hildegarde, 6, 16, 36
Domingo, Placido, 368
Donders, Cornelieus, 135
double-blind studies, 392
Dresden, 378, 379
Dreser, Heinrich, 380, 381, 382
Drinker, Phillip A., 248, 249
Dubos, Rene, 216, 286, 290
Duchesne, Ernest, 177, 179, 221
Duisberg, Carl, 382
Dulbecco, Renato, 260, 269
Dupont, Ethel, 24

Durnev, Valerie, 142
Durrell, Lawrence, 253
Dyes, 7–21, 186, 333, 382
  aniline, 9, 10, 13
  coal-tar or anailine dyes, 9
  Indigo, 9
  methylene blue, 9, 10, 158
  Phenophthalein, 9
  synthetic mauve, 9
  Tyrian Purple, 9
dysentery, 229, 243, 285, 376, 379, 392

eagles, 87
Ebers papyrus, 376
EBV, 259
ECHO, 226
Edelman, Gerald Maurice, 353
Ehrlich, Paul, 2–39, 181, 188, 273, 285, 321, 353, 354
Eichengrun, Arthur, 380
Eisenberg, 15
Elberfeld, 380
electroencephalograms, 388
electron microscope, 233, 237, 266
Eli Lilly, 23, 195, 215
Elion, Gertrude B., 278, 279
Eliot, T.S., 19
Elixir of Sulfanilamide, 24
Emmerich, R, 193
Enders, John Franklin, 227, 237, 240–62
Endotoxins, 341
Enteric Cytopathogenic Human Orphan virus, 226
Environmental Protection Agency EPA, 81, 86
enzymes, 26, 111, 128, 196, 276–287, 310, 315, 352, 385
epidermal growth factor, (EGF), 313, 317
Epstein, Michael, 259
Epstein-Barr virus or EBV, 259
Erythema Infectiosum, 228
erythromycin, 106, 171
Escherichia coli, 280
Eskimos, 134, 135
Euler, Ulf van, 384, 390

excimer laser, 144–148
exotoxin, 341
Exotoxins, 341
eye, 80, 187, 227–276, 312–343
eye derived growth factor, 319

Faraday, Michael, 178
farsightedness, 140, 145
Fascism, 297, 298
Federal Food, Drug, and Cosmetic act, 24
Feeley, James C., 100
fenamates, 389
fermentation, 205, 215
Fermi, Enrico, 51
Ferrell, James, 232
fertilization, 307
fever, 157, 161, 209, 213, 239, 243, 262, 341, 351, 360, 376–89
fibrillar halos, 304, 305
fibroblast growth factor (FGF), 316
filterable virus, 232, 233, 265
Finland, Maxwell, 169
Firemaster, 75, 78
Fleming, Alexander, 15, 39, 177–223, 321. See Penicillin
Fletcher, Charles, 201–3
Flexner, Simon, 233
Florey, Howard W., 39, 149, 173–221
foamy viruses, 234
Food and Drug Administration, 386
FDA, 69, 81, 145–217
Foot-and-Mouth Disease, 232
Ford, Gerald, 283
Fore people, 261, 262, 265
Forssmann, Werner, 333, 334
Fourneau, Ernest, 19, 20, 21
Fowl-pox, 239
fowl-pox virus, 238
foxglove, 376
Fraenkel-Conrat, H., 236
Francis, Francis, 251
Francis, W. Jr., 170
Fraser, David W., 91, 92

420 INDEX

Fred Hutchinson Cancer Research Center, 367, 368
Free radicals, 128
Freeman, John, 191
Friedlander, Carl, 155, 186
Fries, George, 78
Frosch, Paul, 230, 232, 235
fuel cells, 284
Fuller, Albert, 22
Fulton, John Farquhar, 196, 205, 209
Funk, Casimir, 60
Fyfe, Andrew, 48
Fyodorov, Svyatoslav, 136–47

Gajdusek, D. Carleton, 262–65
Galen, 376
Gamble, James, 262
Gandhi, Mohandas Karamchand, 210, 211, 268
garlic pills, 393
Garre C, 193
gas chromatograph, 77, 78
Gay-Lussac, Joseph, 48
Gelmo, Paul, 13, 21, 38, 343
Gerhart, Charles, 375, 377–81
germ warfare, 345
German Measles, 228, 247
Gibbons, Charles, 331
Gibbs, Clarence Joseph, Jr., 264
Gigli, Beniamino, 367
ginkgo-tree extracts, 393
Glasgow University, 377
Glasse, Robert, 265
Gleick, James, 344
globulins, 354
Goeppert-Mayer, Maria, 372
goiter, 44–70
  cretins, 52, 53
  endemic cretinism, 53
  endemic goiter, 52–67
  Himalayan goiter, 52
Gomez, B.A., 8
Gonorrhea, 229
Goodpasture, Ernest W., 237, 238
Gorbachev, Mikhail, 143
Gordon, Harry, 322

Gotschlich, Emil, 171
gout, 217
graft-versus-host reactions, 388
gram negative, 185, 186
gram positive, 185, 186
Gram, Hans Christian, 186
Graves, Robert, 43
Great Lakes, 55, 84
Gross, Robert E., 370
Grove, William, 284
growth factor, 149, 289–326
growth factor receptors, 315
Guignard, L., 193
guinea-pigs, 202
Gunther, John, 65

Hackman, C. H., 288
Hadassah Medical Center, 119
Hadlow, William, 263, 264
Halbert, Frederic, 75, 76, 78
Halsted, William, 58
Hamburger, Viktor, 298–305, 323, 324
Hammon, William, 241
Hannibal, 243
Harada, M., 85
Hare, Ronald, 22, 181–191
Harington, C. R., 50
Harrison, Ross Granville, 238
Harvard Medical School, 64, 241, 242, 262
Harvard University, 113, 290, 336, 387
Hata, Sahachiro, 11
headaches, 201, 262, 381–389
heart, 43, 60, 163, 168, 203, 219, 226, 228, 238, 305, 331, 334, 341
heart attacks, 375, 384, 385–89
Heatly, Norman, 197–215
Hebrew University Hadassah Medical School, Jerusalem, 87
Heidelberger, Charles, 175
Heidelberger, Michael, 2, 23, 38, 168–175, 188, 262, 321, 343
Heidelberger, Philip, 175
Helsinki Declaration, 150
hematology, 344, 346, 347
hemocyanin, 333, 339

hemoglobin, 323, 333, 339
hemolymph, 328, 339, 341
hemorrhaging, 384
Hemphill, Bernice, 365, 369
Hensen, Jim, 37
hepatitis B, 228, 234, 257
Hepatitis B vaccine, 257
herbalism, 393
heroin, 382
herons, 87
Herpangina, 226
herpes virus, 259
Heyden, Fredrick von, 378
Heymann, David, 92
high blood pressure, 333
Hippocrates, 376
histocompatibility, 277
histocompatibility antigens, 355
Hitchings, George, 278, 279
Hitler, 36
HIV, 266
HLA, 277, 352–67
Hodgkins Disease, 277
Hoerlein, Heinrich, 13–21
Hoffman, Felix, 377–82
Hoffman-LaRoche, 31
Hogue, Ruth, 314
Holmes, Sherlock, 63
Holt, Louis, 194
homunculus, 307
Hong Kong Flu, 258
Hong, Richard, 366
Hopkins, Frederick Gowland, 197
hormone, 45–61, 309, 317
horseshoe crab, 327–46
Howard, W. P., 57, 58
Huble, David H., 135
Hudson River, 84
Hudson, Robert, 84, 106
human breast milk banks, 355
human parvovirus B-19, 228
Human Rights Committees (HRC), 150
human sperm, 307
Humphrey, Hubert H., 281–83
Huntington Medical Research Institutes, 345
Hyperthyroid, 49

Hyperthyroidism, 44, 45, 49
Graves' disease, 43
Hypothyroidism, 43, 51–55

I.G. Farben Industrie, 7–31
Ibsen, Henrick, 210
ibuprofen, 389
Ignatowski, 41
immune deficiencies, 226
immunosuppressant, 218
Imuran, 278
inactivated (Killed virus), 250, 255
incubator, 109–120, 229, 239
Indians, 55, 352
indomethacin, 389
infectious mononucleosis, 234
inflammation, 226, 283, 363, 385–389
influenza virus, 259
inhibitor, 309
Institute of Microbial Chemistry, 286, 292
Institutional Review Boards (IRB), 150
insulin, 320, 388
insulin-like growth factor (IGF), 316
Interferon, 258
International Society for Heart Research, 345
iodine, 45–70, 186
iodinated compounds, 51, 65, 68
iodized bread, 64
iodized oil, 65, 66
iodized salt, 48, 54–70
iodoform
  iodine derivative, 48
Iporoniazid, 32
iridium, 283, 284
iron lung, 248, 249
Irwin Memorial Blood Bank, 365, 369
Isoniazid, 31–34, 216
Ivanovsky, Dmitry, 230, 232

Jackson Memorial Institute, 300–302

Jacobson, Leon O., 278, 291
Jacobson, Leon Ores, 277
Jamieson, D., 65
Janus, 271, 274
Japan, 47, 55, 74, 84–86, 134–150, 208, 211
Jarrige, Jean, 382
Jeantet, Dr., 45
Jena University, 13
Jennings, Margaret, 196
Johns Hopkins, 22, 44, 46, 57, 114, 119, 167, 249, 250, 345
Johnson, Lois, 127, 281
Joliet, Frederic, 51
Joliet-Curie, Irene, 51
Jones, Roger S., 87
Jose Carreras International Leukemia Foundation, 368
Journal of the American Medical Association, 145
Jovin, Tom, 324
Juntendo University, 141
juvenile onset diabetes, 363

Kagiwai, Kaz, 325
Kaiser Wilhelm II, 273
Kamen, Martin, 306, 322
kanamycin, 286, 287, 292
Karajian, Herbert von, 367
Kasturbai, 210, 211
Kasugamycin, 292
Keen, Carolyn, 244
Keen, W. W., 253
Kendall, Edward, 50
Kent Community Hospital, 80
Kepler, Johannes, 135
kidney, 46, 171, 217–19, 238, 250, 278, 281,288, 333, 341
kidney transplantations, 362
killer T cells, 355
Kimball, Oliver Perry, 63
King Crab, 338
King Sisyphus, 310
King Victor Emmanuel, 296
Kinsey and Zacharias, 117
Kirchoff, Gustav, 79
Kitasato, 285
Klarer, Josef, 15, 17

Klatzo, Igor, 263
Klien, Bertha A., 117
Koch, Robert, 5, 27, 28–39, 159, 170
Kocher, Emil Theodor, 50, 53
Kolbe, Adolph , 377–79
Kolmer, John, 250
Kornberg, Arthur, 308
Kuratsune, M, 85
Kuru, 261–266

L. pneumophilia, 100–107
La Touche, C. J., 191
Lactobacillus casei, 278
LAL, 328, 333–346. See Limulus Amebocyte Lysate
Lan, Dr., 138
Landsteiner, Carl, 233, 332, 358, 372
Lansing strain, 245, 247
LASER, 145
LaTouche, C. J., 185, 191
Lavoisier, 392
Lederberg, Joshua & Esther, 33, 323–24
Leeuwenhoek, Anton, 230, 307
Legionnaires Disease, 96–107
Lehmann, Jorgen, 28, 29
Leipzig, 377, 378
Leland Stanford Junior, 351
Lenhart, Carl H., 59, 60
lens, 133–143
Leonidas of Alexandria, 272
Leroux, 377
leukemia, 234, 276–278, 353, 367
leukocytes, 158, 353
Levaditi, Constantin, 20
Levi, Adamo, 297
Levi, Giuseppe, 297–99, 302–314
Levi-Montalcini, Rita,149, 295–314, 322–324
Levin, Jack, 328, 333–347
Lew Gehrig Disease, 317
Lewis, Sinclair, 195, 330
Limulus Amebocyte Lysate, 328–341
Limulus polyphemus, 337, 338
Lind, James, 131

INDEX 423

Lindbergh, Charles A., 330–32
Lindenbaum, Shirley, 265
liposomes, 289
Lister, Joseph, 377
locusts, 229
Lode, A, 193
Loeffler, Friedrich, 230, 232, 235
Long, Perrin, 22
Lord Balfour, 182
Lord Beaverbrook, 192
Louis, P. A., 194, 233
Lovell, Reginald, 194
Low, O, 193
Lucey, Jerold F., 123–27
Luciferase, 34
Lund, Erna, 162
Luria, Alexander, 305
lymphatic leukemia, 277
lymphocytes, 234, 275, 353–69
lysozyme, 182–192, 197

Machat, Jeffrey J., 147
MacKenzie, C. G. & J.B., 44
MacLeod, Colin M., 170
mad cow disease, 266
magnesium oxide, 73, 75
malaria, 2, 8–10, 40, 161, 210, 376
Malley, Josef, 31
manganese, 129
Manhattan Project, 221, 291
Mao Tse-Tung, 24
Marine Biological Laboratories, 336
Marine, David, 55–70, 336
Markov, George, 283
Marshfield, Wisconsin, 81
Marston, Robert, 173
mass spectrometry, 78, 79
Massachusetts General Hospital, 64, 249
Massachusetts Institute of Technology, 234, 336
Matsuzaka,J., 85
Max Plank Institute, 324
May, Orville E., 43, 205
Mayo Foundation, 50

MBL, 336, 337. *See* Marine Biological Laboratories.
SeeMarine Biological Laboratories
McArthur, J.R., 261
McCollum, Elmer Verner, 44–46
McDade, Joseph, 97–101
McDermott Foundation, 357
McDonald, Marguerite B., 145, 148, 149
McGill University, 319, 370
McKhann, Charles F., 248
meadowsweet, 376
Measles, 228, 247, 266
mechanical respirator, 249
Medical Research Council, 198, 214, 215
Medicines Commission, 212
menstrual pain, 385
mental retardation, 52, 87
mercaptopurine (6–MP), 278
Merck pharmaceutical Company, 35, 217
Mercurochrome, 1, 182
mercury, 182, 210, 287, 292
Merk & Co., 208, 211
Metchnikoff, Elie, 225, 226
Methe, George, 277
methemoglobin, 333
methicillin, 213
Meyer, Hans, 31
Meyer, Hertha, 304, 306
mice, 14, 16, 40, 95–98, 119, 149, 186, 200, 202–215, 234, 239, 277, 301–312
Michaelson, Isaac, 119
Michigan, 55, 63, 70–85
Michigan Chemical Corporation, 75, 78
Michigan State University, 280
Microbes, 225, 226
Mietszch, Fritz, 15, 16
Millerm H. M., 199
Minamata , 287
Mintz Hittner, Helen, 127
Mississippi Valley Medical Journal, 387
mold, 178, 184–201, 204–23
penicillium, 15, 184–223

424 INDEX

monamine oxidase inhibitors
  MAOIs, 32
mongoose, 242
monkey, 86, 145, 233–56
Montaigne, Michel de, 39
Montalcini, Adele, 297
Montgomery, Bernard Law, 211
Moore, Jonathan, 392
Morgan, Thomas Hunt, 308
mosaic virus, 230–35, 237
Moscow, 136–43
Moscow Research Institute of Eye
  Microsurgery, 137–42
mouse effect, 305, 310
Moyer, A. J., 205, 206
multiple sclerosis, 355
mumps, 233, 244, 245, 261
Mussolini, 298
mustard gas, 275, 276
Mycobacterium tuberculosis, 27
myopia, 133–39, 147

Napoleon, 47, 243
naproxen, 389
National Cancer Institute, 129, 281, 284, 285, 321, 324
National Foundation for Infantile Paralysis, 244, 251, 252
National Institutes of Health, 81, 145, 174, 264, 346, 391, 393
National Research Council, 372
native Americans, 134
naturopathic health, 393
Nazism, 298
nearsightedness, 133–40
negative information, 305
nerve growth factor, 295, 307–24
Neufeld, Franz, 159, 160, 172
neurotactants, 304
neurotrophic factors, 317
neurotropic, 304, 311, 317
New England Journal of Medicine, 387
New Guinea, 54, 65, 66, 172, 261–266
New Orleans, 145
New York Times, 393

nicotinamide, 31
nitre plantation, 47
nitrogen mustards, 276–79
Nitti, 21
Nobel Prize, 50, 177, 179, 183, 188, 192, 212, 216, 220, 221, 225, 234, 240, 247, 258, 260, 265, 269, 273, 278, 279, 295, 299, 307, 322, 323, 330, 334, 370, 372
Nobel, Alfred, 220
Nomura, Junichi, 324
North American Indians, 376
Northrup, Alice, 244
NSAID, 375, 389
nucleic acid, 236, 308
nucleoprotein, 308
Nuremberg Code, 150
Nutrimaster, 75, 78

O'Connor, Basil, 252, 253
Oberlin College, 322
Office of Alternative Medicine, 391, 393
Oklahoma Agricultural and Mechanical College, 357
Olitskey, Peter, 236–37, 245
oncogene, 234, 260
optic nerve, 135, 325, 328, 389
Orphan Viruses, 226
Osler, William, 57, 153, 164, 371
osmium, 284
Owens, William Councilman & Ella Uhler, 114, 115, 126
Oxford Bobby, 201–03
Oxford University, 189, 192–221
oxygen, 16, 42–48, 76, 101–130, 156, 200, 208, 249, 284, 323–43, 388

Pacific yew tree, 391
Palade, 183
palladium, 284
Palo Alto, 351, 361, 362
Pan American Health organization
PAHO, 53
Pane, N, 193

INDEX 425

Papazolu, A., 69
para-amino-benzene-sulfonamide
  Prontosil, 13, 21
para-amino-phenol, 382
Parkinson's Disease, 317, 325
PAS
  para-amino-salicylic acid, 26–34
Pasteur Institute, 20, 229
Pasteur, Louis, 2, 38, 153–55,
  178, 192, 222–250, 377
patent, 203, 206, 207, 215
Patz, Arnall, 118, 119
Paul, John, 232, 237
Pauling, Linus, 39, 222, 262
Pavarotti, Luciano, 368
Payne, Rose O., 349–73
Payne, Thomas, 359–65
PBB, 74–88
PCB, 74–88
Pearl Harbor, 139, 205, 208
pediatricians, 47
Pelletier, Piere-Joseph, 8
Penicillamine, 217
penicillin, 15, 38, 39, 91, 106, 149,
  158, 164–269, 286, 288, 321
penicillin N, 215
penicillium, 178, 188, 191, 200, 206
penicillium chrysogenum, 194
penicillium notatum, 185, 190, 203
penicillium rubra, 185
peptides, 313
peregrine falcons, 87
Perkins, Herbert, 359, 365
Perkins, William Henry, 8, 365
Petrov, Boris, 136, 142
phenacetin, 382, 383
phenol, 378
Phenophthalein, 9
phenothiourea, 45
phleomycin, 288
phosphodiesterase, 308
photoreceptor cells, 325
Photorefractive Keratectomy
  (PRK)., 144
photosynthesis, 109
Phototherapeutic Keratectomy
  (PTK), 144
Piria, 377

Piricularia oryzae, 287
piroxicam, 389
Pitt-Rivers, Rosalind, 51
plague, 243, 285
plagues, 72, 379
plasmoquine, 9
platelet aggregation, 385, 386
platinum, 280, 281, 283
platinum-iridium pellet, 283
Pliny, 376
pneumococci, 23, 153–85
pneumonia, 39, 89–172, 180,
  210–23, 242–62
poison gas, 110, 275
polio vaccines, 247, 253, 255, 256
polio virus, 225–57
  Brunhilde, 247, 249
  Lansing, 249
  Leon, 249
poliomyelitis, 67, 161, 226–55
polybrominated biphenyls
  PBB, 73, 80
polychlorinated biphenyl
  PCB, 70, 74, 77
Polymorphonuclear
  leukocytes, 353
Pontiac Fever, 104, 106
Porter, Rodney R., 353
premature infants, 134
Priestly, Joseph, 42
Prions, 266
proliferative diabetic retinopathy
  (PDR), 320
proliferative vitreoretinopathy
  (PVR), 319
Prontosil
  para-amino-benzene-
  sulfonamide, 13–25
Propylthiouracil
  PTU, 45
Prostacyclin, 385
prostaglandins, 384–390
prostate glands, 384
Prusiner, Stanley B., 266
Pryce, Merlin, 184, 190
pseudomonas pyocyanea, 193
psoriasis, 363
Pyle, Marjorie, 33

pyridium, 15
pyrimethamine, 279
pyrogens, 341

Q Fever, 243
Qian Xinzheng, 24
Queen Charlotte's Hospital, 21
quinine, 8, 9, 161

rabbits, 11, 14, 23, 40, 46, 69,
   137–154, 160, 168, 186, 193, 274,
   342
Radcliff Infirmary, 201
Radial Keratectomy, 144
radium, 272, 273
Raistrick, Harold, 194, 197
Rasputin, Grigorii Efimovich, 384
Ratner, Bret, 25
Rats, Lice And History, 243
rauscher leukemia virus, 234
Ravitch, Mark M., 22
Readers Digest, 283
receptor, 316, 324
receptors, 315, 320, 321, 324
recombinant DNA
   techniques, 314
rectal cancer, 389
Reed, Walter, 232, 262
Refractive Keratectomy, 133, 134
Refractive Surgery Interest Group,
   RSIG, 148
Reid, R. D., 194
replication, 228, 230, 258, 279
respiratory paralysis, 248
retina, 114–35, 316–25
retina derived growth factor,
   319
retinopathy of prematurity,
   ROP, 125
retrolental fibroplasia, RLF,
   114–125
retroviruses, 236
reverse transcriptase, 260
Reye's Syndrome, 383
rhesus monkey, 86
rheumatic fever, 17, 23, 32, 41
rheumatism, 379, 382
rhinoceros horn, 391

Rhoads, Cornelius Packard,
   275–91
rhodium, 284
ribonucleic acid, RNA, 236
Ricketts, H.T., 97
Rickettsia, 97–103, 241, 243
Ridley, Frederick, 187–89, 194
rifampin, 106
Rinderpest, 159, 172
Robbins, Frederick
   Chapman, 240–48
Robbins, William Jacob, 242
Robitzek, Edward H., 31
Rockefeller Institute, 161, 169,
   170–75, 193, 199, 233–39, 248,
   254, 259, 286, 290–97, 345
Roentgen, Wilhelm Conrad, 273,
   274, 289
Roosevelt, Eleanor, 24, 253
Roosevelt, Franklin Delano, 24,
   252, 253
Roosevelt, Franklin Delano Jr., 23,
   24
Roosevelt, Thodore, 156
Rose, 69
Rosenberg, Barnett, 280, 281
Rossiter, Margaret, 371
rous sarcoma virus, 234, 300, 332
Rous, Francis, 233, 234, 259
Rous, Peyton, 332
Roux, Emile, 28, 38,168, 222, 238
Royal College of Surgeons, 384,
   385
Royal Opera House, 368
Rubella, 228
Russell, Frederick, 243
Russia, 137, 150
Rutgers University, 29, 216
ruthenium, 284
Ryan, H., 118

Sabin, Albert, 161, 236–69
salicin, 377
salicylate, 29, 376, 377, 381
salicylic acid, 375–83
saliva, 154, 199, 310
salivary glands, 227, 244, 310–314
Salk Polio vaccine, 257, 268

Salk, Jonas E., 250–69
salmon, 70, 84
Saltykow, 40
Salvarsan. *See* Syphilis
compound 606, 11, 181
Samuelsson, Bengt, 384
San Francisco, 74, 260, 336–69
Sandoz Laboratory, 218
Santayana, George, 85
Sato, T, 138–50
scarlet fever, 209, 213
Schaeffer, Howard J., 278, 279
Schiff, Fritz, 203
Schreus, Hans, 17
Schulemann, 9
Schutz, H. H., 107
Schwab, Ivan, 148
Scrapie, 263–66
Seattle, 357, 367, 368, 393
sedatives, 388
selenium, 129
Selikoff, Irving J., 31, 80, 82
Semmelweis, 97
senile macular degeneration, 320
Septra, 279
Shaw, Louis A., 248
Shepard, Charles C., 99, 101
Sherman, J. Q., 107
Shiga, 285
Shooter, Eric, 317, 323, 324
Sho-saiko, 390
Shumway, Norman, 351
side-chain, 215
Silver Bullet, 36
Silverman, W. A., 115
Slapped Cheek Disease, 228
Slaughter, Frank, 210
Sloan-Kettering Institute for Cancer Research, 282, 290
Smadel, Joe, 262, 264
Smith, J. Brian, 385
Smith, Margaret G., 227
snake venom, 196, 308–11
snake venom anti-serum, 311
Snow, John, 42, 91
Sorbonne, 370
Soviet Union, 137–47
spectacles, 135

Spemann, Hans, 299
Spinal cord injury, 325
spirochete, 210
Sporn, Michael B., 321
Sputnik intraocular lens, 143
Squibb, 31, 170, 194, 206, 324
St. Aspirinius, 381
St. Elizabeth's Psychiatric Hospital, 89, 99
St. Louis Hospital, 272
St. Mary's Hospital, 15, 22, 179–209
Stahlgrenska Hospital, 29
Stanford University Medical Center, 105, 317, 323, 349
Stanford, Jane, 351
Stanford, Leland, 351
Stanley, Wendell M., 230–37
staphylococci, 17, 160, 184–90, 213–15
Steigerwalt, Arnold G., 100
Sternberg, Geroge M., 154
Stone, Reverend Edmund, 376, 377
Storch, Gregory A., 106
Strack, E, 193
streptococci, 16–23, 40, 156, 185, 200–213
streptomyces griseus, 216
streptomyces kasugaensis, 287
streptomycin, 29–33, 171, 216–18, 246, 286–88, 343
Stricker, Franz, 379
strokes, 375, 384–89
Subunit Vaccines, 257
Sulfaguanidine, 44, 45
sulfamethoxazole, 279
sulfanilamide, 2–44, 158, 164, 175, 188, 201–30, 321, 343, 380
sunburn, 388
Sunderman, William, Jr., 96
sunlight, 271
synthetic fibers, 12, 13
syphilis, 2, 11, 20, 100, 181, 210, 273, 285
Syracuse University, 327

Taussig, Helen, 370

taxol, 284, 285, 390
Tchichibabin, 15
tear therapy. *See* lysozyme
Temin, Howard, 234, 260
Temple University, 250
terramycin, 288
Terry, Theodore L., 113–16
Tesla, Nikola, 323
testosterone, 366
tetracycline, 171
Thalidomide, 114, 149
The American Paralysis Association, 326
The Association For Women In Science (AWIS), 365
The Journal of Molecular and Cellular Cardiology, 345
The Laboratory of Cell Biology, 313, 314, 322
the mouse effect, 305, 311
Theiler, Max, 239
Thiersch, Karl, 378
thiosemicarbazone, 30, 31
Thom, Charles, 185, 205
Thomas, Lewis, 170, 176, 282, 336, 337
Thromboxane, 385
Thucydides, 259
thymus gland, 353, 354
thyroid, 43–70, 281, 363, 388
thyroid-stimulating hormone TSH, 46
Thyroid gland
thyroxin, 50, 51
triiodothyronine, 45, 51
thyroiditis, 363
Tillett, William S., 170
tissue cultures, 238, 246, 250, 303–08
tissue typing, 351–73
T-lymphocytes, 353
Tobey. George Loring, 23, 24
Tolstoy, Leo, 225
Tongi University, 24
Toxic Substances Control Act, 86
toxins, 76–83, 168, 257, 341
trace minerals, 393
tranquilizers, 388

transformers, 74
transforming growth factors, TGF, 317
transfusion reactions, 349–60
transplantation reactions, 362
Trefouels, 20, 21, 22
Treponema pallidum, 210
Trichoderma Polysporum, 218
trimethoprim, 279
trypanosomes, 10
Tsar Nicholas II, 384
tuberculosis, 18, 26–42, 216, 221, 286, 290, 357, 379
Tulane University, 145
Turin, 295, 296, 302, 314
Turin Medical School, 297
Tyndall, John, 178, 179
typhoid, 185, 223–43, 351, 379
typhoid vaccine, 223
typhus, 241, 243
Tyrothrycin, 216

U.S. Children's Bureau, 112
U.S. Department of Agriculture, Ames, Iowa, 76
U.S. Food and Drug Administration, 69, 103, 129, 279, 281, 383
U.S.Department of Agriculture, 205, 366
ultraviolet, 79, 144, 146, 271, 388
Umezawa, Hamao, 273, 285–93
Umezawa, Sumio, 288
UNICEF
United Nations Children's Emergency Fund, 64
United Nations World Health Organization, 257
University College in London, 323
University of Alabama School of Medicine, 345
University of Alberta Medical School, 127
University of Bishops College, 370
University of California at San Francisco, 260,266, 347

University of Chicago, 277, 291, 323
University of Delaware, 98
University of London, 119
University of Michigan, 79, 241, 252, 268, 322
University of Muenster, 6
University of Paris, 382
University of Pennsylvania, 167, 176
University of Rio de Janeiro, 304
University of Southern California, 175, 345
University of the Pacific, 357
University of Tokyo Medical School, 291
University of Washington, 295, 356, 357
University of Wisconsin, 85, 234
urethral secretions, 343
uric acid, 218
urine, 343

vaccine, 2, 14–32, 96, 153–76, 223–59, 267–72, 342
Van Camp, Loretta, 280, 281
Van Leeuwen, Stormm, 191
Vanderbilt University, 237, 306–23
Vane, John Robert, 384, 385, 390
Varmus, Harold, 260
Varon, Silvio, 324
Vat fermenters, 212
Veterans Administration, 88
Vidarabine, 258
viral pneumonia, 242
Virchow, Rudolf, 50, 58
virion, 230–36
virus, 76, 159, 269–73, 355, 383
vitamin A, 115–29
vitamin C, 129
vitamin E, 115–29
von Bayer, Adolph, 9
Von Behring, Emile, 38, 168, 285, 353
von Ossietzky. Carl, 36

Wagner-Jauregg, 3
Waksman, Selman A., 29, 39, 216, 287, 343
Walshe, John, 217
Waring, George O., 139
Washington University, 227, 298, 306–24, 345
Wasserman, Marcus & Dora, 87
Watson, James, 39, 222
Wayne State University, 345
Weaver, Robert E., 100
Weaver, Warren, 199, 268
Weichselbaum, Albert, 155–59
Weigert, Karl, 10
Wellcome Institute, 263
Weller, Thomas Huckle, 240–47
Wells, Percy A., 205
White, David, 219
Wilde, Oscar, 131
Williams, W. W., 59
Willis, Anthony L., 385
willow, 376, 377
Wilson's disease, 217
Wisel, Torsten N., 135
Witebsky, 69
Witthauer, 380
Wollstonecraft, Mary, 371
Woodruff, Alice, 238, 239
Woodruff, Charles Eugene, 237, 238
Woods Hole Oceanographic Institute, 336–46, 356
Woods Hole squid, 337
World Health Organization WHO, 54, 56, 63, 64, 67
World War II, 51, 150, 170, 207, 243, 275–79, 286–98
Wreed, F, 193
Wren, Christopher Sir, 358
Wright, Almroth, 14, 15, 42, 162–84, 192, 194, 223, 224

x-rays, 65, 80, 164, 177, 212, 272–74, 289, 333
x-rays, 272
Xu Haidong, 24

Yale University, 205, 346

Yamagiwa, Katsusaburo, 274, 285
Yamaguchi, A., 85
Yasnaya Polyana, 225
Yellow Fever, 232
yew tree, 284
Yoshimura, T, 85
Yosho, 85, 86, 88

Yu-Cheng, 86, 88

Zeide, 15
Zhou Enlai, 25
Zigas, Vin, 262
Zinsser, Hans, 241, 243
Zovirax, 279
Zyloprim, 278